The Creative Animal

Roberto Marchesini

The Creative Animal

How Every Animal Builds its Own Existence

palgrave
macmillan

Roberto Marchesini
Etologia
Scuola d'Interazione Uomo Animale
Bologna, Italy

ISBN 978-3-031-07413-4 ISBN 978-3-031-07414-1 (eBook)
https://doi.org/10.1007/978-3-031-07414-1

This Palgrave Macmillan imprint is published by the registered company Springer Nature Switzerland AG.
The registered company address is: Gewerbestrasse 11, 6330 Cham, Switzerland

Contents

About the Author

Roberto Marchesini is Director of the School of Human-Animal Interactions and the Center for the Study of Posthumanist Philosophy, both based in Bologna, Italy. His main publications include: *Over the Human. Post-humanism and the Concept of Animal Epiphany* (Springer, 2017), *The Philosophical Ethology of Roberto Marchesini* (Collected essays edited by Jeffrey Bussolini, Brett Buchanan, Matthew Chrulew (Routledge, 2017)), *Dialogo Ergo Sum. My Pathway into Posthumanities* (University of Virginia Press, 2018), *Beyond Anthropocentrism* (Mimesis International, 2018), *Philosophical Ethology and Animal Subjectivity*, in *Animality in Contemporary Italian Philosophy* edited by Felice Cimatti and Carlo Salzani (Palgrave Macmillan, 2020), *The Virus Paradigm* (Cambridge University Press, 2021), with Marco Celentano *Critical Ethology and Post-anthropocentric Ethics* (Springer, 2021).

CHAPTER 1

In Praise of Improvisation

Premise

Observing the adaptive behaviour of animals, it is difficult not to be fascinated by the complexity and aptness of their actions: they match their environment and their chosen lifestyle down to the smallest detail. The various species express their skills spontaneously, almost nonchalantly and yet precisely, without requiring any additional training. They show that they have built-in knowledge, something that does not require masters and apprenticeships and comes even before experience. The close correlation between morphology and behaviour as a whole, between an animal's appearance and the challenges it has to face, has always struck the human imagination. When Rudyard Kipling (1865–1936), in his *Just So Stories* (1912), imaginatively illustrated the origin of the morphological characteristics of various animals—the whale, the camel, the kangaroo—he provided us with the clearest example of how, since the earliest times, man has recognised functional predicates in the characters of the zoological universe, judging them to be extraordinary in their predisposition to the performance to which they were called.

Both shamanism (Eliade 1964) and natural theology (Paley 1836) are based on the recognition of an intrinsic virtuosity of living beings, whereby nature becomes the first teacher also for human beings. We marvel at the care with which a weaver builds its nest, a swallow prepares for migration

R. Marchesini, *The Creative Animal*,
https://doi.org/10.1007/978-3-031-07414-1_1

or wasps build their hive, all with geometric and precise perfection. There are therefore countless testimonies, widespread in mythology and religion, of the feeling of *thaumazein*, which is equally admiration for animal wisdom, but also religious fervour and nostalgia for a spontaneous dimension that is irretrievably lost. The Hellenic founding myth of the two titans also shows us how animal qualities are based on somatic expressions—one might say intrinsic to the body—resulting from Epimetheus' bestowal, as opposed to humans, who have to call upon Prometheus in order to receive the technical equipment they need to survive. Animality shows off an embodied competence, which made Giacomo Leopardi (1887) say (1798–1837) "di natura è frutto ogni vostra *vaghezza*". And that word embodies admiration and longing for the inexplicable wisdom of nature, which is now precluded to man.

On the animal stage, the spectacle of nature's wisdom is on display: the techniques and performances, the precision and foresight, but above all the correlation is stupefying, stunning and sublime. Every challenge finds the right structure in the animal's body, every opportunity finds the right behaviour to take advantage of it, and the whole system runs like clockwork, dictating precious synchronies. We are thus led to think of the work of an omniscient architect, the only one who could have arranged such perfection (D'Aquino 1970). The wisdom of nature appears as much in the harmony of the whole, that is in the reciprocal dependence of living beings and in the organismic texture of the various ecosystems, as in the performative adequacy of the individual, which shows a manifest balance between needs and capacities, a functional specialisation to which we attribute precision and lucidity of judgement. Everything seems to be part of an organic design and the animal appears to be called upon to bear witness to the pervasive ingenuity of nature.

The manifold strategies of the living are not only masterfully tailored to the typical challenges of each habitat—that particular niche in which each species is embedded—but often display such precision and extravagance in their details and peculiarities as to leave naturalists speechless. The great competence deployed by animals in their activities is baffling and suggests minds capable of actively raising various problems and delving into them so as to find the best solution every time. Human beings find themselves in front of this set banquet and cannot help but be astonished by the order it displays, so they are led to seek the explanation of such perfection in the here and now, attributing it all to the work of a designer and assigning expressive finalism to the individual whenever it manifests particular adherence between the action and the result achieved.

Human beings also exhibit innate behaviours, some of which are particularly evident in children, precisely because they have not yet been modified by first-hand experience of the world, the relationship with their parents and cultural conditioning. The presence of a substantial number of hereditary patterns contradicts, moreover, the humanistic assumption that considers our species different from other animals insofar as it lacks natural endowments (Gehlen 1988). On the contrary, ethologists, starting with Konrad Lorenz (1903–1989), have challenged this claim: the great researcher Irenäus Eibl-Eibesfeldt (1928–2018), by studying similar behaviour in populations of very different cultures, has been able to describe in detail a large number of innate human patterns (Eibl-Eibesfeldt 1989). The reason for such skills lies in the history of the species: phylogeny, in addition to providing morphological solutions that are strikingly suited to a species' adaptive challenges, is a process of information build-up through standardised patterns of expression. These patterns provide the subject with "operational tools" already designed to deal with the most common survival problems. It is clear that the organisation process of such displays takes much longer than an individual's ability to learn, but for this very reason these skills appear extremely refined.

Innate knowledge gives individuals an initial advantage in facing the challenges of life, and this will then be followed by the experiential processes that they will develop through the particular circumstances with which they will be confronted. This knowledge is a fundamental asset for all living things: it results from natural selection, the mechanism highlighted by Charles Darwin (1809–1882) that represents the cardinal explanatory principle in biology. If we consider each character as the result of a slow refinement process, lasting millions of years and carried out over many generations—going beyond the species dimension—then we can understand the origin of such skills. We fall into error, however, when we attribute relevance and performative effectiveness to the occasional and specific response of the individual's ingenuity. The reason for this distortion is easily explained: we are accustomed to believing that intelligence is manifested in performance responsiveness and success, and that every solution results from individual work.

There is no doubt, moreover, that the performative collection of the animal world is a veritable Guinness Book of Records: think of the surgical meticulousness of actions, the inflexible sequence of flow charts, the precision with which time frames are respected, the convergence of organs, functions and practises. It is no surprise that we should feel wonder and tend to attribute these masterpieces either to an omniscient deity or to the skills of the individual. The Darwinian theory is counterintuitive in many respects, because man struggles to understand the idea of a knowledge built up over millions of years through competition in the reproductive process (Dawkins 1986). We find it hard to relate to biological timescales, so we look for a comprehensive explanation in the behavioural manifestation itself. In reality, the individual is nothing more than a drop of water in the great river of life and its qualities are the result of a process that transcends its brief passage through the world.

We are wrong because we persist in believing that knowledge can be attributed to individual experience and that, at birth, the subject is a kind of *tabula rasa*, to be filled from scratch (Locke 1690). In reality, every animal comes into the world with a wealth of knowledge. Its sensory organs are already predisposed to perceive those stimuli that are useful to it, organising them into meaningful *Gestalts*, and its cognitive system has expressive and orientational tendencies from the earliest moments of its existence—such as chasing a moving target for predators or gathering for omnivores. Puppies already have communicative and interactive patterns for parental request behaviour, for example. All this results from an accumulation of selection processes that bring together millennia of adaptive evidence in the individual. The great wealth of past generations, subjected to the grindstone of articulated selective pressures, brings forth existential geometries from below.

The human being is excited by the adaptive responses of animal behaviour and tends to attribute to individual ingenuity a wisdom which, conversely, results from a long build-up of successful attempts to overcome life's challenges. We delight in the thoughtfulness of parental tasks, we admire the foresight and orientation skills, and we are enraptured by the variety of hunting strategies we find in animals. On the other hand, we can see that these competences tend to recur almost unchanged in different individuals belonging to the same species, so it is evident that these behaviours cannot be assigned to the intellectual work of a single subject. There is a kind of experience that precedes the individual, a learning that does not develop in the instant of the subject's attempt at a solution, a functional adequacy that precedes it.

For naturalists, the animal world, where morphology and behaviour add up to a style that reflects the context, is a source of interest and passion, precisely because of the large number of strategies involved. The great explorers of the eighteenth and nineteenth centuries, such as Alexander von Humboldt (1769–1859) and Alfred Russel Wallace (1823–1913), could not help being fascinated with the multiformity of the living universe and the adherence of morphological and behavioural strategies for survival adopted in different geographical areas. The ability of some snakes and scorpions to conserve water in arid areas, the deep trophic chain links in complex environments such as rainforests, the ability to avoid thermal dispersion in arctic or high-altitude areas—all these things inspired the creation of new intellectual figures who, alongside their profession as geographers in the service of different countries, inaugurated a growing interest in biodiversity.

It is easy, on the other hand, to fall into the trap of assigning to the species itself a title of cognitive individuality, personifying a process that, conversely, is simply one of accumulation. The apparently simple Darwinian explanation actually contradicts certain intuitive heuristics, so that we struggle to understand this process of adaptive adherence. This process takes place over a period of time that exceeds our comprehension, and if we already struggle to reckon with a millennium, let alone millions of years. Life may appear to us in an individual form, but we must not forget that it is more like a flow, a great river, where the various generations follow one another through small modifications which, in the long run, change the morphology and the overall rules of the corresponding ecosystems. Performative adherence is therefore only the result of a momentary balance between the different actors.

But this correlation between character and function, or between challenge and performance, is undoubtedly what most strikes our imagination, in the apparent cyclical stillness of a background world: the seasonal migrations of many birds, the springtime awakening of mating, the time of hibernation, the winter diapause of the insect world. Animal life seems to impart teachings concerning the appropriateness of actions, in the right ways, at the right times and in the right contexts. Human beings, in admiring the behaviour of each species within their own niche, assign to each individual a wisdom that is unique to them and deserves to be imitated. Thus, from the vast performative panorama of biodiversity, man has drawn inspiration for his instrumental translations, crediting birds with the role of auspices, incorporating the virtues of bears and wolves in shamanic rites, relying on animals as spirit guides and double identities (Marchesini 2017). There is a strong sense of enchantment that humans feel towards animal mastery, because it is spontaneous and does not require any additional teaching, but for this very reason it is all the more amazing.

Praising animal skills therefore becomes a veritable narrative *topos* in human tradition, which often magnified the richness and properties of animal ingenuity. Each species has precise expressive habits together with a morphological profile that is consistent with them, so that form and behaviour make up a whole corresponding to an existential style. A species becomes a way of inhabiting the world and lends itself to taking on symbolic values or being used as a metaphor. Animal shrewdness was easily used as an analogy for vices and virtues, qualities incorporated into nature that can be accessed through liturgies, models to be imitated through techniques and tools, properties to be represented in avatars or heraldic representations. The form itself thus embodies a content, because anatomical features and expressiveness overlap: a structure becomes, therefore, fully manifest in the corresponding behaviour. It is said, then, that the animal is intelligent insofar as it is sensible in the execution of a task, admirable in the effectiveness and efficiency of the results achieved, and capable of correlating the body to the function.

Phylogenetic Knowledge

We cannot fail to be amazed at the precision with which a bird builds its nest or a mammal copes with the task of giving birth, at how a tiny wasp is able to create complex architectures or a whale feeds on plankton, how a cicada knows how to camouflage itself on the trunk of a tree or an ant is busy supplying its community with food for the winter. The knowledge that each animal deploys to achieve its objectives of survival, foraging and reproduction, in the multiplicity provided by biodiversity, leaves us astonished and fascinated, conveying the image of a foresighted intelligence filled with dexterity. And there is no doubt that the interplay of animal interactions, in the balances of ecological homeostasis as well as adaptive functions, is as close as it gets to what we consider perfect. How many times have I heard the virtues and expertise of different species listed as proof of their intelligence, their mastery of ordinary tasks as unequivocal evidence of their mental acuity! The carnival of animal performance cannot but leave us astonished, and it is understandable that we should try to explain it through the individual's ability to see into the problem and give a well-considered answer.

However, this is a big blunder, because it leads us to believe that there is an overlap between performative property and intelligent evaluation, whereby we focus on animal competence in order to obtain evidence of its subjectivity. The mistake lies in looking for an overlap between competence and the individual's ability to find a solution to the problems of everyday life. The skill with which birds build their nests and the diligence they show during hatching, the oxygenation of eggs that fish achieve by beating their fins, the early obstetrical care we observe in many mammals on the part of their mates, the precise and meticulous attention paid to the young, which we find even in insects—all these features cannot fail to amaze us and testify to the profound wisdom of animal instinct. Parental attention astounds us with its lyricism, a poem of parental love that on the one hand moves us, and on the other seems to suggest implicit intelligences projected towards the goal of reproduction. Similarly, survival skills, from mimetic capacities to various predation processes, speak to us of the wisdom of the subject, which seems to knowingly camouflage itself in order to escape or surprise its adversary.

Thus, the architecture of a termite mound is reminiscent of a homeothermic construction project devised by the master insect, while the use of a stone to break the shell of a mollusc seems to have been devised by the ingenuity of an otter. There are countless examples of animal expertise, such as beaver dams, parasite tricks, the cryptic disguise of an octopus or the Batesian camouflage of a hoverfly. The animal world is a kaleidoscope of qualities: from the reproductive rendezvous of fish, turtles and toads, excelling in synchronism, to the chemical battles of moths based on pheromones of engagement or repulsion. While sexual contests between the males of some species turn into exhibitions, in others courtship consists of psychedelic manipulations. Interspecies interactions are also extraordinary, such as the allelochemical phrasings and alliances between plants and insects—for example, when aphid-infested shrubs attract ladybirds—or the dissuasive stotting of gazelles towards cheetahs (Caro 1986). It is a world of multifaceted and precise messages: from the topographical dances of bees, hymns to collective harvesting, to the predatory mimicry of some fireflies, which turn love invitations into lures. This whole display of ingenuity cannot leave anyone indifferent.

But on closer inspection, these behaviours occur in all subjects of a given species, as choreographies whose individual units follow rigid and predetermined schemes, so that the final performance is accurate and exact precisely because the whole is detached from the variables of experience and individual will. Our projective interpretation misleads us, attributing an intentional character to something that in reality derives its competence not from the individual's planning and decision-making, but rather from phylogenetic knowledge. We are misled by the anthropocentric perspective that tends to assign to every event a finalistic and subjective explanation, as also used to be the case with meteorological phenomena. On the other hand, the teleological explanation is the most intuitive way to account for what we see, so we say that fish have fins to swim and birds have wings to fly. And in this objective-based interpretation any behaviour that shows adherence to the goal, whatever the functional chapter it is supposed to serve, seems to support an operational judgement, that is the evidence of intelligence in action. But we have to think again.

Since I was a child, I have been fascinated with the plural articulation of insect life, which in a certain sense represents the youth seminar, one could say the propaedeutic apprenticeship, for anyone who wants to try their hand at the subtleties of ethology. In the phantasmagoria of entomology, in fact, one can find such a vast catalogue of life strategies as to go beyond any science fiction or horror movie, with correlations and appointments so precise in their space-time coordinates, despite every degree of improbability, as to leave one breathless. Consider Hymenoptera, these mind-bending miniature masters of cruelty, far removed from the harmonies of butterflies or beetles or the flying feats of flies and dragonflies: their refined homicidal skills cannot fail to astonish an enthusiast such as myself; their life strategies have something demonic, but also sublime, about them.

Ichneumonids, who anaesthetise a larva to enclose it in a cell together with their eggs, which once hatched will feed on their immobile host, seem to possess truly incredible foresight (Fabre 1879). On the other hand, the multiplicity of parental strategies of solitary wasps is no less impressive than the complex constructions of social wasps in the articulation of community life. We are thus faced with a large catalogue of examples of animal foresight. Ants are portrayed as the ultimate hard workers, an example that seems to need no further commentary to show that animals can think about the future. And yet, I ask, is this really the case? Is the ant busy filling the chambers of the anthill because it is thinking about the future? Undoubtedly, when one compares its never-ending toil with the cicada's presence in the festive *hic et nunc*, one finds it hard to argue otherwise. We see it scurrying for food, cooperating with its sisters for the good of the community: it is inevitable to explain this with the yardstick of intentionality.

We are led to believe that an ant's behaviour reflects clarity about the need to forage, insight into the problem of supply and projection into the future, as if to say "the fate of the community" depends on this. The ant therefore becomes a counterpoint to the bacchic lightness of the cicada. The latter, however, with its hymn to carelessness, *chi vuol esser lieto, sia* (De' Medici 1991), certainly does not seem to lack wisdom, given the short number of days left to the insect once it has left its exuvia. Each behaviour is coherently embedded in an overall style, without smears or contradictions. It is no coincidence that the classical tradition of Phaedrus and Aesop handed down wonderful fables based precisely on the qualities of animals, where each species is an embodied idea, a predicate that can be extracted in its essential and translatable qualities. Animals seem to be directly responsible for the ingenuity they display, the latter being the result of their single, active presence at the interface with the world. Nevertheless, we must not be misled by this project-based epistemology, by this eagerness for finalism.

The aerial image of nature offered by natural theology claimed to extract proof of the divine creator from the harmony of the whole, where every correlation took on the title of perfection, mirroring the omniscience of its designer. The metaphor of the watchmaker, formulated by William Paley (1743–1805), was intended to justify the presence of a divine design. This would later give way to the counterargument by Thomas H. Huxley (1825–1895) who, by entering into the details of life, in the ruthlessness of the struggle for survival, could find, on the contrary, evidence that there was no beneficent author. For Darwin's Bulldog, it was not permissible to attribute the deadly harmonies of living things to a benevolent deity, but rather to the lack of one, so the dictates of natural theology could even be reversed (Huxley 1893). Once again, an attempt was made to identify a universal through a personifying projection: the harmonic design was contrasted with the struggle for existence.

Darwinian evolutionism has provided us with a key to explain how animal traits respond to the challenges of the world: the principle of adaptation, that is the gradual emergence of the most advantageous traits (fitness) within an animal population. Once we understand the evolutionary mechanism by which a better performing character is more represented in subsequent generations, the slow process of adaptive correlation becomes clear. What is so extraordinary about living things? The answer is simple: they reproduce. But not everyone succeeds in doing so during their lifetime, so that over time a kind of accumulation of advantageous traits is created in a species, depending on the lifestyle and environment it adopted. The reproductive differential strongly depends on several factors that affect the population of a species, inducing a selective pressure, which can be more or less powerful depending on the available resources and which includes different selection levers according to the characteristics of the species. The behavioural wisdom of animals can be treated in the same way as the information that chisels the adapted body of different species, regulates cascades of physiological functions and keeps the system in a state of metabolic homeostasis.

The risk, however, is to fall back into the error of finalism, attributing to natural selection the same teleological matrix previously assigned to the omniscient designer. On the contrary, taking Darwinian thought to its extreme consequences—a task which, in my opinion, was carried out by the palaeontologist Stephen J. Gould (1941–2002)—angels and demons no longer have space in the world of the living, because life on Earth neither followed an a-priori design nor can be compressed into the selective algorithm (Gould 1989). Phylogeny has generated events, rehearsing its morphopoiesis in a chaotic manner within the constraints of necessity. Only a generous flowering of forms from below, subjected to the grinding of fitness along the score of time, that is of its incompressible historical dimension, could have achieved such an admirable concert of adaptive adherence. Every form of finalism and predetermination fails, and with them fails the essentialist claim of an idea embodied in the morpho-function of a species. At the same time, thanks to evolutionary dynamics, it is easier to understand the process of information build-up in animal predicates, producing skills that do not require individual dexterity, because they are the fruit of phylogenetic wisdom.

Without this clarification we would risk confusing intelligence with performative competence: but this is precisely the mistake I want to avoid. In order to search for intellectual subjectivity, it is necessary not to be misled by the carnival of adaptive respondence of multiform species-specific performances. However much we may continue to be amazed at these competences, once we understand the morphopoietic guidelines of evolutionism, we no longer need to invoke individual wisdom to explain precision and adherence. We could say that the performative marvels we observe, even when they act with perfect adequacy to the problematic context they have to address, should be read in the light of innate information, built from below by accumulation. In other words, these are standardised abilities, the result of pre-established schemes and sequences, which in reality have very little to do with the process of *intus legere*, that is looking inside a problem and formulating a relevant answer *ex-novo*.

This knowledge diluted in time, a knowledge that belongs neither to the individual nor even to the species—since there are countless skills acquired from ancestors common to other animals—should make us reflect, if anything, on the relative importance of the individual in the great river of life, placing a sort of *epoché* on the narcissism that elevates individual experience to the sole engine of knowledge. We could say, then, that the biosphere learns through the generational repetition of its creatures, bringing out morphological and expressive features calibrated to the needs of the lifestyle and environment of each species. On the other hand, these forms, constantly tested and changing due to the transformation of surrounding conditions, do not pre-exist the great palette of possibilities, but are simply enabled by it. Life itself is a process of knowledge: this can be seen in its colonisation of the Earth, in the priority given to development and reproduction, in its resistance to thermodynamic decay, in its capacity to accumulate information within the aperiodic crystal of DNA, in its modification of the environment and thus the matrix of selective pressures.

It is marvellous to study these plural variations on the theme of replication, to discover how one law, that of mutation with differential reproduction, has created a multitude of strategies and relationships that exceed our capacity for understanding. This process transforms the individual into a small portion, an infinitesimal fraction of a magmatic whole that travels through time and brings back its traces as resonances: the individual is a fossil of a lost world. The parable of the individual is always fascinating in its budding, opening up to the world and building an existential singularity. And I continue to be amazed by the memory of the experiences that came before me and that still live inside me. We think of ourselves as corpuscles, totally disjointed monads, thrown into the world in a state transitory Being-there, yet we have already been in the world. We are an organic part of a flow that transcends us, and the innate memory that each of us has received from phylogeny speaks of this "having-been-there-before". It brings back fragments of non-human ancestors who preceded us, so that even our presumed identity as a species has blurred boundaries: we resonate with ancient worlds that, though past, have never really come to an end.

The individual is the fruit of a process that is metamorphic and sedimentary at the same time, a sort of epistemology of the living that, in the body, bears the stigmata of the useful cognitions that have evolved and been maintained in the species. The river of life changes over time, collecting, albeit partially, the succession of remote experiences, since, as Konrad Lorenz suggested, the a-prioris are nothing more than phylogenetic a-posterioris (Lorenz 1973). Traces of distant memories cross the species threshold so that, in spite of the essentialist view, the individual carries within itself endowments that evolved in its progenitors, so that even in the behavioural sphere we can speak of homologies, that is of traits of trans-specific similarity through common inheritance. Each species develops its own competences because the complex dialogue between population and environment leads to the emergence of a niche and a rank, acting as a chisel on useful traits. But at the same time each species does not evolve *ex nihilo*, so it retains other skills that evolved in its ancestors: each vertebrate tetrapod retains in some remote corner of itself echoes of a liquid world in which it swam in its distant past as a fish (Shubin 2008).

Today we know that different geological periods have followed one another on the face of the Earth, with equally different players, creating ecological balances that cannot be superposed on the one that finally took place in the Tertiary era, which saw the emergence of mammals and birds. On the one hand, evolution is constantly producing new forms, and on the other hand it is just as voraciously dissipating them, following the biological timeline, so that what we are seeing today is not an Eden prepared by an omniscient mind, but just a point in time of life on our planet. This, however, is an important stage, whose biodiversity has been achieved over millions of years. Incidentally, this awareness should make us reflect on the absurd way in which human beings are destroying it at a rapid pace, having known no more than an infinitesimal fraction of it. The innate reminds us of our participation as individuals in something greater that develops over time and has its most important expression in this diachronic dimension. We cannot, in fact, understand the individual, in its orientations and impulses, if we try to detach it from the generational flow.

Expressive Singularity

It is not my intention to belittle the innate dimension, that performative articulation of living beings which does not descend from the creativity of the individual, but from the legacy received from phylogeny. I do not wish to downplay the elegance and complexity of instinctive strategies, but, on the contrary, to place them in the dimension they deserve. What I am proposing is to distinguish this form of creativity of the living—which is diachronic and diluted in the succession of generations through processes of natural selection—from the creativity of the individual. In fact, I would like to point out that it is not precision or competence that should be called upon to bear witness to animal cognition, because in very many cases—obviously not all—this performance does not belong to the individual, but to an entity dispersed over time and transversal to the various forms of life. Phylogenetic wisdom has always fascinated me, precisely because of its transformation of the individual into a drop in the great sea of life, so that its presence is far from being annihilated or alienated, but on the contrary finds comfort in the fact that it participates in a realm of knowledge that goes beyond its temporal presence. Each living being emits a resonance that lingers in the great flow of life, associating itself with all the others. However, each can leave an imprint on this flow, in which the individual is both moved and moving. And that is what I intend to talk about in what follows.

The subject of this book, therefore, is not the innate wisdom of animals (although this is fundamental to understanding the identity of the various species), but their creative capacity and therefore also their ability to improvise in the great lottery of life. The fact that an individual does not start from scratch when entering the world but can make use of skills developed by natural selection over generations is certainly an important argument that should not be underestimated by assigning every expression of the individual to the manifestation of its ingenuity. On the other hand, this heritage should raise at least a second question, namely: what is the relationship between this heritage and the individual? We can believe that the individual is nothing more than the passive executor of innate neurobiological algorithms, thus adhering to the model of the "animal machine" professed by René Descartes (1596–1650), or we can consider this heritage as a set of endowments that the individual actively uses. Choosing this second option would certainly lead us to a different order of ideas, one that values and focuses on the individual in the expression of its existence, because we would assign the role of protagonist to it.

This perspective is precisely what I am going to develop, trying to show how the model of simple repetition and passivity in the exercise of phylogenetic endowments cannot stand and how, on the contrary, the singularity of individual action is the only way to enliven phylogenetic memory. What I am trying to suggest is that innate competence should be distinguished from the use the subject makes of it, adapting it to several situations and often adjusting it where necessary. Without being a passive executor or even a puppet moved by the strings of the innate, in reality the individual most often uses its innate competences in a rather flexible way, but sometimes even in an innovative way if not as a solution to unexpected problems. In other words, what I am proposing is not to question the innate, but rather the interpretative model usually adopted to outline phylogenetic competence: that concept of automatism which claims to be the only key to explain animal behaviour.

If phylogenetic information is thought of as the constitutive basis of automatisms, it is evident that there is no room to speak of expressive subjectivity. If, on the other hand, we consider the innate as a set of endowments available to the individual, a totally opposite explanatory question appears: the protagonism of the individual becomes, then, an inescapable explanatory factor. In this case, it is true that the individual does not start from scratch—we could say that the competence it manifests is not all its own making. However, the innate does not give rise to automatisms that passively drive the subject, but to endowments that the individual uses: (1) to build tools that are even more adherent to the specific situation in which it finds itself; (2) in a fairly free manner, like any other tool. When it comes to animal subjectivity, the alternative between these two models is the real issue, not the appeal to consciousness or the appeal to the astonishing abilities of the animal world. In fact, the problem does not concern performative extraordinariness or awareness, but ordinary behaviour—the very foundations of animality.

The question brings us to a central issue, namely whether we can speak of animal subjectivity or whether we must recognise that the animal is nothing more than a puppet moved by its own innate strings (Marchesini 2016). Tradition seems to continue to prefer the deterministic model based on automatisms, where it is the innate pattern that drives the individual. This explanation merely refers to the trigger, whether a drive or a stimulus, maintaining the direct and analytical relationship between input and output, that is reading behaviour as the expression of an automatism. It is clear that as long as we continue to explain animality in this way, there is no point in talking about subjectivity. I have referred to abilities, because any performance, even the most complex or extraordinary, can be carried out by a machine on the basis of adequate information. We have seen this with the development of extraordinary software capable of acting as an expert system, of neural networks capable of learning and modifying their functions, of robots that have become increasingly interactive in the environment (Hutter 2004).

Similarly, I questioned the appeal to consciousness because any intentional gradient is a reference to something, an illumination of a certain event or process, which must therefore precede the reference or explication. If the individual is an empty room, it is not enough to turn on the light to furnish it, since it is its automatisms that animate it. I believe that the study of animal consciousness is very important and should be developed more fully, and yet it cannot be consciousness that sustains subjectivity. Consciousness can only make subjectivity explicit—in the human being, it can make it declarable. In other words, it can only shed light on subjectivity, but not create it from scratch. If endowments were automatisms, subjectivity would be mere appearance; it would then make no sense to attach consciousness to something that does not exist. To appeal to consciousness is to avoid the issue of subjectivity, that is to use it as a kind of deus ex machina.

The automatism model is based on the direct relationship between trigger and performance (in the behaviourist view, the stimulus-response nexus), which therefore can only repeat the performance or place it within pre-established cascade chains, domino style. Being a subject means having a certain freedom of action in relation to the world, having agency over one's own endowments and not being a slave to them. The individual never repeats, even when implementing similar behaviours: rather, it is reminiscent of a craftsman who uses his own tools, with competence of course, but with the power to use them as he wishes. Subjectivity calls us to give meaning to the individual, a meaning that, if it were merely to repeat itself, would certainly be negligible. Indeed, this meaning is given by the singularity of individual action, which is first and foremost presence and not simply the echo of a past that imposes a certain behaviour. The singularity lies in the individual act, in its emergence from the here and now to impose an orientation and thus give rise to something that did not exist before its action. Reading heritage as automatism denies any space to subjectivity, because it deprives the individual of ownership of its endowments.

The automatic and repetitive conception of animal behaviour, associated with random expression in the attempt to find a solution, has been the prevailing view in twentieth-century epistemology, even if, in my opinion, this explanation has more than one flaw. The most macroscopic error is the pretence of directly and exhaustively linking behaviour to endowment when, on the contrary, it is evident that the individual would not be able to manage the multifaceted everyday situations it encounters if it had no free use for its endowments. Being able to use them freely means: i) being able to make them more and more adherent to the specificity of one's life context, by producing new endowments; ii) being able to use them in a flexible way when faced with situations that do not require the simple repetition of a competence, but a careful work of adaptation. The innate, then, rather than a repetitive automatism, is more like a "tool" available to the individual, which can evolve, generate new endowments and be used flexibly. Subjectivity, therefore, is realised through the biographical singularity with which it arranges its tools and thanks to the possibility of using its tools in a flexible and creative manner.

In this model, innate competence alone is no exhaustive explanation for a given behaviour, at least no more than a helm can account for the steering of a boat. The subject is therefore the helmsman, to keep this metaphor, who uses and supervises the equipment. Phylogeny sets up patterns of expression that are in some ways very much in tune with the particular challenges that the species faces, based on the given lifestyle-environment. However, the world is not a stable stage, not even in the short time span of an individual's life, so it is necessary to adapt one's heritage through new acquisitions. Innate endowments are therefore essential, but not sufficient to enable the individual to rise to the challenges of the world. At the same time, the individual must be able to use these endowments in a flexible manner, so that they are more like tools than automatisms: the individual uses them, it is not they who drive it.

In adaptation, the innate represents the starting material through which individuals construct their adaptive biography to match the specific conditions in which they find themselves. The innate provides the individual with an initial syllabary of behaviours that have already been experienced by previous generations, displays that realise the fundamentals of the species and support the individual in its first steps into the world. But the innate does not only have this "tutorial" function towards the young individual without first-hand experience of the world. It also represents the developmental substratum for the construction of new endowments, those which we define as individual. The innate is an interpretative tool, one could say a theory, or an operational tool, a heuristic, which allows the individual to construct new theories and new operants, as in the genetic epistemology developed by Jean Piaget (1896–1980). The subject, therefore, always represents a singularity on the stage of life, both because it modifies its endowments and because it uses them freely.

What, then, of the particular kind of competence that appears to us as a kind of species uniformity, whereby we detect a kind of species-specific repertoire? Undoubtedly, different animals exhibit behavioural recognisability, an identity of expressive traits, as well as a certain morphological structure and a well-defined physiological regime. When we observe the behaviour of individual members of a species, we find uniformity of expression, so that individuality seems to disappear, to be nothing more than background noise. Yet that noise should be taken into account far more than we do. It is one thing to detect common traits among individuals of a given species, quite another to claim that the innate is manifested through automatisms and that individuality must be considered background noise. This tendency to homologate is the result of an interpretative bias, the perspective distortion which leads to zeroing out the differences between various entities when we place them at a certain perspective distance. It is the same effect that, for the ancient Greeks, turned all non-Hellenic peoples into barbarians, and it is the same effect that makes us speak of animals as opposed to human beings. It is how we observe the population that generates our standard of reference.

If, on the other hand, we deal with the individual, and therefore use a different lens of observation, we realise two things. The first is that the species dimension, rather than a precise standard of behavioural patterns correlated with different situations, appears as a range of possible variations within the same theme. In other words, the innate generates a field of virtuality of development, within which we find the singularity, that is, the protagonism of the subject, which achieves a unique and unrepeatable identity. This field of virtuality of development and singular expression paves the way for significant differences between one individual and another and between one moment and another in the life of the same individual. This highlights the importance of subjective experience, which is capable of transforming the innate not into a mere echo of the past, but into a resource that the individual brings to life and evolves through its interaction with the world in the course of its existence. The innate therefore becomes available as a useful resource for the subject and not as a source of determinism.

In fact, the individual receives its phylogenetic endowments or what I have called innate memory, but then, through its individual experiences, adapts this heritage, developing an expressive identity that distinguishes it from all other subjects. On the other hand, the phylogenetic endowments, precisely because they constitute the "building" material of the learned, define constraints and possibilities, giving rise to the range of variations on the theme that characterises individuality. The animal, therefore, incorporates two different experiential matrices: the phylogenetic one, relating to the experiences of previous generations, and the ontogenetic or biographical one, relating to the experiences of the individual. However, these are not two juxtaposed worlds because, as we have said, the innate is the basis of learning, so that the two identities actually form a single dimension. This means that, precisely by remaining within the canon of the species, the individual manifests a distinctive expressive personality.

The second interesting thing is that when the same individual faces similar situations, we would often expect a certain behaviour to manifest itself in the same form, yet we observe, on the contrary, a further variation on the same theme. In other words, the individual never automatically repeats the same pattern, even when in situations that present overlapping problems. There is always a small difference: it is as if we were always making a slight diversion when returning home from the same place. In this sense, we can say that, while learning produces individual competence, giving rise to a well-defined personality, in this case the use of the endowment does not resemble the triggering of an automatism. Rather, it is a singular expression that inevitably produces a variation on the theme of the acquired competence. To simplify, we can say that while learning also produces competence, the individual never simply repeats what it has learned in a standardised way, but always adapts it to the specific circumstances.

This second aspect of singularity sits alongside the biographical composition of the individual, whereby singularity is a matter of the utmost importance. Just as the speciation event represents an emergence of singularity within its starting taxon, so the individual is singular within its species, and its here and now defines a subsequent moment of singularity. Singularity indicates that there is always a creative protagonism in the living that can never be dismissed (Bergson 1907). Standardisation is therefore a deliberate result, the outcome of filtering out individual differences and expressive singularities. In reality, an animal cannot simply repeat itself if it is to deal with the singularity of reality because, in adhering to a particular situation, it must always manage a degree of novelty, however small it may be. In fact, situations present themselves in terms of similarity, but never complete identity, so competence is not enough: one also needs creativity, that is the free and particular use of one's endowments.

Creativity, in fact, is a far cry from competence; more often than not, it is opposed to it, resulting in sketchy works, functionally clumsy DIY, a misplaced use of tools and juxtapositions bordering on the paradoxical. Creativity is often uncertainty, contradiction, it appears as a lightning and unrepeatable flash showing the solution or as a long reflection often lost in suspension or failure. Creativity increases when the degree of novelty to be managed is considerable; it manifests itself most clearly in challenging situations. This is where a mind in action is revealed: in a moment of difficulty, when competence alone fails to grasp the opportunity or the danger of the situation, in a situation which shows a gap between the stimulus and the response. Yet even when the degree of novelty is small, there is still something unforeseen to handle, that is, there is a need for the helmsman to use its resources in an innovative way. Not blind repetition, but the animal's ability to attend to its here and now: that is what can lead to adaptive adherence. This adaptive intervention, even in failure, reveals the intellectual work of going beyond the appearance of habit and the repetition of the acquired. The package of information possessed by the individual is necessary, but not sufficient to explain the expression.

But then, what is the relationship between competence and creativity? It is not easy to answer this question, because any behaviour expressed by an individual is always a mix between these two components. We can say that in certain situations the contribution of creativity is very modest, limited to adapting competence to the singularity of the situation being experienced at the time, managing a narrow margin of novelty. The level of creativity in these cases is not even remotely comparable to those situations in which the animal lacks endowments predisposed to that type of problem: then creativity peaks and manifests itself in hesitation, partiality and improvisation, which often lead to a failed attempt to find a solution. An intermediate situation can be identified in the process of assimilation-accommodation exposed by Jean Piaget, where it is possible to find a bridging endowment that allows the development of a new competence (Piaget 1971). However, even when the degree of creativity is small and limited to modifying the expressive volume of the response or generalising its scope, even in these cases it is not possible for the animal to simply repeat a pattern: a touch of creativity is always necessary.

We could summarise this statement by saying that "it is not possible for an animal not to be creative". In fact, even innate competence requires reapplication, a touch of creativity, because when faced with the fluctuations of the world, repetition is of no help. What, then, of species-specific competence? As I said, it is difficult to deny that there are traits of behavioural identity within a species and that some expressions of competence are not directly acquired by the individual. There is no doubt that: (1) it is possible to find general patterns in the behaviour of species, repetitive patterns that individuals of a certain taxon have in common; (2) there are innate behavioural patterns that adhere very closely to the adaptive problems faced by that species, what I have called "phylogenetic knowledge". It is necessary, however, to place the apparent uniformity and repetition of patterns within a particular observational perspective: one needs to be aware of using the lens to find patterns, not the lens to capture fluctuation. The skills we detect in the animal world are far from being as precise and standardised as they appear at first sight, and far from being mere repetitions.

There are species-specific patterns of expression: these are configured through hereditary contributions but also through social learning and interactions with the niche to which they belong. So, if we focus on these patterns, the behaviour of various individuals seems standardised. In reality, each pattern is nothing more than a basic script that the individual interprets with a range of expressive freedom, which may be greater or lesser depending on the situation. But looking at what generically appears to be uniform—identical among individuals of a particular species—the more we go into the details, the more it exhibits marked differences. In reality, species behaviour is not a standardised canon of expression, but a range that allows each individual to position itself in a singular way within experience. And it is not simply a matter of admitting individuality in the declination of a pattern, but of accepting that there is situational freedom in the expression of a pattern.

The homologated reading of phylogenetic competence stems from the model called upon to explain its application, a model based on the de-subjectification of the non-human animal, such that competence itself—understood as automatism—could be exhaustive in the explication of the behavioural phenomenon. The model was designed precisely to do without a subject as the helmsman of its own competences. In accordance with the Cartesian paradigm of the automaton-animal, it was essential to transform the innate and learned endowments into automatisms capable of explaining behaviour, without involving any entity delegated to supervising the individual's expression in the here and now of the situation. This meant constructing a full overlap between accumulated competence and compliance of the implemented behaviour. Automatism, in fact, does not require presence, because it is free from control and not subject to calibrated application. But if this were the case, we could not speak in any way of animal subjectivity, because once the ownership of use of the endowments is lost, the presence and, with it, any attribution of inherence in the animal also necessarily vanish. We can, therefore, speak of animal subjectivity only if we admit a present individual that is capable of applying its innate and learned endowments freely in every situation.

The Art of Improvising

Competence can be discussed in various ways. At times one can note the correspondence between the structure of the body and the animal's expression—the stick insect, for example, not only has the shape of a twig, but also sways as if moved by the wind, just as the cat, in addition to its plantar pads and retractable claws, also exhibits stealthy behaviour. At other times one can stress the adherence of innate behaviour to the adaptive challenges of the species. There is one aspect, however, that is before the eyes of any observer: the here-and-now aspect of animal behaviour, the fact that it can never be assigned to the domain of mere repetition of prior skills.

The animal is present in the moment it is living, it is not directed towards a past of automatisms to be triggered, but is always alone in front of what is happening to it. Animal expression is always a work in progress, a phenomenon that takes shape at the very moment it takes place, according to the situational here and now in which the subject finds itself. The animal has to invent its behaviour whenever an opportunity presents itself, gathering everything at its disposal and trying to produce the best possible response. Most of the time it fails, but it does not have the easy option of simply reiterating the skills it has acquired: every time an animal expresses a certain behaviour, it has to apply a little creativity.

More often than not, competence, whether innate or acquired through experience, provides the individual with endowments that are sufficiently suited to the challenge it faces, sticking to an adjustment-based approach, that is a partially innovative use of the instrument. In these cases, it is sufficient to find new operational paths or try other ways of using the endowment itself. At other times one has to rely on the art of improvising, like a trickster capable of shaking up the cards in order to find an entirely new or unusual solution; the individual must try and make attempts, after having understood the problem. The term "improvising" may seem far-fetched, and it is obviously necessary to clarify very well what is meant by this definition. Indeed, although it may seem unproblematic and intuitive, in reality it requires explanations that are anything but simple. When we say that an animal is also called upon to improvise when going about its ordinary business, we are primarily saying that competence is never sufficient in the difficult task of adapting to the specific nature of the situation.

Of course, gaps can always be filled through the acquisition of new information or by modifying or adapting innate endowments. This is often what happens. Yet even learning is not sufficient to explain the development of behaviour, which is never a linear or algorithmic sequence of sedimented information, but rather a composition that takes place in a singular way in the here and now of the subject's expression. In other words, behaviour is more like a dialogue between the individual and the world than a linear sequence of already structured expressive units. When I say that a behaviour is never a repetition of a pattern, but is always an interpretation, that is, a singular way of using an endowment, I mean to say that the individual always has to put in place a creative surplus to manage a margin of situational novelty. The particular condition experienced by the subject is singular, and the ability to respond in tone is always urgent and immediate.

In order to do this, it is not enough to make use of previous responses, be they innate or learned: it is necessary to accept the challenge of the ad hoc responsive predisposition. This of course may have different degrees of complexity, but is always an emerging product in the animal's interactive here and now. In other words, the individual must use its own previous endowments as useful material for developing responses that adhere to the particular nature of the situation. This means that the model of the innate endowment, however related and schematic in its structure, as mentioned above, cannot be thought of as ready-made, so that it only needs to be applied. On the contrary, it must be viewed as an instrument that presents a multiplicity of functional outcomes and is also endowed with internal plasticity such as to allow an immediate reconfiguration when necessary.

Admitting a certain amount of creativity in all animal behaviour—that is believing that the expression occurs at the very moment the animal manifests it and is not a simple repetition of a previous pattern—means questioning the model with which we are used to explaining animal behaviour. Let me be clear: the problem of subjectivity lies primarily in this, that is whether we think of behaviour as a product freshly made by a mind in action or whether we consider it the expression of a preset mechanism. In arguing for this inevitable "expressive improvisation" of animal behaviour and this poietic role of the mind, which does not limit itself to repetition but continually produces content, I do not want to diminish the importance of the past, whether innate or learned. In fact, it is one thing to consider this background in the form of fixed schemes, which are activated when stimulated by triggering factors, and quite another to think of these endowments as construction material, which allows the mind to produce the expressive result in a specific way, that is in the moment.

In the mechanistic paradigm the animal is a puppet not because it possesses endowments of competence—I do not wish to question this in any way—but because it simply has to obey them, it does not own them. This means that, in fact, it lacks presence in the here and now (it cannot attend the event), because its actions are nothing more than the repetition of something already set. In this model, the individual is absent, so that animal behaviour does not require the work of a helmsman called upon to adapt the instrument to the situation. In this model, the endowment, in its trigger-sensitivity and productive automatism, is considered exhaustive in explaining behaviour. If, on the other hand, behaviour is the instantaneous fruit of a mind-in-relation with the event, however corroborated by all the endowments of competence it possesses, it is evident that expression always requires a touch of improvisation and never arises from mere repetition. The subdued and local nature of behaviour, its being improvised—*in provisus*, that is, not fully predictable—means that it is impossible to explain animal experience through a static model, which defrauds the individual of creative capacities and of ownership over its resources.

Clearly, the explanatory model needs to be modified in order to understand animal creativity, and it is not enough to appeal to superior cognitive faculties—perhaps it is not even necessary. It is not a question, therefore, of carrying out a simple operation of retouching, adjusting or fixing up the traditional model, perhaps by adding this or that cognitive endowment, as was attempted in the last decades of the twentieth century. In so doing, due to the multiplication of falsifications and exceptions that escape every single model, we have gone so far as to produce an explanatory tripartition of animal behaviour. We use (1) the psycho-energetic model for the innate, through the drive motive; (2) the associative model for learning processes, by virtue of the concept of conditioning; (3) the elaborative model for all behaviours that do not fall within the previous two. This is, of course, a case of epistemological redundancy, justified, like the famous Ptolemaic epicycles, by how difficult it is for the traditional paradigm to explain animal behaviour.

But what is wrong with this approach? At first glance, we note that the use of the explanatory tripartition depends on the inability of each model to subsume the others, or rather, to explain the different expressions that an animal presents. I will deal with this subject when discussing the interpretative model. With regard to the question of creativity or expressive adherence, two basic difficulties can be identified. The first (1) is the tendency to consider the animal as an automaton, with the impossibility of admitting its actual presence in the situation, that is the ability to look into reality (*intus legere*) and produce a singularly correlated behaviour. The second (2) consists in placing the expressive motives outside the individual, in the heritage of the past for the innate and in the environment for the learned, operating a de-subjectivation of the animal. It is clear that, conversely, in order to speak of animal creativity it is necessary to go beyond both these positions, granting a creative faculty to the individual (1) and placing it at the centre as the subject of its own actions (2). A puppet cannot improvise, so we must imagine the mind as an operating system capable of producing its expressions as and when occasions and events arise.

As I have said, if we look for standard patterns of animal behaviour, reporting the variations—between individuals or in the same individual on different occasions—in terms of "noise" and expanding the scale of evaluation, even if only numerically, what we will get is a canon. Thus, we will have the impression: (1) of a deficit of expressive individuality, that is of a uniformity of outcomes, through the mechanism of expressive fashion; and (2) of a lack of freedom of interpretation of the endowments on the part of the individual, as the subject didn't express itself but were driven by its endowments. What is most misleading about the current discourse on creativity is precisely this last point, namely the idea that it is not the subject that uses the endowments to achieve its own ends, interpreting them to make them fit the situation, but that it is the endowments that mechanically and repetitively drive the animal-puppet.

We must put the mind back at the centre of the expressive process and think of it not in a static way, as a recapitulation of the past, but in dynamic-productive terms: it is an emergent condition capable of producing an experience expressed in a behaviour that is always singular and improvised. In conclusion, I do not think it is arbitrary to look for common patterns within a species, because this helps us better understand its ecological and phylogenetic position. Similarly, it is correct to identify a biographical identity, built on individual experiences: this is why we talk about the individual's tendencies, knowledge and abilities. But we are mistaken if we forget: (1) in the case of the species, that we have found patterns because we have subjected our research to a precise focal point, capable of mitigating variations between individuals; (2) in the case of the individual, that we have given importance to recurrences, propensities and common expressions, leaving out the variability within the individual itself.

Research on common patterns, able to extrapolate innate schemas and laws about learning, does not equal the assertion that an animal is nothing more than the automatisms that the innate and the learned place on its presumed expressive passivity. Claiming otherwise is the real obstacle to recognising animal creativity. If we do so, we fail to notice the animal's production of solutions in the moment, through DIY adaptations and strategies (Jacob 1977). Improvisation, in fact, admits the animal's ability to have presence and a leading role in its life, so it is not simply a matter of adding new schemes to the innate endowment through learning, but rather of maintaining a constant flexibility in regulating responses. Of course, this improvisation does not come out of nowhere, since every burst of creativity rests on the shoulders of the giants of the past, that is to say on the phylogenetic knowledge I have mentioned, but also on all the experiential events that sum up the biographical history of the individual.

On the other hand, the animal does not live in a condition of passivity with respect to its surroundings: rather, it is the protagonist in its interactions. Likewise, the animal cannot limit itself to repetition, because any endowment, even the most complex—and I would say, especially if it is complex—cannot be based on a simple application, but requires an interpretation that makes it fit with the specific situation. It is necessary to start from an important consideration, namely that reality always gives itself to the individual in a singular way, setting up situations that are in some way similar—certainly!—but never identical, so it is not possible to simply rely on repetition, because margins of novelty must be managed at every moment. It is interesting, then, to note that creativity is anything but an exception in the animal world: it is only a question of understanding its characteristics without falling into clichéd anthropomorphisms. However one must also avoid, for fear of this, rushing headlong into the antithetical hypothesis of the random solution, because when faced with a problem an individual could never reach a solution through random attempts.

There is a fundamental difference between the concept of random solution and that of improvisation, and it is useful to address it right away. It is clear that when faced with a problem, not having the solution at hand, an animal must make attempts to solve it. However, the question is a different one: how are we to interpret this trial-and-error mechanism? In the traditional view, the animal reaches the solution by chance, and this assumption is understandable. If this were not so, we would have to hypothesise the animal's ability to understand the problem and formulate answers that are in some way coherent with the structural requirements of the problem, but above all based on the here and now. To deal with a problem, one has to be present and able to give an answer referring to the present moment: in other words, one has to improvise. Obviously, in a model that denies the animal even the possibility of formulating goals, or at least considers the concept of mind useless for explaining behaviour, it is clear that such operations cannot be taken into account. However, what disproves the effectiveness of the random solution is none other than the second law of thermodynamics applied to the concept of order (goal) and disorder (failure). The reason for this is straightforward: the chances of a random attempt reaching a solution are incomparably lower than the chances of failure. It would be like hoping to reach a destination by getting into a car and driving blindfolded.

Improvisation does not deny that the animal makes attempts, it simply considers these attempts as solution recipes coherent with the characteristics of the problem, that is "heuristics" or targeted, and not random attempts. When we talk about the solving abilities of animals, we immediately fall into the trap of puzzle-making and consider the problem as a condition of obvious checkmate. Now, the question remains whether an animal is constantly confronted with problems or whether these are very rare occurrences. I would like to stress that even if the hypothesis were very remote, it would still be possible to explain animal behaviour with the automatism model, that is with the mere repetition of what has been acquired. But in reality the animal must constantly manage degrees of novelty and therefore must raise and face countless problems, so we must break out of the bias of puzzle-making, which considers as a problem only something that is an objective setback. An animal, in fact, is faced with continuous challenges, so the art of improvising is one of the main strategies it has to address the subjective problems of everyday life, because it changes all the time. It is therefore necessary to hypothesise an interface with the world that is creative and non-repetitive.

Works Cited

Bergson, Henri. 1907. *L'évolution Creatrice*. Paris: Presses Universitaires France.
Caro, T.M. 1986. The Functions of Stotting in Thomson's Gazelles: Some Tests of the Predictions. *Animal Behaviour* 34 (3): 663–684.
D'Aquino, Tommaso. 1970. *Summa Theologiae*. London: Blackfriars.
Dawkins, Richard. 1986. *The Blind Watchmaker*. New York: W.W. Norton.
De' Medici, Lorenzo. 1991. *Trionfo di Bacco e Arianna. Canti carnascialeschi*. Roma: Salerno editrice.
Eibl-Eibesfeldt, Irenäus. 1989. *Human Ethology*. New York: Aldine De Gruyter.
Eliade, Mircea. 1964. *Shamanism. Archaic Techniques of Ecstasy*. Princeton: Princeton University Press.
Fabre, Jean-Henri. 1879. *Souvenirs entomologiques*. Paris: Delagrave.
Gehlen, Arnold. 1988. *Man, His Nature and Place in the World*. New York: Columbia University Press.
Gould, Stephen Jay. 1989. *Wonderful Life*. New York: W.W. Norton.
Hutter, Marcus. 2004. *Universal Artificial Intelligence*. Berlin: Springer.
Huxley, Thomas H. 1893. *Evolution and Ethics*. London: Macmillan and Co.
Jacob, Francois. 1977. Evolution and Tinkering. *Science* 196 (4295): 1161–1166.
Kipling, Rudyard. 1912. *Just So Stories*. New York: Garden City Pub. Co.
Leopardi, Giacomo. 1887. The Lonely Sparrow. In *The Poems of Giacomo Leopardi*. New York and London: The Kinckerbocker Press.
Locke, John. 1690. *An Essay Concerning Humane Understanding*. London: Eliz. Holt, for Thomas Basset.
Lorenz, Konrad. 1973. *Die Rückseite des Spiegels*. München: R. Piper & Co.
Marchesini, Roberto. 2016. *Etologia filosofica. Alla ricerca della soggettività animale*. Sesto San Giovanni: Mimesis.
———. 2017. *Over the Human, Post-humanism and the Concept of Animal Epiphany*. Cham: Springer.
Paley, William. 1836. *Paley's Natural Theology*. London: C. Knight.
Piaget, Jean. 1971. *Genetic Epistemology*. New York: W.W. Norton & Co.
Shubin, Neil. 2008. *Your Inner Fish*. New York: Pantheon Books.

CHAPTER 2

On Subjectivity

Premise

At first sight, living organisms differ from abiotic phenomena in that they are projected towards achieving their own goals: they have an intrinsic orientation that ensures that their turning-to-the-world is never passive, but always proactive. Animals, for the most part, add a further distinguishing factor to this: they organise their own motion, moving a multi-part body in their environment, with the help of sensory compasses. Other forms of *bios*, for example protozoa, are also endowed with motion, but animals, even those with the simplest neurobiological systems, take things to a further level. Even in some nematodes, which only have a few hundred neurons, somatic movement requires sensory adaptation systems. In this sense, animals have developed nervous networks which are responsible for processing incoming information. In fact, it is necessary to distinguish the landscape modifications produced by an animal's motion from those that derive from changes attributable to external events. This is why, when we look at different targets scattered over the horizon, we do not have the impression that the world is also moving.

This ability to separate the effects experienced when interacting with the world (*sensibility*) from the effects produced by one's own action on the world (*experience*) can be considered the basic requirement of a Self that reacts to phenomena and actively projects itself onto the world. Both sensibility and experience are mediated by the sense organs, that is an

R. Marchesini, *The Creative Animal*,
https://doi.org/10.1007/978-3-031-07414-1_2

animal's faculty of accessing present stimuli through species-specific receptivity. It is evident that if one can move one must be able to distinguish between different effects. The disjunction between the sensation that something is happening to the individual, as interference of the world, and the experience the animal obtains by moving or manipulating entities is achieved through complex mechanisms of neurobiological reaffirmation. On the one hand, this dichotomy transforms the animal into an entity capable of promptly detecting external phenomena, and thus into a *sensory subject*; on the other hand, it allows it to correct the feedback it gets when it moves around or grasps an object, so that it does not have the sensation that the landscape is moving or that it is touched *by* the object—that is, the animal is a *subject open to experience*.

The ability to differentiate a Self that receives and produces effects on the world from a non-Self that undergoes what happens to it can therefore be considered the first form of subjectivity. It is a basic condition that does not require complex neurobiological structures to be achieved, but needs them only to create a clear separation in the domains of receptivity. As we shall see, this property represents the basic core of subjectivity, a kernel capable of being enriched in more complex organisms through new predicates, such as affective states and biographical memory. Subjectivity manifests itself in the animal's fierce protagonism in action, which gives a sense of ownership to the individual in charting its own course, developing strategies and doing the necessary activities. We often say that animals are subjects, but we do not always pause to reflect on what subjectivity is. In my view, animal subjectivity is the consequence of the fact that these organisms are called upon to move an articulated body in an external milieu and to perform actions on the world.

Talking about subjectivity is never easy, since it is usually done by referring to the higher levels of cognition, namely to complex functions such as free will or self-consciousness. On the contrary, I would like to start from its simplest expressions, such as the property of distinguishing what happens in the world from what the individual produces. Such a predicate brings out one's Self: we can see it in a fly dodging our attempts to swat it, in the way a spider waits for prey, or in an earthworm wriggling to sink back into the soil. Subjectivity, in the most basic sense, is the faculty of distinguishing the Self through the effects produced by its own action. For example, consider the hummingbird hawk-moth (*Macroglossum stellatarum*): it can coordinate its flight with the motion of a flower, in order to insert its proboscis in the part where it can find nectar. It is evident that there is a need to calibrate the different afferences at play and not only its own motion. Animality is therefore feedback about one's own action: being able to reconstruct the perceptual returns when performing structured activities, such as spinning a web, digging a hole, weaving branches for a nest, making an ambush, learning, solving problems and so on. This requires a centre of confluence of information, that is an internal organisation.

Let us take another example. In order to identify prey, the praying mantis uses the sensory heuristic of the synchronous movement of its parts, which allows it to detect a grasshopper as a figure against the background, despite the latter's cryptic camouflage. But in order to follow the grasshopper's movement and coordinate its raptorial legs, the mantis must also move the axis of its head, directing its eyes on the moving target. Hence the clear need to differentiate the sensory feedback due to its own head rotation from that linked to the grasshopper's movement. The ability to receive sensations from events in the world and to distinguish afferents due to one's own movement is a primary requirement to be a successful predator. It is a basic faculty, indispensable for those organisms that must both carefully monitor any external interference and move about the world, manoeuvring a more or less complex apparatus of organs and readjusting sensory feedback. These are fundamental qualities in an environment where opportunities are never within reach and danger is always lurking.

Moreover, it is evident that this distinctive core leads to the emergence of a Self—understood as a self-referential dimension—endowed with its own register for processing incoming information. Therefore, an animal is subjective insofar as it can distinguish, from a perceptual point of view, between what it receives from the world and what it produces. This dichotomy creates the prerequisites for a Self to take charge—albeit not necessarily consciously—and interact with the outside world: a Self capable of emphasising both receptivity and activity. We can say, then, that the role of perceptual wiring is central to subjectivity, because it lays the foundations for sensibility and proactivity. Moreover, these aspects are intimately linked to owning inherent and self-defined purposes, such as defending oneself from danger or managing to feed oneself—traits that appear to us as the most authentic and explicit characteristics of being a subject.

In this chapter I do not want to talk about consciousness or the different realms of awareness. The reason is very simple: I do not consider it necessary to talk about subjectivity. The subjective condition is illuminated or made explicit by consciousness but, precisely for this reason, it must come *before* it. I think it is much more useful to look for the basic characteristics that make up what I have defined as *one's own*. Obviously, in more complex neurobiological systems, increasingly magnifying predicates, such as affectivity or consciousness, can be added to this primordial nucleus. However, the first form of subjectivity must come before these functions. Subjectivity rests on owning a specific perspective on external reality, which takes shape in the dialectic between feeling and acting. The coordinated movement of the body requires, in fact, a continuous work of sensory correction that represents the first rudiment of a sense of Self. And it is precisely from this faculty that one must start in order to understand subjectivity.

Emerging between feeling and acting, the individual transits through the world, bringing its—innate and acquired—demands and adapting them to the possibilities unfolding before it. The animal's actions display its protagonism, in that dialectic between neurons and muscle fibres from which it develops its own perspective, its own determination. Behaviour, in fact, is characterised by an internal motive that, though it can be influenced by external factors, seems to impose its own will or arbitrariness of conduct. Subjectivity and protagonism are linked together by the emergence of this perspective. When we observe ibexes ascending extremely steep walls or bears climbing snow-covered peaks, constantly at risk of slipping down, we are amazed not only by their obstinacy, but also by their iron resolve that, having set a goal, fights furiously against what could be the foregone conclusion of passivity.

Life, more than any other natural event, is an expression of the improbable; this is true of its own internal organisation, which is thermodynamically unstable, and of its energy-hungry functions and catalysts. It is only by being an open system, feeding on the sun's radiant energy and carrying it like a river along the trophic chain, that the biosphere can subsist. Life is a history of events whose echo continues through the generations thanks to the mechanism of heredity; it is the protagonism through which individuals adapt the environment to their own needs, through processes of active modification. It is autopoiesis, as we read in Humberto Maturana and Francisco Varela's *The Tree of Knowledge* (1992), and, equally, it is heteronomy or sympoiesis, as Donna Haraway suggests in *Staying with the Trouble* (2016). Animal behaviour is certainly the most explicit and evident expression of this darting into the territories of the improbable. The composite movement of an animal organism that flies, jumps, swims, crawls or walks shows us this will to oppose the outcomes of passivity, whereby subjectivity is always marked by activism.

Animality is endowed with agency, that is *industrious proactivity in relation to the surrounding context*, in the sense of being able to actively exploit the resources and possibilities offered by the environment. An animal does not bend to the laws of the world but uses them to achieve its own ends. To attain these functions it is not enough to have a membrane-like sensibility, a chain of reflexes or a system of tropisms, because it is necessary to coordinate in space-time, correctly processing the dialectic between external interferences and the consequences of one's actions. The concept of agency identifies the capacity to monitor the contingent situation in the here and now, to judge one's own actions, to anticipate future events, to set goals time after time, to make choices consistent with them—given the boundary conditions and the feedback received—and to evaluate one's own performance. These properties give the animal a proactive, selective and transformative role in relation to its environment. Agency therefore combines two basic functions: on the one hand, sensibility, the primary faculty of receptivity, and on the other, experience, the capacity to cognise the object of one's actions.

Subjectivity as a Perspective on the World

According to the Scottish philosopher Thomas Reid (1710–1796), it is useful to distinguish *sensation*, what happens to the subject, for example when an object touches us, from *perception*, what happens out there, when we reach out our hand to touch something. Changes due to one's actions are felt as one's own and not due to the world. Neurophysiological research today shows that perception is linked to the active movement of the subject, through the phenomenon of reaffirmation: it is clear that the functions of agency are realised through integration processes. Movement is also present in unicellular organisms, but it is in animals that it is configured as *peripatesis*—from the Greek *peripateo*, to stroll—as the ability to organise different parts of the body through the correlation of perception and proactive coordination. Even in the simplest organisms, it involves a neurobiological configuration that puts the somato-motor system in dialogue with the sense organs. The property of agency thus translates into the animal's *sense of self*: the ability to have one's own perspective on the world during activities performed for oneself. A subject is therefore the (not necessarily conscious) owner of one's own actions, sensitive to external phenomena.

The ability to act requires a dynamic reorganisation of the sensory feedback, capable of recalibrating the image of the landscape in the course of one's action, both in its becoming and in its specificity. The reason for this is easily explained: every action is carried out by setting one's motion according to precise implementation plans, such as grabbing an object on the fly or chasing it along a trajectory. For this reason, the sensory correction in animal motion is related to the activities typical of that species. Therefore, subjectivity is a subjective construction of one's perspective. Somatomotricity, even in the simple act of moving the body or articulating parts of it, such as the head or the eyes, imposes mechanisms of sensory adjustment concerted with the peculiarities of the movement itself. Having said that, it is clear that even in the simplest animals we never observe a mere translation of the body into the environment, but the performance of complete actions—picking a fruit from a branch, seizing prey, engaging in play, exchanging food—which consequently require correlated and specific sensory coordination mechanisms: namely, a primordial sense of a Self agglomerated around a perspective.

If we observe a single-celled organism, an amoeba or a paramecium, for example, we will notice a sense of direction—approaching or moving away from a source—regardless of the fact that the amoeba does so through pseudopods or the paramecium through the blink of an eye. In animal behaviour, on the other hand, we see actions being carried out according to precise coordinates, reminiscent of *verbal predicates*: think of a cat chasing something or a spotted nutcracker picking up a nut. These verbalities, which define the type of act produced by an animal, are sense-motor units that involve the individual in different circumstances, influencing the sequential unfolding of its perspective. Action is therefore an expression—from *ex-primo*, that is to bring out—that dialogues with the external environment but cannot be deduced from it. The animal's action takes on the aspect of a subjective event insofar as it transforms the environment into a space of agency, that is, into a sort of theatre where the individual plays its part according to directives *in fieri*.

Each of these activities requires a specific sensory reaffirmation and an equally specific motor organisation, developed in the very becoming of the action. In fact, the horizon of constraints and possibilities, of opportunities and risks, changes every time, thus making it necessary to correct the course of action on the go. The simple, albeit fundamental, mechanism of reaffirmation is not enough, because the individual must constantly review its route plan: it is peripatetic in that it is always on the threshold of feeling and perceiving. It is not enough to detect an opportunity, it is essential to assess its scope, to relate it to any possible risks, to define the timing of its achievement, to compare its advantages with other profitable activities. Very often an animal is faced with two or more possibilities, both of which are fundamental from the point of view of survival, such as escaping a predator or reaching a food source. The individual, therefore, is required to be able to act in such a way as to best negotiate the interests at stake—it is clear that this requires ownership of the action.

In an ever-changing context, this negotiation cannot be given a priori, but needs to be achieved on the spot. As John T. Bonner reports in his essay *The Evolution of Culture in Animals* "these flexible responses to the larger components in the animal's environment are of special significance because they will permit the animal to invade new habitats more readily and take on new sources of food" (1980, 147). Therefore, we must not consider subjectivity as a pure sensory distinction between a Self—or one's own sensory and perceptive dimension—and an external world. Subjectivity is protagonism in action. It combines: (1) the ability to interact with the environment, differentiating what I receive from what I produce, and (2) ownership in the choice of useful actions, defining the here and now in a singular way. Movement is subjective in that it indicates a certain way of interpreting the situation through a complex action that reflects both the context and the identity of the individual performing it. It is not, therefore, a question of merely approaching or moving away from an entity-event, nor is it a question of taking advantage of a resource at hand, which are very rare events in nature. Rather, it is a matter of ensuring, through an active contribution, that certain conditions occur.

Subjectivity, therefore, is characterised not only by one's generic protagonism in one's journey, but in presenting specific coordinates of action and being able to carry them out, in their complex sense-motor organisation, according to directives established in the moment. The environment is therefore interpreted, bending its characteristics according to particular angles and thus establishing a singular perspective: one's own *being-in-the-world*. The interference of external events and presences is thus perceived through a sensibility that not only registers the events taking place and the consequences of one's actions, but also projects their course, enabling anticipation. While consciousness is not required to do all this, it is certain that this would not be possible through simple automatisms. What is required is a supervisory—not necessarily centralised—function that correlates the various needs and afferences, bringing out a criterion of conduct.

Having a perspective also means anticipating events and, consequently, being able to project oneself into a condition that is not yet present and get an idea of what is going to happen or could happen. We know, for example, that animals that are very distant from each other from a taxonomic point of view can recognise the approach of an object through a comparison of the images projected on the retina. As Giorgio Vallortigara writes in his book *Pensieri della mosca con la testa storta*, "Animals as diverse as children, fiddler crabs, chicks, and rhesus macaques, placed in front of a computer screen showing a simple disc that suddenly expands, pull back as if to dodge it" (2021, 132). Animality is manifested, therefore, not only as a reaction to external stimuli present in the here and now, but as a capacity to anticipate events in progress according to parameters of probability, with physical integrity being an absolute priority.

Thus, what emerges is an entity with intrinsic values, supporting on the one hand the faculty of caring, that is of assigning meaning to the situations encountered, and on the other hand the propensity to make choices based on a future perspective, that is to inhabit a time not yet present. According to Peter Godfrey-Smith, in his essay *Other Minds* (2017), a nervous system evolves to cope with the coordination of the organism, but the result is such neural complexity that eventually other capacities emerge as by-products or relatively simple additions to what has been constructed in response to action-shaping needs. The peripatetic condition of the animal cannot therefore be superimposed on its being heterotrophic or on movement in the generic sense, but is rather due to the need to organise perception and motion by modelling the whole body in view of the goals to be achieved. Just as the animal is able to cancel or modify a sensory afference in order to recognise its own movement, so it can anticipate the effects of an event in progress.

The ability to attach value to present or probable conditions shows that we do not stop at immediacy but rather worry about the future. Whether to anticipate an answer or to make an event possible, subjectivity offers a new image of protagonism: owning a perspective not only as a peculiar angle, but also as the ability to project oneself into something. Subjectivity therefore means not stopping at the status quo, but rather thinking ahead, occupying a time to come and, therefore, not being totally encompassed in the here and now. We have the feeling of a subjective movement whenever we witness foresight in action, namely the act of escaping from the present condition, going beyond it through a projection towards the immediate future.

However, this propulsion into the future is not a simple unfolding of triggers, a cascade of mechanisms whose result is already given by the starting conditions, but is literally the creation of a becoming. By prospecting, the subject invents its future. It imprints a precise direction on the course of events, implementing a succession that cannot be deduced from the intervening variables, but is only influenced by them. If a subject has a purpose, its moving about in the world always has a value for-itSelf, that is, it does so in such a way as to achieve what it wants. The individual thus prescribes its own coordinates in its wandering about the world. As John Z. Young puts it in his book *Philosophy and the Brain*, "This incessant pursuit of aims is the essence of the maintenance of life. The search is not random but is directed towards the achievement of particular immediate targets that are pre-set for each creature" (1987, 3).

An animal is constantly faced with the need to make choices in the face of situations that are always presented in a singular way, that is with wide margins of novelty: both in the sequence of events and in the systemic organisation of the phenomena involved. In *Your Brain Is (Almost) Perfect: How We Make Decisions*, Read Montague writes: "The ability to evaluate different choices in the world emerged because it was an efficient solution to fundamental problems of survival—the most important problem being the need for flexibility in the face of uncertainty" (2007, X). Autonomy in constructing one's own future in the moment—an electivity that comes from the individual's perspective on the horizon of possibilities—brings out the animal's protagonism in the here and now. This feature, which perhaps more than anything else expresses its subjectivity, is a careful evaluation of the alternatives, with reference to a future into which the animal projects itself; it is a choice of the most appropriate attitude, or of one that is consistent with the projection, but also a correction of course. There is certainly a very close relationship between active expression—the animal's ownership in making choices and therefore its ability to assess the specifics of the situations encountered—and the concept of subjectivity.

In animals, even those with a simple neurobiological system, the ability to attend the here and now, that is to have experiences and therefore to behave subjectively, is widely demonstrated. Giorgio Vallortigara, in the book mentioned above, gives many examples of how it is not necessary to have particularly complex brains in order to have correlative behaviour, that is adaptation to the situations, through specialisation processes such as lateralisation and co-option, that is the reuse of certain areas for new functions. Thus, we read about bees that convert their ability to differentiate flowers into the ability to recognise human faces, antlion larvae that learn to anticipate the arrival of prey after just a few sessions and cockroaches that remember experiences they had in the larval stage. The reuse of behavioural modules in new contexts demonstrates that there is flexibility in the use of operational tools, so that it is possible to adapt them to the situation and release an expression from its canonical context of use.

Subjectivity is manifested in this protagonism of action, as the individual cannot avoid making choices and developing strategies to deal with problems that always include margins of unpredictability. Walking about in the world means embarking on a path that requires openness to the uncertain. It is necessary to inform one's own project by identifying the operational space available: acting in the context involves *thinking* about the environment and not simply adapting to it. An animal must necessarily take risks, and bring something new or non-experienced to its behaviour, if it wants to fulfil its needs. It is essential to try, but without falling into the combinatorial boom of possibilities: this inevitable consequence of acting based on random choices would be a strategy doomed to ineffectiveness. A subject's activity refers to a for-Self that guides not only its competence or adherence to previous information, but also its innovative capacity. A subject must be able to use the ability to bend situations to its own advantage, often through performative DIY and, sometimes, by coming up with an entirely new strategy to achieve its purposes.

The ability to have goals and therefore inherent values—what Henri Bergson (1859–1941) calls intrinsic teleology—is a fundamental quality that cannot be disregarded. Animal subjectivity appears to us, therefore, as a new level of the intrinsic teleology that characterises the living. In agreement with Geoffrey-Smith, I believe that the peripatetic property, that is the function of sensory and motor organisation, has brought forth a different plane of inherence: the sense of a Self endowed with a singular perspective. I do not think that this faculty should be ascribed to a particular neurobiological constitution, nor do I accept tautological simplifications, such as that of the *homunculus*. I rather lean towards an emergentist explanation: from my point of view, subjectivity emerged when living beings began to organise their bodies in space-time, trying to anticipate events and modify their development to their own advantage.

Subjectivity and Agency

As mentioned in the previous chapter, an animal cannot limit itself to repeating a previous instruction: it must monitor the specifics of the context it is experiencing and, at the very least, adapt its behavioural pattern to the novelty of the situation. The real world is made up of singular moments which, while appearing similar, have specificities that are most often crucial in the survival game. The margin of operation is extremely narrow in the biological context, and the slightest error can cost an animal its life: for an animal it is essential not only to identify the features that relate to innate mechanisms or previous experience, but also to analyse the novelty of the situation, in order to construct implementation schemes that adhere strictly to the particular circumstances. It goes without saying that this process cannot be based on the simple reproduction of preset skills but requires an inventive effort and above all a proactive presence in the here and now as it is experienced. Subjectivity and presence therefore seem to be two intimately linked features.

The capacity to attend the situation at hand, like a helmsman who is always alert and ready to correct the course, is what we call agency. Differentiating the afferences produced by actions from those concerning interfering phenomena produces a fundamental effect in this predicate of subjectivity: it gives rise to the present, as a condition based on a here and now marked out according to the species-specific rhythm of sensoriality. We can say, then, that the present—the animal's being—is a construct of subjectivity itself. We must not imagine the present as a prison for the individual, but rather as a treadmill that allows it to combine the past with the future, in an existential diachronic dimension. Presence in the moment is perhaps the most difficult ontological predicate to explain, because it means immersing oneself in a certain condition while also transcending it. The diachronic coordinates that create the present through a close dialogue between past and future concern the becoming of external conditions, the unfolding of the individual's actions, its past and anticipations. It must be said that the physiological state is also a variable of singularity, which influences the choices the animal makes in the moment, since its internal condition produces a fluctuation of priorities.

The combination of these singularities requires creative presence from the animal because, as Bergson points out, "As soon as we go out of the encasings in which radical mechanism and radical finalism confine our thought, reality appears as a ceaseless upspringing of something new"

(1911, 47). In order to construct an adherent behaviour, it is necessary to interpret the situation, to reorganise the present conditions in order to make them converge where the individual's projection places the project. As mentioned above, somatomotricity requires distinguishing between the reception of external phenomena, which intersect the individual, and the landscape modifications produced by the animal's motion-action. The mechanisms of reaffirmation, studied by Erich von Holst (1908–1962) and Roger Sperry (1913–1994), show that the animal's movement retroactively affects the perceptual systems by reshaping its image of the world. This produces not only a sense of Self, but also a *Dasein*.

A system based only on a membrane reaction or reflex produces a simple response to a stimulus, albeit sometimes organised along associative chains. Instead, when the subject's active movement develops, we find the distinction between feeling and acting which lays the foundations of presence. The condition of agency is to be in the moment by looking at it from the outside, that is through the resonance of the past and the projection into the future. An animal looks at the background world that surrounds it, and it is not just a landscape, but a field in which its action takes place. It is only then that we can speak of a clear distinction between feeling between the Self and the non-Self, the first sketch of a subjectivity which experiences the intersection with the world through an acted present. The tendency to interpret the moment in a prospective way therefore adds another character to animal protagonism: that of bearing a project or prospect, being capable of grasping the moment by subsuming different times in the here and now.

If perception is knowledge of external reality and not simply the disposition to react to stimuli, and if this property is the result of action and not mere exposure, it becomes clear that subjectivity is intrinsically linked to the construction of an operational project in the world. The echolocation of microbats does not only say something about a specific way of constructing an image of the world through ultrasonic echo, but also shows a type of hunting, a reconfiguration of auditory properties, the difficulties regarding the type of prey and the nocturnal context. Similarly, the ability of bees to perceive polarised light indicates the importance of always knowing the position of the sun as a mapping marker to reach food sources. The character of peculiarity is given by the animal's orientative endowments, whether they are innate or the result of experience, but always in a dimension where perception and operativity are linked by rings of recursiveness, as already pointed out by Jakob von Uexküll (1864–1944) through the concept of *Umwelt*. The animal's perspective relates the projective character of experience, that is being already preset on becoming and *being for the future*, and that of partiality, which can be traced back to what is one's own or referred to oneself. The subject's actions or, better, its agency makes the individual the bearer of a temporally prospective view of the world, capable of bending becoming itself, including the resonance of the past, to its own service.

This ownership of agency should not be considered in terms of absolute arbitrariness, but as an informed interpretation of one's own script. A cat playing with a ball follows its singular bounces around the furniture of the house, organising a specific prospect of becoming and adapting its responses to this, but it does so by interpreting its own predatory canons in a particular way. A dog is able to communicate by touching a target with its nose: again, the interpretation is a modified reuse of an innate behaviour; the same applies for tricks, or herding sheep. It is clear, therefore, that we cannot exclude the innate from the field of subjectivity, because orientation and becoming are based on previous proactive and reactive tendencies, the result of one's specific species identity. However, we should not consider these a-priori aspects as limitations to subjectivity, but rather as means of defining fields of agency. The species dimension informs canons of agency. Consider, for example, the engagement behaviour of wolf pups, the predatory sequence of wild cats, the feeding request of estrildid chicks: most often, these are not presented as fixed patterns, but as open ones, with ample room for operational flexibility or reuse in new contexts.

We are wrong if we believe that having innate instructions limits or hinders an animal's creativity and expressive uniqueness. We have seen how subjectivity manifests itself through perspective specificity in interacting with the world, which therefore defines the ownness of one's action, that is to say, being the bearer of values. An animal comes into the world with the dictate of bending reality according to peculiar self-referred perspectives: a wasp transforms the world into a landscape of waspness, that is, of values referring to its being-a-wasp, and the same goes for any other animal. It is evident, then, that subjectivity can express itself and develop precisely thanks to the pre-existence of a partiality of perspective. The innate is what enables the individual to take the first steps in the direction of subjective partiality. Its evolutionary heritage induces the wasp to consider the world through its own perspective, but it is the world that confronts it with the difficulty of being-a-wasp and the challenge of singularity.

Therefore, it is the uniqueness of each situation that allows and requires presence in the interpretation of agency and makes the real wasp emerge from the script of the virtually possible wasps. An animal is like an actor who only partially follows the script they have been given, and certainly does not passively adhere to the reactions of the audience: while taking both factors into account, they nevertheless add a touch of their personality and improvisation to their performance.

Subjectivity in the animal world is intimately linked to this operational condition that we have defined as agency. This endowment is affected to varying degrees in the various animals by higher-order mental functions. If reafference shows that a sense of Self emerges even in very simple animals, thanks to the fact that movement creates a clear distinction between the individual's protagonism and the background world, it is equally evident that, with the development of a brain equipped with nuclei responsible for affective states, subjectivity takes on new meanings. The affective condition has a direct link with feeling, in the sense of *feeling something*—such as yearning, fear, restlessness, anger, frustration—that is to say of experiencing becoming dispositions, correlated to the physiological states occurring in the body and to the events encountered by the individual. Affectivity, that is the capacity to be in an affective condition, is an even stronger sign of the presence of a Self that intersects the world, translating it into a subjective declination. In this case, too, I believe that affective states precede awareness, that is, they are such even when unconscious.

As the brain becomes even more complex, the database of information it possesses—we might say the overall memory of the system—grows in parallel, and that is when there is a further gap in functionality. Memory gives rise to an inner landscape of reminiscence, thoughts, maps and procedures that reproduce, always in deference to perspective partiality, what is present in the external world. Interiority thus translates into a kind of internal theatre where ideas and concepts can interact, allowing for processes of simulation and imagination. Mnestic functions are fundamental in animals that have to achieve individual adaptation to environmental fluctuations, because they are characterised by a low reproductive regime that does not allow for strict population selection. In these animals we notice that the a-priori aspects also become more complex, both in terms of evaluative and operational endowments, demonstrating that the innate does not limit or obstruct subjectivity, but is at its service. In fact, the greater the innate endowments, the richer and more articulated—that is expressed in terms of one's personality—the experiential life and learning of the individual will be.

The animal's affective and cognitive evolution shows that subjectivity is based on operative-disjunctive functions, bringing out a Self that reproduces the dialectic between inside and outside through the capacity to be present in the moment. Both affective states, which translate this dialectic into feeling something, and cognitive states, which give a mapped representation of it, are therefore functions of agency. Interiority is often confused with consciousness, and experience is defined on the basis of this condition. As I have said, this is a mistake. Consciousness, in its various intentional stages, is by definition the explication of something. In a sense we can say that, in its reference to something, consciousness recalls the act of illuminating a process, for example an emotional condition, a desire or a thought.

It is then evident that affective and cognitive interiority precedes its illumination: from the simplest to the most complex forms, it is not consciousness that produces the subjective condition, because all its expressions—from evaluation to decision-making—can be illuminated only insofar as they are already underlying this process. We must conclude, therefore, that subjectivity is the set of all the conscious and unconscious processes that bring out a Self as the author of its own becoming, insofar as it is endowed with agency. We can identify different degrees of subjectivity, such as (i) the differentiation between sensation and perception linked to reafference, (ii) the presence of affective states that characterise a psychological condition endowed with a for-Self value, (iii) the presence of a complex database that allows the mnestic function and the construction of an interior theatre reproducing the world. Subjectivity, therefore, is a basic quality of animality, an ontological dimension made possible by activity.

Evaluation as an Expression of Subjectivity

We have seen how the animal dimension is influenced by the fact of possessing an articulated body that moves in space-time. An animal must select sensory afferences, distinguishing the interferences produced by the world from the changes in landscape that result from its own actions. Motor coordination must also be expressed in specific verbal predicates, such as the chasing of a predator or the gathering of an omnivore—activities that require an articulated unfolding and, therefore, a projection or propositionality. Even in its most elementary forms, passing through the world and acting on external reality raises the need for a prospection in the near future to be monitored in the course of the action itself. We can say, therefore, that the property of agency is expressed in the prefigurative capacity and in the maintenance of evaluative attention, step by step, comparing the consequences of one's actions with the responses of the world.

Evaluation does not simply precede action, inspiring the choice an animal will make, but rather a process that accompanies all action. This is why I have defined subjectivity as attending to becoming, being present, presiding over the here and now. Considering, then, the complexity required in the operational performance of common behaviours, it is clear that simple course correction mechanisms based on feedback are not sufficient. It is always essential to enter into the specifics, so a percentage of creativity must necessarily accompany the operational becoming. In acting on the world, an animal never follows a linear trajectory, which is a simple synthesis of the variables present, but imposes its own course on becoming. Protagonism therefore indicates the unpredictability of the choice, or rather its taking shape at the very moment it is made. This implies moving away from simple mechanistic evolution. As Henri Bergson points out in his essay *Creative Evolution*, "The essence of mechanical explanation, in fact, is to regard the future and the past as calculable functions of the present, and thus to claim that all is given" (1911, 38).

For an animal, there is never a condition where everything is given, ready-made and linear. Action is like a dialogue, where every argument that is put forward also requires proactive presence: being ready to counter the counterpart's assertions and, if necessary, to take advantage of its divergences to improve one's own point of view. Behaviour, when deployed through complex verbal predicates—predation, in its various forms, is the best example to understand this—is confronted with a chaotic whirlwind of intervening internal and external variables. This is why the individual cannot avoid being present in the here and now and coming up with corrective measures for its project on the spot. Becoming conditions must be set up as field functions, that is as events presenting a range of possibilities with different probable outcomes. On this basis, even the project is never given, but is rather realised, taking into account the opportunities and risks that unfold in the course of the action. An animal, therefore, cannot avoid being present, that is building up a specific framework of possibilities in each situation, keeping the horizon of options open.

One of the fundamental aspects of subjectivity is, therefore, the ability to translate external conditions according to a self-referenced perspective and consequently to attribute values to the external world on the basis of one's own state. In order to do this, an animal is required to weigh what is in front of it on the basis of changing cost-benefit dynamics. Evaluation is primarily a matter of readiness to bring the development of the situation back to oneself, following a logic of relevance to internal needs. In other words, the subject must be equipped with what Montague defines "value labels" (2007). Evaluation is an unavoidable necessity in steering one's course of action and requires a set of interface conditions which are the more complex the greater the horizon of choice available to the individual. In principle, we can say that the ability to assess any situation in a precise and detailed manner can be considered a metacondition of selective pressure: in fact, only high-precision detectors have been able to overcome the adaptive challenge in the course of phylogeny. The evaluation of external conditions, of course, requires specialisations related to the environment and the lifestyle of the species, but the assessment capability should be considered a precondition of subjectivity.

An organism has many devices for detecting external factors—think of the sense organs—but it is equally sensitive to internal variables, through the endocrine system, the immune system, and the selective and dynamic activation of neurobiological centres. The entire external surface is studded with contact sensors, capable of reporting mechanical, chemical and electromagnetic afferences, thanks to subsequent nervous transduction. Sensibility is perhaps the most important quality in the first assessment of what is happening to the animal, and nociception allows it to promptly withdraw from sources of harm. In the different animal taxa, the neurobiological system has its own specificities: think of the ganglionic chain that characterises arthropods, the diffuse system of cephalopods or the encephalisation of vertebrates. However, the capacity to promptly detect damage to the integrity of the body by retreating shows us a particular character of animality, that is what the philosopher Jeremy Bentham (1748–1832) had already defined as the ability to suffer.

There is a very close relationship between subjectivity and suffering, so it is commonly said that, unlike objects, subjects have self-ownership precisely in relation to this being exposed to suffering, to use the words of Jacques Derrida (1930–2004). On the other hand, animal receptivity is not limited to nociception, but influences in various ways the individual's orientation and the type of response it will put into action. In this sense, the ethological school speaks of key signals, each appointed to open a certain behavioural lock. External stimuli can act directly to provoke a physiological reaction: for example, many pheromones induce orientation and readiness for copulation in insects. Some species, such as the monarch butterfly (*Danaus plexippus*) or the silk moth (*Bombyx mori*) are able to detect sexual pheromones and respond selectively even to infinitesimal concentrations. The Oriental fruit moth *Grapholita molesta* has abdominal glands that produce sex pheromones which are released into the air by flapping its wings, while the grayling (*Hipparchia semele*) not only does this but also traps the female's antennae in its wings to reinforce the effect. In these cases, it is more appropriate to speak of stimulatory effects than of a true evaluation.

A famous example, reported by Jakob von Uexküll (1864–1944), is the smell of butyric acid emitted by the skin of a mammal, which stimulates the tick to leave its bush and drop onto the host to feast on its blood. Direct stimulatory effects, in some cases even of a physiological order, are also active in mammals: for example. The pheromone of a male rat activates the prostaglandin mechanism in pregnant females, causing them to have an abortion. In reptiles, mammals and birds, but probably also in many other animals, alongside these mechanisms of direct induction, we see processes of influence mediated by affective states of an emotional or motivational order. In these cases it is not correct to speak of a direct induction, but rather of a psychological predisposition that brings together several influencing factors—what Konrad Lorenz called the parliament of instincts. Hence the individual's evaluative plan and the emergence of its own for-Self.

Subjectivity, therefore, also comes out in this ability to make choices based on an assessment of the various factors involved. It is not only a matter of weighing up the elements of the situation—for example, of considering possible risks and opportunities, or of calculating costs and benefits—but also of comparing them with internal predispositions, all in a dynamic overview capable of bringing becoming to the fore. Let us take an example. A moving ball undoubtedly has an influence on a cat's predatory motivation, but in order for the behaviour to become active it must meet the cat's appetitive condition. However, a cat that is initially uninterested might very well wake up from its torpor and start playing, just like a cat that is very excited at first might tire later on. We can say, then, that evaluation arises when the external stimulus is not able to produce behaviour directly but becomes an influencing factor in a wider range of elements to be taken into account—including the animal's predispositions. Here subjectivity is charged with a new meaning: the display of an electivity of state that regulates the individual's response. In other words, behaviour is subjective because it refers to the individual's momentary dispositions.

Having affective states, that is a motivational and humoral condition that changes over time, shows us subjectivity as self-referential dispositionality, that is non-homogeneous responses to events. Subjectivity is therefore antithetical to the condition of impartiality and objectivity that makes a course predictable and repetitive. Indeed, animal behaviour must avoid this condition, keeping within canons of competence—that is the species-specific repertoire and biographical identity—but always illuminating the evaluative control and operational definition through presence and creativity. All the influencing factors must necessarily enter a process of convergence or, if you like, of overall elaboration, in order to give rise to a state synthesis. However, this condition is anything but the achievement of a balance of forces, since the psychological condition of the subject is marked by the principle of becoming and cannot be fixed within a framework of persistence. Evaluation is an interpretation of the spectrum of possibilities in existence and not simply an objective photograph of the here and now. It is achieved through a creative act of projection into an unspecified future, certainly related to the influential variables but not limited to them, because it is conducted by the animal in a singular way at the moment it is living.

What remains to be defined is the psychological device that—like a compass—guides the subject in this creative act. The most qualified for this purpose is the pleasure system, which can be considered a multilevel centre, that is a system with several state conditions, in terms of both quantity and quality. There are, of course, more than one pleasure condition, but it is equally evident that they all contribute to defining the value of the different options available. On the one hand pleasure indicates the state of the subject, in terms of yearning or satisfaction, and on the other hand it signals the value of the consequences compared to the choices made. A pleasurable consequence of an action performed confirms the action and is therefore an orientation signal, that is an element that informs the subject about the direction of its aims. As Montague (2007, 118) reports, these signals consist "of two basic parts: a piece representing immediate experience (feedback) and a piece carrying long-term judgments about the future". We could say that the latter measures the gradient of possibilities (evaluation map) with respect to the value of different subsequent options.

A subject, therefore, is an entity capable of experiencing pleasure, and is led to orient itself through the hedonic feedback that follows its actions. Reversing the Benthamian principle, we can say that a subject is capable of feeling pleasure and therefore instructs its action to obtain the maximum possible hedonic feedback. If the judge of phylogeny is fitness, that is the reproductive value of a certain morphological or behavioural character, it is the pleasure principle that establishes the individual's orientation compass. The relationship between these two values—fitness and pleasure—is perhaps the key to understanding subjectivity. Following Darwinian dictates, a character asserts itself in a species because it is able to enhance the replication of its carrier. The result is very simple: in the next generation there will be more carriers of that character. But the presence of a certain trait produces a hedonic need in the individual: if you are a predator you enjoy chasing things, if you have wings you enjoy flying, if you are an explorer you love having no boundaries, if your diet needs sugar you cannot resist the scent of a ripe fruit.

Thus the fitness value achieved by a character is not the cause of behaviour but its consequence. An animal does not eat to feed itself, but *because it eats*, it feeds itself; it does not mate to reproduce, but *because it mates*, it reproduces; a bird does not have wings to fly, but *because it has wings*, it can fly. It becomes clear, then, that it is not fitness that causes behaviour, in its proximate meaning of unfolding, because that judge's only task is to flag that character as compatible and advantageous. Once a certain character is established in the species, it is the hedonic meaning that regulates its expression, in the sense of activating, modulating or terminating it. I speak of a hedonic need because pleasure always has antithetical counterparts, such as yearning, restlessness, appetite, discomfort, boredom and unease. Behaviour is therefore not necessarily driven by the search for pleasure, since most of the time what determines it is the attempt to extinguish these antithetical counterparts and obtain pleasure only as a consequence.

Some orientation signals are related to the physiological state of the individual, who can define its list of priorities on that basis: for example, whether to get food, rest or look for a partner. In this case, we note that any supposed fixed hierarchy of needs is arbitrary, because priorities change during the different phases of an animal's life. Moreover, a priority may simply emerge on the basis of unexpected events, such as the presence of a predator, a competitor or an unexpected source of food. An animal's condition is, therefore, one of standing at the window of the worldly horizon and constantly assessing the situation. The lived world is more reminiscent of a succession of occasions than a mere tropism directed by variables: an animal can never be absent from its factive presence, so that evaluative protagonism is a basic necessity rather than a high-profile function. The individual cannot act mechanically, in deference to already defined automatisms, but must attend the moment like a sentinel, acting with extreme flexibility in reference to the pleasure feedback.

Other orientation signals concern external phenomena with meaning, which can influence behaviour in line with innate legacies—such as *Gestalt* reconstructions of perceptions, key signals and emotional stimuli—or learned ones, such as search images, primes or emotional markings of particular entities/events. In animals that possess a central nervous system, the hedonic system is correlated with an affective state, so pleasure is not only an immediate feedback, but an expectation. As Jaak Panksepp and Lucy Biven point out in *The Archaeology of the Mind* (2012, 95–144), one of the most important factors in the affective state is the seeking system, which influences many motivational propensities through the dopaminergic system. When in the search phase, an animal does not simply exploit the resources present, but actively explores the surrounding environment trying to bring out, if not create, opportunities.

In a state of high motivation, the individual is more susceptible to the calls of the world, but it is not just a matter of evocation. Rats are gripped by a sort of exploratory frenzy, and also by a state of expectation that leads them not only to increase their attention, but also to interact with the environment, modify it and engage in playful behaviour. According to the authors, the activation of the seeking system is able to generate "expectant behaviors and euphoric-enthusiastic affects" (2012, 127) and to "go to new places where they are apt to learn" (2012, 135). As can be seen, evaluation can never be dissociated from the psychological state of the subject, so that, the authors reiterate, "the affective neuroscience vision is that all mammals are born with an urge to engage the world in various ways" (2012, 136). A similar situation can be found in the parental care system, governed by the epimeletic motivation and supported by oxytocin, which makes the individual much more self-confident and more willing to evaluate the social context as favourable. A completely opposite situation occurs with the activation of areas such as the amygdala, related to fear, or the insula, related to disgust, which involve a narrowing of the explored field.

Evaluation is therefore always subjective; we could say that it is a dialogue between internal conditions and potential opportunities in the environment, to which the subject must add that "director's cut" that allows it to interpret its singularity. Competition, which plays a powerful role in the everyday life of every living organism—to access food or escape from predators, in to defend one's territory as well as to reproduce—puts the animals face to face with the urgency of interpreting the situation at hand in a specific way, through the evaluation of all the factors in play and their significance with respect to their hedonic value. Assessing means attributing a different value to the circumstances and to the options available; we could say that it is a matter of identifying the gradient of coherence and adherence with respect to hedonic needs and of assigning a weight to the different possibilities through a cost-benefit analysis. Evaluation obviously implies the animal's ability to choose, that is its elective orientation, but it is also linked to the fact that existing situations constantly present margins of novelty that the animal must be able to manage.

The singularity of reality, and the possession of hedonic needs—which, moreover, fluctuate or contradict each other, for example when an animal is faced with the choice of whether to feed or flee from a predator—means that assessment takes place at the very instant it is carried out and that it continues while the action is taking place. Moreover, not only does an individual have to assess what is in front of it—if, for example, the risk of being preyed upon is so high as to outweigh the benefit of finding a food source—but it also has to judge which behaviour presents the best level of efficiency. Judgement is the act of weighing one's actions, also through simulation and counterfactual processes, considering the possibility of tackling the problem through several tactics presenting not only different levels of probability, but above all different degrees of efficiency, that is of resource saving. A choice is therefore the convergence of assessment and judgement, but in any case the animal can never operate by mere repetition.

The Subject's Protagonism

Directionality, self-organisation and autopoiesis reveal qualities that clearly differentiate the biological condition from other physical and chemical phenomena. Indeed, many authors have gone so far as to hypothesise the presence of a vital substance, or at least something that transcends matter, to explain this radical gap. In fact, however, it is not necessary to add external factors in order to admit that there are supervening qualities: this is demonstrated by the properties of emergent systems, as discussed in Steven Johnson's *Emergence* (2001). According to the emergentist theory, when matter takes on particular configurations—by introjecting organisational information and a systemic structure—and when it reaches a certain critical mass and number of interactions between units, characters arise that are not present in the parts that constitute the system. In the phenomenon of life, subjectivity can be considered a second level of emergence.

Among other things, I believe that the assumption of a vital principle to explain biological phenomena risks leading us to circular reasoning or petitions of principle. All expressions of the living world are based exclusively on physico-chemical factors and there is no need, therefore, to imagine supplementary elements. However, as Konrad Lorenz argued in his *Intervista sull'etologia*, if it is true that all vital phenomena are physico-chemical, it is also undeniable—and this is what determines their originality—they are something different and much more complex than that (2016, 108). As we shall see, it is more coherent to consider protagonism as an emergent property rather than related to some hypothetical intangible components. To quote Montague, "Biological systems possess layer upon layer of physical structure and hence layer upon layer of information-processing levels" (2007, 30). Therefore, we evidently cannot ignore these qualities that show the protagonism of the living.

A phenomenon appears to us as objective when it passively follows, we might say, the laws of physics: it does not change its trajectory on its own, it is not endowed with autonomy, it does not struggle to achieve a goal, to give just a few examples. If we compare inanimate objects to the acrobatics of birds in flight or the pirouettes of dolphins—that is their ability to exploit physical phenomena to impose their own operational project—we cannot fail to notice a significant difference. That is what Lorenz calls their originality. Yet it is never so simple to identify a precise threshold where subjectivity begins and ends. The concept of arbitrary decision-making is also not easy to define, even if we abandon the chimerical image of free will, because the animal world is much more articulated and complex than our claims to identify universal laws. The presence of inherent purposes certainly shows the animals' partiality in their intersection with reality, which more than any other character strays from the objective domain. An animal is a subject insofar as it is the bearer of a-priori interests that transform the world into a field of action for its protagonism, interests that are translated into hedonic requirements that the individual must meet.

Having the faculty of agency, and thus being the bearer of inherent values, presupposes a compass in the animal's peripatetic trajectory which must obviously be based on innate or learned contents, that is a database of references; it also requires the ability to use them not in a repetitive way, but by constantly referring them to the pleasure principle. This orientation has found many explanations over time, such as the theory of tropisms by Jacques Loeb (1859–1924) and the chain of reflexes by Charles Scott Sherrington (1857–1952). However, animal behaviour cannot be explained exclusively as an orientation determined by gradients or by a cascade of triggers, otherwise we would not be able to account for phenomena that contradict the domino model, such as the pre-activation highlighted by Karl Spencer Lashley (1890–1958). The American neuropsychologist demonstrated, in fact, that during an expressive sequence some behaviours can precede the normal order of unfolding. Therefore, if subjectivity manifests itself in the faculty of orientation, we would be wrong to superimpose the former on the latter, because a direction is not enough to define this attribute: an operational plan is necessary.

The structuring of behaviour is based on two main requirements: (1) the need to conform in the unfolding of the behaviour and (2) the maintenance of an irreducible margin of unpredictability in the course of the action. The first is given by the need to adapt the action, step by step, to the specific singularity and to the equally singular evolution that the scenario presents and assumes under the blows of animal operation. The latter is a process of correlated unfolding, which takes into account the consequences arising during the operation and is achieved through systems of correction and redefinition. A cat playing with a ball cannot simply set up its action to the initial condition of its target via predefined behavioural patterns but must constantly reconfigure its action on the basis of the consequences that unfold during the game. The construction of the operational plan is therefore an ongoing process that requires a system of superintendence.

In addition, in order to adapt the action to the margin of novelty of the situation, and to dynamically and accurately change the course of the action, it is necessary to rely on endowments which proportionally provide for a certain degree of virtuality, so that they can be useful resources which are not bound by their canon of use. In this respect, we must recognise that creativity can take many forms: not only inventing a new intervention strategy, but also reusing an endowment and applying it to a new context. Of course, this process of co-optation is possible if, and only if, we presuppose the presence of a subject, that is, of a function capable of supervising action and evaluating the operative field as a place of action. We must not, however, consider the subject as a sort of *homunculus* that observes behaviour—such an explanation would condemn us to an infinite regression—but rather as a synthesis function, capable of evaluating the consequences of actions and of using endowments as useful tools.

Creativity can also be developed by using previously unexploited potentialities of the operational equipment: in this case we are not talking about co-opting the resource to new operational areas, but rather about new ways of using it in the same area. Here evaluation acts as a reconnaissance of the potential of the available resources. On the other hand, if the experience is presented through situational-evolutionary singularities, the action must also conform to expressive singularities. This can only be achieved by assuming the presence of open programmes, that is programmes endowed with operational redundancy (there are more options available) and with implementation degeneracy (more ways of achieving a given result), the combination of which can yield new ways of dealing with the situation. This is what happens, for example, when using a topographical map, which allows a plurality of itineraries to reach the destination. The itinerary, that is the procedure related to the margin of novelty that the situation envisages, is inherent in the structural characteristics of the equipment (the map) but also by virtue of the presence of an operational function: the user.

When we speak of adaptation and correlation we are referring to the unpredictability of the resulting action, given the need to conform the behaviour to the specific nature of the situation. Animal protagonism lies in the making—that is not being already given—of the operative scheme, so the individual does not merely repeat but invents, though not from scratch, its own intervention model. Subjectivity appears to us, in this case, as the capacity to conform one's state on the basis of both the dynamic condition of one's aims and situational novelties. On the other hand, there is also a need for singularity that does not have a conformational explanation, but concerns the animal's ability to deal with highly critical situations. I am referring to the second point, namely the need to maintain an irreducible margin of unpredictability in the course of the action itself. To take a good example, in the competition between prey and predator, it is essential to acquire skills and behavioural strategies, but it is equally necessary to keep these operational programmes open, that is unfinished, not only to adapt them to specific circumstances, but also to endow them with irreducible unpredictability. In fact, in order to be valid, any strategy must maintain a margin of operational uncertainty even on the part of the executor, who may otherwise unwittingly reveal any pre-established moves.

Animal protagonism in operational practice therefore involves four fundamental aspects, which are always present even in very simple situations that do not require particular solving skills. We have: (1) the ability to *evaluate* different situations in terms of coherence-adherence to the individual's hedonic needs; (2) the ability to *judge* one's actions through simulations, comparisons, counterfactual projections of consequences; (3) the capacity to *conform* behaviour on the basis of situational specificity and evolution; (4) the capacity to maintain an irreducible margin of *unpredictability* in the programme, which makes it possible to conceal the individual's strategy against a competitor. Subjectivity is thus shown to be a space for arbitrary decision-making that the animal holds in its movement in the world, in the protagonism it manifests in carrying out its activities, in being endowed with an agency that allows it to make choices.

There is no doubt that phylogenetic inheritance, contained above all in the genetic code, overrides constitutive and functional processes capable of defining the specific directionality of living organisms—the teleonomic aspect proposed by Jacques Monod. On the other hand, it is undeniable that hereditary information cannot be explained as a simple series of learned processes because, as Charles Taylor pointed out in his essay *The Explanation of Behaviour* (1964, 57), the agency of living beings should be considered in its purposeful value, that is where the action is not preconstituted in its development but takes shape in the moment in such a way as to achieve its goal as circumstances change. The animal, then, explicitly shows its title of agent by being endowed with movement, expressing its ability to move closer or further away from a target, to deviate and choose its own trajectory, to bring out hidden opportunities and to modify its environment. The animal's agency does not manifest itself in simple correlative aspects, given by the correction of its behaviour when environmental conditions change, but also concerns anticipating opportunities and often making them possible, through expectations and niche constructions.

Clearly, being endowed with purpose and presenting a kind of intrinsic teleology are characters that help us describe living organisms and animals in particular. However, these same predicates risk leading us into dangerous explanatory circularities when we try to use them to explain characters that, conversely, deserve in-depth analysis. The search for a framework for the meaning of agency in the animal world, going beyond the genetic determinism of the first formulation and the environmental determinism of behaviourism, has characterised ethological research, especially through authors such as Niko Timbergen and Robert Hinde. However, this research has not always been able to create a consistent framework comprising both approaches. A third path was taken by Konrad Lorenz when, in the second half of the twentieth century, he enacted a profound revision of the innate/learned relationship through the concept of the innate instructor, that is phylogenetically adapted teaching mechanisms that allow the individual to build its own identity and a subjective style of behavioural expression through experience.

Ernst Mayr (1974, 650–659) takes up this subject by distinguishing the mechanisms of phylogenetic inheritance into two categories: closed programmes and open programmes. The latter allow the animal to actively and selectively search the environment for certain types of information that will influence the formation of its identity repertoire. In other words, we could say that the innate is not binding but, on the one hand, plays the role of first teacher in accompanying the individual into the world, and, on the other, acts as a pupil ready to grow through the teaching that follows. Inheritance ensures that interaction with the environment does not turn into a passive being exposed to the world, but makes experience possible. At the same time, the evolutionary condition of open programmes allows for knowledge expansion via modification of the past—a process that Jean Piaget called genetic epistemology. As can be seen, according to these authors, agency cannot be dissociated from the characteristics of the individual's identity development, that is the ability to avoid both genetic and environmental determinism. Animals' ability to use hereditary and environmental information to construct a self-centred singularity is what most suggests the presence of a state of subjectivity in them.

Animals show that they know how to be present, moment by moment, in their existence, through their protagonism in the interpretation of both the phylogenetic canon and the environmental conditions: in the first case, they do so by avoiding the simple repetition of closed instructions, in the second by selecting and reconfiguring the contextual variables. At the same time, animals manifest a full ownership over themselves, a sort of self-ownership that can be essentially traced back to the pleasure principle. There is more than a simple release from determinism and therefore a certain degree of operational freedom, but rather a principle of self-revolving power, which animals use to maintain and extend their range of ownership.

In Conclusion

To conclude this line of reasoning, subjectivity does not have a specific quality that is capable, by itself, of defining its character. If we analyse the different connotations outlined above, we realise that none can ensure the status of subjectivity on its own. For example, having an orientation or the capacity to reflect the existing situation does not necessarily testify to a condition of subjectivity, or at least it does not unequivocally prove its presence. The same can be said for having responsive properties or producing functional specificities or, again, showing a state partiality. Only the combination of these characteristics developed in an operational key is capable of producing the dose of creativity that sustains the animal's protagonism in interpreting its behavioural script in the here and now. At the same time, it is becoming increasingly clear that subjectivity is not a property found exclusively in higher animals or in highly complex neurobiological systems, but a basic condition of animal life. In this sense, it is necessary to distinguish subjectivity from consciousness, with the latter being called upon to illuminate the states and functions of subjectivity itself.

Works Cited

Bergson, Henri. 1911. The Evolution of Life. Mechanism and Teleology. In *Creative Evolution*. Trans. Arthur Mitchell. New York: Henry Holt and Company.

Bonner, John Tyler. 1980. *The Evolution of Culture in Animals*. Princeton: Princeton University Press.

Godfrey-Smith, Peter. 2017. *Other Minds: The Octopus, the Sea, and the Deep Origins of Consciousness*. New York: Farrar, Straus and Giroux.

Haraway, Donna. 2016. *Staying with the Trouble*. Durham: Duke University Press.

Johnson, Steven Berlin. 2001. *Emergence: The Connected Lives of Ants, Brains, Cities, and Software*. New York: Scribner.

Lorenz, Konrad. 2016. *Intervista sull'etologia*. Ed. Alain de Benoist. Sesto San Giovanni: Oaks Editrice.

Maturana, Humberto, and Varela Francisco. 1992. *The Tree of Knowledge. The Biological Roots of Human Understanding*. Boston: Shambhala Publications.

Mayr, Ernst. 1974. Behavior Programs and Evolutionary Strategies: Natural Selection Sometimes Favors a Genetically 'Closed' Behavior Program, Sometimes An "Open" One. *American Scientis* 62: 650–659.

Montague, Read. 2007. *Your Brain Is (Almost) Perfect: How We Make Decision*. New York: Penguin Group.

Panksepp, Jaak, and Biven Lucy. 2012. *The Archaeology of Mind: Neuroevolutionary Origins of Human Emotions*. New York: W.W. Norton & Company.

Taylor, Charles. 1964. *The Explanation of Behaviour*. New York: Routledge & Kegan Paul.

Vallortigara, Giorgio. 2021. *Pensieri della mosca con la testa storta*. Milano: Adephi.

Young, John Zachary. 1987. *Philosophy and the Brain*. Oxford: Oxford University Press.

CHAPTER 3

The Role of Heritage

Premise

As children of a very long history that has forged precise characters of adaptive conformity in each species, we animals come into the world not only with a body predisposed to live in a certain context, but also equipped with the expressive endowments needed to combine our somatic structure with the challenges we will encounter. Just as the gills of a fish speak of oxygen dissolved in water and the wings of an eagle of the swirling thermal currents, so the behaviour of an animal is an echo of its ancestors' past experiences. The environment is inscribed in the intimate fabric of the body: in the morphology of the organs and the rhythm of their functions; in the unfolding of proteins in a tertiary structure as well as in the sequential unfolding of germ layers; in the orderly migration of neurons to reach their sites and in the activation pathways of neuromodulators. Thus, even the body's behavioural expression involves information that precedes experience—a having-been-there-before state—and the animal's responses to the stimuli and opportunities presented to it follow preset grammars, in terms of both the peculiar sensitivities and the choreographic structures put in place.

We are talking about more or less complex instinctive reactions, such as the grasping reflex in the human infant or the rooting reflex in the puppy dog, but also about an evolutionary chronoprogram that includes periods of particular sensitivity to experience, such as the imprinting phase in

R. Marchesini, *The Creative Animal*,
https://doi.org/10.1007/978-3-031-07414-1_3

anatids or the socialisation period in carnivorous mammals. It is fascinating, and undoubtedly moving, to observe how a hatchling immediately looks for a reference figure to follow, to see a blind puppy reach for the mother's teat, to witness the first flight of a nestling, the emergence of a lepidopteran from the nymphal cocoon, the first flicks of the tail of a fish or the mad dash to the sea of a little turtle. There is something magical about the spontaneous, yet immediate, expression of one's own nature, but all this is easily understood if we embrace the evolutionist explanation. The body bears the signs of a very distant past that speaks of common heritages (i.e. of more or less close relatives and ancestors still present in the depths of each of us), and of adaptive convergences (i.e. of refinements brought about by selective pressures), which reveal the challenges and opportunities with which our lineage has had to contend in the uninterrupted timeline leading up to us.

The body is a serial expression of belonging: being a wolf, belonging to the canid family and then to the carnivore order, being a mammal, a tetrapod that took its first steps on land as an amphibian and then as a reptile. The body somehow subsumes this story, even if parts of it have been lost and certain passages have been emphasised. Behaviour is nothing more than the unfolding of this memory hidden among various somatic textures, a being-present, suddenly in the world, through confused echoes that yet perfectly match the environmental fabric in which we are thrown. The memory of our phylogenetic past rests on a thousand layers, so intertwined that it is never possible to pinpoint the main source, because our presence is essentially being a body. Its sequential and inclusive development, whereby each phase instructs the next, not only determines the development of a uniform morphological organisation, such as to characterise the members of a particular species, but also implements specific physiological performativity that then translates into specific behavioural patterns.

Phylogeny is a long history of meticulous refinement of the characters of different animal populations, capable of correlating a species to external host factors, making it dependent on certain environmental presences-variables. Specialisation, in fact, can be interpreted, at the very least, from two points of view: (i) the ethological one, which highlights the acquisition of adaptive skills, focusing on the performance shown by that population in the difficult game of survival; (ii) the ecological one of conjugation, which conversely shows us the dependencies that a species assumes by binding itself to particular environmental conditions and lifestyles. The evolutionary history of an animal taxon, on the other hand, does not see the population as a passive entity subjected to the environmental grind: animals move around and colonise new territories, develop sexual orientations, modify their living context and adaptive milieu and invent strategies or tools by shifting selective pressures onto technomediated performance. Phylogeny, therefore, is undoubtedly a path of mediation with respect to environmental challenges, but also a path of animal protagonism and accumulated experience.

In some ways we animals are born old, because the conformation of our vital and expressive systems—for example the structure of the sensory system, the endocrine and immune apparatus, the basic neurobiological organisation—reveals and transposes past experiences into the present. These are resonances of events that have not been experienced individually but are nevertheless vivid, memories that are not ours but are grafted onto us by phylogeny. This wealth of a-priori experiences makes us such that we open up to the world as if waking up from a dream: certainly not the Heideggerian being thrown into the world without any anchor or connection, but rather the confused recollection of a shared past. If we observe the physiological and behavioural development of a body, we realise that marking a starting date is as fictitious as it is arbitrary, because the body travels through time. Each individual is part of a larger story that goes beyond the individualised and particularised image of the self. Perhaps we should revise our merciless judgments of the anamnestic theory expressed by Plato in the *Menon*, obviously disregarding the implications about metempsychosis. Reminiscence is, in fact, the fruit of heritage and the shared history of the living.

The difficulty in interpreting this heritage lies in the meaning we often attribute to it, considering the innate as a diminution of expressive ownership that limits the subject, a sort of destiny or determinism that would condemn it to the blind repetition of preformed automatisms. And yet, if it is true that some instinctive displays leave little room for protagonism and resemble reflexes—no more and no less than the perineal reflex in puppies or the knee-jerk reflex in humans—it is equally undeniable that a large part of innate competence should not be considered in terms of reactive automatisms. This is an error of judgement due to numerous distortions. I will only mention two of them: (i) the prospective bias, whereby the further away an entity is, the less detectable are the differences that its representatives present; (ii) the epistemological assumptions of the research itself, because if we look for uniformities, we will classify variations as noise and we'll adopt a rating scale suited to bring them back within the average.

It is more correct to imagine heritage as construction material—interpretative and choreographic schemes that must be applied, enriched and subjected to specific refinements—available to the individual. Thus the innate is presented in the guise of a useful resource, a tool at the subject's disposal to create *ad hoc* behaviours suited to the situation. On the other hand, when we observe the behaviour of an adult animal and find a uniform repertoire, for example the dog's submissiveness, we often overlook the fact that such choreogram is due to the mother's teaching, which is why puppies that have lost their mother do not show it. In this case the heritage requires finishing through parental modelling. The same can be said for the song of the zebra finch (*Taeniopygia guttata*), for predatory behaviour in cats, for sexual imprinting in anatids and so on. At other times, the innate component comes across as a tool with a wide range of operational possibilities and therefore open to reuse in new areas, as a solution recipe or as a creative redefinition.

Phylogenetic inheritance does not, therefore, limit protagonism: it allows it. This subjective capacity is the basis for the local adaptation required of the individual, provided that one does not believe that subjective singularity equals expressive indeterminacy and requires starting from scratch. All that has preceded us is still *in* us, but it does not act as a constraint or predetermination: it is the germ that leads us to be protagonists in our own intersection with external reality. Thanks to heritage, we do not passively adapt to the world, but we rather bend it to our needs for conformity. The innate precedence of the individual, who already expects a given scenario at birth and deals with it by selecting, modifying, responding to or facilitating certain phenomenal inflections, distances it from any passive adaptation to external forces, be they persistent or circumstantial and incidental. Precisely because they carry a proactive legacy about the world, animals escape environmental determinism.

This, moreover, does not mean that the innate imposes an authoritative condition, since its task is mainly copulative. The innate is a call to action, languishing in the world, the urge to dialogue with external reality as soon as possible in order to update the system. It is thanks to heritage that an individual enters into dialogue with everything around it. Of course, this dialogue is developed through electivity, but still requires a welcoming attitude. The innate dictates our first steps but, immediately afterwards, it urges us to walk—through processes of variation, completion, adaptation—so as to bring out an expressive thirdness capable of escaping both hereditary and environmental determinism. Evolutionary heritage is only the script of what is to be staged but, then, it is the actor who brings it to life through its own protagonism. This is the process of *subjective interpretation* that the individual makes of its species canon and of the environmental situations it will encounter on its journey through life.

Being a Body

Phylogenetic inheritance allows the individual to open up to the world thanks to a set of basic endowments that enable it to play the dual role of belonging and individuality. An animal can thus benefit from the comfort of having a behavioural *residence*, that is a predefined dimension of living, a sort of welcoming nest capable of filtering its relationship with the world. At the same time, having a heritage means having endowments that enable it to be expressive: to be able, that is, to construct new adaptation events. Heritage is therefore the collision of different times in the individual, a process that does not hinder protagonism but makes it possible precisely by multiplying the actors of influence. The individual, in fact, is at the centre of several directives—linked to metabolic, motivational, emotional, elicitative, stimulative, representational and other aspects—and the very fact that it serves a large number of potential masters actually gives it control over its choices. The body is the substance of this dialogue between influences, so we would be wrong to consider it a mere theatre of the self. We animals do not have bodies: we *are* bodies, capable of constructing a present by collapsing the past onto the unpredictability of the future.

The body is hunger for the world and for time, it is wonder and excitement. Animality is manifested, in fact, in being a body: a structural and functional organisation that owes its internal order and its continuation in time to external contributions that oppose thermodynamic degradation. A body, therefore, is not a closed system, driven by preordained cascade automatisms, something autonomous and autopoietic, that is "finished" from a functional point of view. A body can never be enucleated from the context and the flow of occurrences. The somatic dimension is a field function: it is certainly part of a matrix of recursive and self-organised relations, but it presupposes a continuous exchange with the external world. In fact, it owes the maintenance of its order to dissipative processes, which stabilise it without, however, fixating it in an invariable condition. The body is first and foremost a metabolic function, that is the transformation of chemical elements that change the very conformation of the organism, and do not merely consume energy. Precisely for this reason, the body traverses time, it is not static, and every somatic expression reverberates on the system and changes its structure. The body, therefore, absorbs what is external to it and internalises it in the form of organisational information.

This being-in-time of the body, the impossibility of fixating it (keeping it) in a synchronic condition, represents the distinctive aspect of the somatic dimension, which is always a projection towards a possible future. The animal is projected towards something to come, almost magnetised towards the future: it is expectation and aspiration. The protagonism of the body lies in the dynamic dimension of its constitution; its becoming transforms the past into a working hypothesis that continually requires to be expressed through functional correlation. The brain is perhaps the clearest expression of this metamorphic dynamism or *structural metastability*, since every psychological condition that is voiced in a behaviour also modifies its synaptic organisation and neuromodulation setting. That neuronal set, the activation of which produced a certain behaviour, is no longer present in its original configuration after it has been actualised. The world is schematically reproduced and constantly redefined in the organisation of the somatic matrix.

On the other hand, even this prevalence of actualisation—the feeling that its existence is primarily a celebration of proceeding by converting spaces of virtuality—does not grasp the animal's diachronic condition and the retrospective gaze that dictates its path. By restricting the focus to its here and now, we cannot understand it, because every animal expression is also a resonance that takes on new life through the advent of the individual in dialogue with the world. The body is fulfilled thanks to a formative process that vivifies the information coming from the past, making it flesh that acts on the world, probing its possibilities. The animal dimension is, therefore, like a time machine that cancels chronological distances. Even the stringent position in the present is only an appearance: a body is an entity that emerges by exploiting the adaptive experiences that preceded it in order to cross over into the proximate dimension.

The emergence of the individual would not be possible without this continuity with the past, since all learning must be based on a past and cannot be constituted *ex nihilo*, being the revision of a hypothesis and not a simple association between a stimulus and a response. As Stanislas Dehaene points out in his essay *How We Learn*, "Any learning (…) contains, in one way or another, a set of assumptions (…) The more innate assumptions there are, the faster learning is" (2020, 52). The body is therefore an information-recycling system that allows the individual to construct a framework of adherence to the world, that is, its own existential style. Starting from a set of tested assumptions that define the probability gradient of an event, and therefore a predictive compass, an animal constantly compares the innate component with the results obtained and uses it as a picklock on the bolts of reality. In this sense, it is clearly thanks to the innate that the subject can express its creativity.

Life in its general aspect is, therefore, a process of production and accumulation of information, which circulates on multiple planes of storage, such as the structure of the genome, the mechanisms of epigenetic translation, the cascading sequences of developmental processes and finally the subject's learning. The body is the seat of this flow, which runs uninterruptedly from the past to emerge like an underground river in the resurgence of the individual. The subject is therefore only a moment in this trajectory, the instant in which history becomes flesh to produce new events in the unfolding of its tale. On the other hand, it is precisely somatic translation that makes sense of phylogenetic content, thanks to processes of evolutionary self-organisation. If we consider the development of a complex brain, as found for example in mammals, we discover that DNA cannot in fact contain the planimetric information to detail the highly specialised circuits of the central nervous system. Dehaene speaks of "brain highways" constructed through the metabolic function itself in the embryo-foetal period, thanks to axon exploration of the neuron. This exploration is "guided and channelled by chemical messages, molecules whose concentrations vary from one region to another and which act as spatial labels" (2020, 106).

The somatic dimension is, therefore, only apparently defined by a packet of heritage information, since it is not possible to separate form and function, also because its matrix is constantly made through metabolic expression. The body self-organises through proximity gradients and topographical arrangements. The emergence of a structural plan also takes place according to certain rules of complex systems, such as the construction of hexagonal cells, pyramidal structures and fractal compositions. According to Alan Turing (1912–1954), this self-organising phenomenon can be explained as the outcome of local amplification and inhibition at a greater distance. Hereditary information already predicts certain rules and processes inherent in the world and likewise relies on the functional consequences of somatisation itself. The body is therefore the condensed product of a complex set of dialogues between inside and outside, and this is true for all functions, including behavioural ones.

Behavioural development in mammals and birds, as well as the corresponding growth and wiring of the different neurobiological areas, is affected by the individual's experiences, especially during the sensitive periods of development. If a mammal has one eye permanently blindfolded from birth, the corresponding occipital visual area will be invaded by afferents from the other eye. The structure of the body therefore strictly depends on the unravelling of its functions and not on a mere translation of genetic information. Somatisation is a process which, assessed on the level of its autopoiesis, appears miraculous, so improbable are the architectural directives it follows. But, as said, if it is true that everybody is realised according to a structural plan encoded by the heritage information, this information is neither autonomous nor self-sufficient to reach the phenotypic dimension which, in the end, is responsible for fitness itself. In this sense we must admit a heteronomy of the body, a condition of being already implicitly predisposed to external norming.

On the other hand, it is indispensable to clarify the body's reliance on the influences of the world, avoiding fluctuations and occasionality, so that we cannot simply speak of an additional process or of completion: heteronomy is something very specific, like the receptive niche of a species. An animal relies on external content because the heritage information itself expects this input. Therefore, somatisation is nothing but the emergence of a very special dialogue between the body and the world. And in the condition of being a body there is no hierarchical directionality, because every developmental or expressive process always makes use of circularity, recursive effects and emergencies coming from the various apparatuses. It is difficult, for example, to say whether it is the endocrine system that controls the neurobiological one or vice versa, but it is just as difficult to separate the two systems physiologically: think for instance of the relationship between the hypothalamus and the pituitary gland. Behaviour is an expression of the body and fitness is the result of this performativity.

The genocentrism of the second half of the twentieth century turned the body into a simple container for replicators—see, for instance, Richard Dawkins' essay *The Selfish Gene* (1976). This led to phylogeny being considered as a kind of algorithm, as in Daniel Dennett's *Darwin's Dangerous Idea* (1995). On the contrary, the new readings of evolutionism, based on the evo-devo plan and on the theory of niche construction, show us that the body has a central function in outlining evolutionary events. Also, time can no longer be thought of as an irrelevant constant, because as Stephen J. Gould (1941–2002) has already pointed out, life on Earth has a historical dimension. The life that the individual undertakes inevitably develops its own story, albeit starting from the legacy of experiences that preceded it, and this story, which the being carries out with biographical protagonism, has more or less extensive repercussions on the phylogeny of the species and often even on ecosystem balances.

For a long time, attempts have been made to remove anecdotal evidence from the study of animal behaviour, and there is no doubt that individual accounts have more than one flaw when it comes to defining species-specific behavioural repertoires. A good rule for the scientific validation of research is, in fact, to check the intervening variables (internal validation) and the correct sampling of the group (external validation): all this is impossible in anecdotal reporting. This, however, should not make us forget the importance of the individual who, in drawing new coordinates, when successful, modifies the repertoire of the entire group. Protagonism appears *prima facie* in the proactive tendency of the animal, which does not passively submit to the world, but rather bends it to its own needs through non-predetermined actions. The body is therefore an opening towards the improbable, an event that cannot take place without a dissipative process, so we must not consider heritage as a repetitive constraint, but as a resource to be dissipated through a narrative that is completely in the hands of the individual.

The improbability of somatic events is first overcome through the organisational information that a body possesses from the earliest stages of development, thanks to phylogenetic inheritance. However, we would be wrong to consider somatic phenomenology as the outcome of an already completed morpho-functional programme. From conception onwards, development is based on an external contribution that is in some ways specific and in others open to possibilities. A mammal embryo expects a womb to nourish and protect it, just as a bird's egg would be meaningless without a parent to brood it. On the other hand, predatory behaviour in a carnivore only comes with very general hints, such as recognition of the *Gestalt* of the probable prey and the key signal of movement. The actual hunting experience will then produce a seeking image, which is very often circumstantial, that is linked to the fact that that prey is abundant that year.

The relationship with the world is therefore always twofold—confirmatory in some respects, illustrative in others—but necessarily implied by the somatic dimension. The animal, in the final instance, in its heterotrophic dimension, could not subsist outside a biosphere, that is without the outside world. The body is, therefore, the product of an openness to the world already implicit in the heritage that each organism receives from the past. We could say, then, that the task of organisational information—an inheritance that, as we shall see, is not only genetic—is mainly that of catalysis, that is of allowing improbable events to take place under the various formative or expressive profiles.

In the course of development—as in adaptation, attachment, immunisation, endocrine regulation and learning—what in one phase is an external contribution in the next is presented as an internal element. It is clear, then, that the body's main function is to somatise what it encounters: this occurs in lymphocyte selection following antigenic exposure as well as in post-experiential synaptic wiring. The inclusive internal-external dialectic is repeated fractally on the somatic multiplane, so that any essentialist approach that claims to offer an exhaustive explanation of the functions fails miserably in its attempt. The body is the expression of a relational ontology, so it is useless and incorrect to search for the essence of life or the *primum movens* of biological phenomena—be it the gene or the protein—or to establish some hierarchical linearity in the morpho-functional process. The expressions of the body are stratified but do not have a core: the effect is reminiscent of peeling an onion.

The condition of being a body—and not a disembodied mind—shows us an interweaving of demands and needs spread over a multiplicity of physiological layers, an inextricable network that cannot be separated into an operational centre coordinating a multitude of internal variables. The body is functional incompleteness, a reference to an external note capable of making that project possible. There is no such thing as a self-sufficient body and every autopoiesis is always a recalling of, or dependence on, completion factors. Just think of the importance of serotonin in influencing the emotional condition of mammals: this neuromodulator is produced largely by the intestinal microflora and therefore is not really under the individual's autogenous metabolic control. Inflammation also has a significant effect on the production of cytokines and thus on the possibility of significantly influencing animal behaviour. And the same goes for food metabolites, exposure to sunlight, altitude and oxygen gradients, trace elements in water and radiation. Every event in the body is therefore an opening to the singularity of the world and every inheritance is a weft to be placed in the fabric of the whole.

Heteronomy

It may seem paradoxical, but the body owes its prominence to a kind of ontic virtuality, which we might identify as an implicit "call to the world"—a veritable footnote that serves to organise and in many ways to understand the somatic text. The synaptic architecture of a mammal owes its design pattern to hereditary information, but then it is the embryo-foetal development process that produces the organisational consequences, each phase informing the next. Once the process is set on the right track, what follows is the result of movement itself. Virtuality at birth is not a lack, but a redundancy of possible paths in the construction of a particular wiring. It will be the experience of the subject, its dialogue with the outside world—and not a mere exposure to external stimuli—made up of questions and proposals, inventions and stimuli, that will define the final outcome: it is much like a statue being carved out of a marble slab. Learning is not the act of filling an empty jar, but the act of actualising a virtuality or imposing a certain organisation on the system, ultimately choosing a path among many.

To live in the world is to be an integral part of it through one's individuality, to be hungry for it and equally projected into it, wishing to include large portions of it in the intimate texture of one's personality. At the same time it means dispersing oneself in the environment, scattering it with traces of oneself that have profound consequences in the paths of others and produce resonances that do not end with the death of the individual. The vicissitudes of a subject are like ripples on the water: its experiences are never repetitive and sometimes produce profound upheavals in the history of the species or in the matrix of ecological relationships, while at other times they are mere breezes that confirm a collective path. Interactivity is a somatic prerequisite that is repeated in a fractal way in all of the animal's biological planes, right up to the most subsumptive level of behaviour.

Functional needs and psychological dispositions therefore have a copulative role, in the sense of enabling the identity organisation produced by the interaction between subject and world. An animal is born with a package of appetites that connect it to the outside world, which refer to presupposed contexts and identify, on the whole, a horizon of interests to be pursued. The structure of any bird's body and its motivational disposition are shaped for flight, but its dialogue with the air currents and with the challenges of soaring, gliding and ascending will shape the aeronautical identity of *that* particular bird. Inheritance speaks of an existential aerial dimension, translates celestial rules into biomechanical connotations and creates not only the possibilities of flight but also the need to fly. This call, therefore, goes beyond the simple need to replenish resources (heterotrophy), as it is a multi-layered connection between the animal and everything around it. The gills of a fish envisage a liquid world with oxygen dissolved in water, just as the wings of an albatross imply upward currents in the air: the body is modelled on an organism's needs and these define the package of interests that drive the subject as a protagonist of its own life. The somatic dimension of the animal is therefore constitutively open to continuous external contributions.

The animal is characterised by two openings to exogenous contributions: (i) the need for energy to avoid entropic decay; (ii) the use of organisational information coming from the phylogenetic pathway. If it is true that an animal owes its persistence to external energetic contributions, capable of acting in a negentropic way, it nevertheless dissipates information and history. For example, the structure of the gene, which is responsible for the primary structure of the protein, is not exhaustive in relation to the three-dimensional conformation that makes it a functional entity. The intergenerationally transmitted genome, as it were, is a recipe book thrown into a kitchen where expert cooks and precise environmental boundary conditions will take it upon themselves to translate it into courses. The social behaviour of wolves, for example, is only minimally inscribed in their inheritance because it is assumed that they will learn the basic rules of social interaction through practice, through their relationship with their parents and play with other pups. Thus, phylogenetic legacies are increasingly open to external contribution, and the final result of animal expressiveness follows a cascade process of inclusions or finishes that depend on its interaction with the outside world.

In mammals, for example, most behaviours have innate displays or pre-experiential expressive units that, like the pieces of a jigsaw puzzle, are then organised into species-specific behaviours through interaction with the environment or with conspecifics, especially through play and maternal relationships. The emergence of a structured, functional and expressive body therefore follows stages of introjection of exogenous elements, a process already foreseen in the conformation of hereditary information. The relationship with the outside is not really a completion or a juxtaposition, but a predicative relationship, that is, an interaction capable of bringing out the qualities of the entity itself. We can say that the relation is predicative when the predicates that the entity manifests descend from the relation itself. It is the pruning of neuronal redundancy that brings out the subject's identity, so the subject's experience is a predicative relation. In other words, the body cannot be understood by looking for its qualities exclusively within. It cannot be explained by internal reconnaissance, on the basis of an essentialist conception, because all the qualities it presents are the outcome of its relationality. Still, this should not lead us to believe that the body is a deficient entity without a form of its own, but rather the opposite: it is somatic redundancy that makes external norming possible.

The body is the bearer of such a luxuriant complexity that it contemplates many possible manifestations, so that every predicative expression is but one of the many declinations it can assume. The innate, in this sense, does not rigidly prescribe a domino sequence of behavioural pieces, but rather a range of possible choreographic interpretations, so that behaviour, unfolding in a choreography, relies on occurrences of singularity. There is no mere repetition of pre-constituted automatisms, but a reinterpretation of the heritage in its real expression. Moreover, any particular interpretation is always the result of protagonism in the subject-world dialogue: it would not be feasible without the choice of one among many possibilities. The relationship is, therefore, predicative in that it "actualises" certain connotations that are not given a priori except in their potentiality. The relationship brings out specific opportunities present in the world, but at the same time it reflects on the body by assigning it a certain connotation.

This is not a simple reflection, but a reciprocal organisation process: the foliage of an oak tree actualises its context, just as the latter takes shape in the oak tree. The predicative relationship, in fact, is not a simple assumption of the form that the world imprints on a presumed animal clay, as if the individual were a passive entity that made an objective copy of the developmental environment. Rather, what happens is a mirroring of the outside through the being's heritage, thanks to previously introjected contents. Just as the canopy of an oak tree does not merely develop following the light, but does so through the specific leaf pattern and trunk arrangement that characterises its taxonomic belonging, so an animal places its inherited contents as mirroring bricks in the morphopoietic relationship. The canopy of an oak tree quite clearly mirrors the environment, though not neutrally and passively, but by using its inheritance, which in turn was the result of previous predicative processes attributable to natural selection. I have called this reciprocal process *heteronomy*, which is never a pure assumption of the form of the other, but a re-presentation of the other through one's own morphopoietic devices.

In other words, we can say that the predicate, which makes the body recognisable, already implies the relation, as if the body foresaw it a priori, on the basis of a copulative pre-constitution. On closer inspection, what in animal behaviour highlights the role of the innate in the context-referred dialectic of expression can already be found in the phenomenon of the living. Most biological dialectics are based on a presupposed relationship, already assumed from the outset, and the reason for this is simple: every characteristic of the body is the result of a previous relationship. The past, that is the set of previous relations, is not really past, but casts its life-giving shadows on the body's present. In accordance with this, countless somatic relations are established and determine phenotypic qualities by virtue of externalisations introjected into the body. For this reason, the relation does no more than actualise a predicative event that was already presupposed, albeit in a virtual way—a predicative event that must be, so to speak, "awakened" by the relationship. The relationship is already implied in the very constitution of the body. Therefore, it is not just about the living being's need to receive energy from the outside, but rather about the fact that the being has constituted its complexity by relying on an external informational contribution, that is by referring to normative contents that are taken for granted.

Unfortunately, when it comes to defining what is meant by "external", things become increasingly ambiguous, especially in the early stages of the body's emergence. Are the epigenetic contributions of methylation or the cytoplasm of the egg cell to be considered external? Are gravitational laws defining the embryo's developmental axis, the proximity of growing cells defining the germ layer organisation or metabolic temperature external? Clearly, it is not enough to derive a genome to make a dinosaur, because the qualities of a body are scattered across relationships that develop sequentially between the evolutionary core and its surroundings. What is previously external then becomes an integral part of the developing entity, in a fractal sequence of inclusive relationships. On the other hand, it is evident that the relationship always has an occasional and singular character, even when set on very rigid environmental directives, which leave little room for improvisation or variation because they are kept on a precise register by strict selective factors. In other words, if the relationship is predicative and at the same time it is influenced by protagonism, the role of the individual in its existential trajectory becomes clear.

When evaluating the characteristics of the so-called higher animals, especially birds and mammals, we dwell on the striking individual qualities that these species present, as if they had to be attributed to their essence, forgetting that the greater the complexity, the more the system owes its structure to the predicative contribution of the relationship. On the contrary, we must recognise that from a behavioural point of view, the parental and social interactions implicit in these classes of vertebrates tend to increase their ethographic dependence on external contribution. This leads to the development of traditions within the species, such as hunting instruction in female cats, and cultures within particular populations, such as the use of objects in different chimpanzee communities. Individual variability becomes all the more pronounced the greater the virtuality of the system and, consequently, the greater the dependence on external organisational input. This does not mean that the condition of virtuality is a quality exclusive to homeotherm vertebrates, because individual characteristics are also evident in invertebrates and not necessarily only in cephalopods.

In assessing the construction of the paedotrophic nests of solitary wasps, Guido Grandi, in his essay *Gli insetti. Un mondo occulto di dominatori* (1968), underlines the individual differences and often the creativity shown by these insects in trying to adapt the nest to the peculiar characteristics of the environment, using various material for construction or for encryption. The author emphasises this aspect: "Insect behaviour is often very plastic and its stereotyped manifestations are liable to change according to the circumstances, as well as to the contingent needs faced by individuals. It is evident that, in many such circumstances, insects demonstrate a concrete understanding of the situation" (30). It is not possible, therefore, to understand the predicative relation aside from the protagonism of the subject, who has to adapt its copulative drive—to build a nest, to reach a prey, to court a mate—to the contingent situation it encounters. Different situations can only be superimposed on each other to a small extent: if the relationship is presupposed, the terms of the relationship itself are not. The animal relies on a context that constantly changes and requires a DIY attitude from the individual.

Therefore, it is not possible to understand an animal by referring exclusively to its presumed essential completeness, responsible for the traits it manifests: its predicates, in fact, are not expressions of intrinsic qualities, that is, of traits that manifest themselves autonomously, as emanations. It is the animal's predisposition to relate to the world that gives rise to relationships resulting in the characteristics that make it recognisable. The body is a flywheel of relations and at the same time a point of convergence of hereditary contributions and environmental factors. What is external to it makes it possible, in a process of inclusive relationships. The world is not, therefore, the animal's dwelling place, but rather the revealing element, that which brings forth those contents which are indispensable for the individual to take shape. The subject's reference to the outside world does not simply mean that it is *modelled by* the world, but rather that its individual protagonism *dialogues with* the world.

We have the feeling that the body is a self-sufficient entity, simply drawing raw material and energy from the world, but this reading is the result of an epistemic prejudice. The animal is never thrown into the world like a castaway, but is connected to the world through retrospective references, that is bearing the experiential resonances of the phylogenetic past capable of activating dialogic flows. The body is a time portal, translating into flesh the dialogue between different epochs: the animal is the product of relationships that take place on multiple chronological levels. On the one hand, the dissipation of previous information allows for the personalisation of its dialogue with the world—trivially, the specificity of the questions it asks or the type of copula it proposes—and on the other hand, it achieves an openness which allows the subject to take the lead in its path. This creativity is often almost imperceptible—subtle rehashes, generalisations, reuse of operational strategies in new contexts, targeted attempts—but nevertheless the result of individual ingenuity.

It is not infrequent, however, that the predicative relationship takes on traits of marked subjectivity, especially when hypotheses about the world are disproved, situations become critical, or the environment is invested by profound mutations: that's when the subject is no longer able to count on the canonical questions and must make new copulative proposals, adapting the innate or, in any case, the experiential past to unusual or completely new coordinates. The animal thus becomes an explorer treading on unknown ground, and everything it knows seems to be at odds with the challenges in front of it. But this is not the case. Once again, its heritage comes to its aid to provide it with the necessary tools not to repeat automatisms in a disconnected way, but to express its own creative flair. The animal's protagonism takes the form of a narrative sustained by continuous flashbacks. Events that are epochally very distant thus fall into the same plot: selective processes mix with problem-solving activities, species-specific heritage becomes available to build local adaptations. The intrinsically copulative nature of the body makes the subject-world dialogue "the co-factor" *sine qua non*, in a seriality that runs through time.

The cub's relationship with its mother, for example, becomes an integral part of its affective-relational identity and this, subsequently, will predispose the subject to the relationship with the other members of the group to build its social identity, and so on. In the hybridisation process that causes the mother to be introjected into the affective identity, which in turn is part of the relationship with others, what is external at first becomes an internal quality, which makes itself available to a new hybridisation relationship. Looking back over the individual's development, we will therefore find that what we define as intrinsic to the individual is nothing more than the result of a previous hybridising relationship. Leafing through the phases of this introjection of the world, we will find no essential core, but only relationships: the main feature of the animal character is this copulative predisposition, this opening up to the world to take on a form that arises from the coordinates of the relationships it entertains.

Morphopoiesis is thus a connection between the animal's phylogenetic past, the previous predicative relations of that species and the subjective dialogue that takes place in the here and now. When we analyse, for example, the swallows' autumn preparations for migration, we discover a convergence between remote past, the adaptive value of unsettlement and the present of the world's events, in the seasonal change in daylight: combined, these factors give rise to the individual's predisposition for tomorrow's flight. It is therefore not possible to explain what the individual does by referring only to what happens in the here and now. In fact, one could attribute those behaviours exclusively to changes in environmental luminosity or to the cascade of endocrine transformations, but the ultimate reason why birds migrate eludes us if we stick to proximate causes.

As Niko Tinbergen (1907–1988) rightly pointed out, in order to understand animal behaviour it is also necessary to refer to two remote causal sources: (i) the progeny of that behaviour, that is the line of transmission that led it from the progenitors to the species, and (ii) its adaptive significance, that is the selection events that shaped it. While it is true that an animal adapts in a thousand ways to the characteristics it encounters in the world, it does not do so by going against the innate or by juxtaposing it with new skills, but rather by using phylogenetic resources to achieve its local adaptation. It is in this temporal chiasmus, that is in the use of the innate to realise the learned or, conversely, in bringing the innate dimension to fruition through experience, that we discover the diachronic nature of the animal, where connecting different times also means opening up to the co-factorial contribution of dialogue with, and need of, the world.

Belonging and Individuality

Phylogenetic heritage indicates the coordinates of the predicative relationship, so that both the outcome of ontogenesis and expressiveness in the here and now conform to the necessary dialogical condition. However, this doesn't happen by betraying the innate but rather by using it to effect the mirroring. Expressive morphogenesis is necessarily hybrid: it gives rise to a form that subsumes the situation into the heritage and vice versa, as it introjects the characteristics of the context, making them its own, and assuming them in its own phylogenetic form-function. In mammalian ontogeny we recognise two essential needs: (1) on the one hand, the need to refine the species repertoire through maternal and social learning, which enables the development of the communicative and interactive codes of belonging; (2) on the other hand, the need to adapt the open innate schemes, in order to tailor the subject's behaviour to the specific challenges found in the environment, thus creating an individual, necessarily singular, profile. In these animals, the relationship with the world is what provides a predicative structure on the basis of (i) expected contributions, that is external elements that correspond to phylogenetic expectations, giving rise to apprenticeships of belonging; (ii) occasional contributions, that is variable and unforeseen elements, enabling ontogenetic variations and producing effects of singularity.

Both belonging and individual singularity are variations on the theme of heritage. In the first case, we can picture heritage as a jigsaw puzzle, where the innate is the set of pieces that the individual has to put together in a preset configuration. In order to succeed in this difficult compositional task, the following are helpful: (1a) the characteristics of the species niche, which, through the uniformity of its variables, produces a homogeneous evolutionary direction among individuals, because they are all subjected to the same growth environment; (1b) first the parental relationship and then the social relationship which, through predominantly modelling actions, suggests, just like the puzzle box, how to put together the different behavioural pieces. In the second case we can imagine the heritage as a Lego set, which allows the subject to put the pieces together in its own way in order to build what is actually needed. To achieve this singular interpretation of the heritage, the animal will make use of: (2a) compositional modes, linking motivations and emotions to target and circumstantial elements, for example learning to chase lizards in a meadow; (2b) assimilation and accommodation processes that will allow it to modify its predictive and operational tools through experience, much like a system update.

Evolutionary heritage, therefore, certainly defines a range of behavioural expressiveness that cannot be surpassed: a dog will never be able to assume the ethographic characteristics of a cat or a dolphin. However, within this range there are many possible variations that allow both belonging and individuality. And if individuality seems to be the character that has most to do with protagonism, we must not forget that parental and social learning are what yield animal cultures, namely styles, habits and particular forms of use of the environment typical of a certain population *within* the species. We must always start from the hereditary component if we are to understand protagonism, avoiding the erroneous superimposition of expressive ownership with the lack of innate education, which involves linking individuality and learning and above all considering ontogenetic variability as inversely proportional to innate education. In fact, the opposite is true: the greater the phylogenetic heritage of a species, the wider the learning horizon. The reason is very simple: it is called the virtuality dimension.

A complex and articulated (redundant) innate is like a network made up of many nodes and possible connections. It is obvious that such a network offers many possible wirings, so the more complex it is the greater its virtuality. What does this mean? It's not difficult to understand. If a network can be potentially wired in many ways, that is if it has a high degree of virtuality, the combinations will be directly proportional to its complexity, so that on the one hand there will be a powerful space for the intervention of experiences, and on the other there will be a greater emergence of singularities. Cephalopods and hymenoptera, which have a highly articulated innate repertoire among invertebrates, also excel in learning, in building individual profiles, and in the innovative use of tools. Inheritance therefore needs to be redefined, because we are inclined to see it as a repetitive mechanism, a constraint on free expression, a form of determinism that cancels out the individual, a state of stupor in the performance of a function: these are the misreadings that have led us, for example, to reify the role of the innate in human beings, in order to celebrate or simply grant them free will.

We must move on from a conception of inverse proportionality to a new reading that emphasises how, conversely, learning and individual freedom—we could say the protagonism of the subject—are *directly* proportional to the individual's innate endowments, that is to the complexity of its heritage. The mistake we make stems from a series of distortions and prejudices. We want to differentiate human beings from other species, recognising their freedom of choice as opposed to the supposed mindless instinct of non-human forms. What is at work here is a dichotomous bias, a distortion that tends to place the two terms of a relationship in an oppositional relationship and in an antinomian structure. Furthermore, we are unable to correctly interpret individuality in the subjects of another species, due to the distortion of perspective—the *oi barbaroi* effect experienced by the Greeks—whereby we tend to homologate their behaviour; hence our preference for explanations that involve instinctive automatisms. And, finally, we must not forget the very assumptions of ethological research, which, wishing to identify a species repertoire, made up of choreograms, sonograms, patterns and uniform displays, necessarily uses a scale that reduces individual differences to zero, thereby stigmatising anecdotal description.

Protagonism is a process in the making, a sort of avalanche effect, which must rest on a starting point, on a source core, in order to develop its full power. There is no doubt that the process is self-implementing, so we must abandon the static-repetitive view of the innate and see it rather as an explosion of vitality that launches the individual into the events of the world, activating its interest and capacity for dialogue. Animal protagonism is born, as I have said, from the collision of various epochs in the here and now, in the dystopian conflagration of powerful expectations, the product of inherited resonances, with events that sometimes respect them and sometimes clearly betray them. This is why phylogenetic inheritance is never a simple piece of equipment but a call to action, a prompt to flex those grafted memories on the basis of new interpretative configurations. But nothing is built *ex nihilo*: hence the importance of being able to lean on that original core and use it as an explosive charge on the world.

An animal is not tossed here and there by the environmental waves like a dry leaf: while respecting the laws of this universe, it soars in defiance of gravity and swims against the current; it does not merely occupy a space but builds its own niche. It makes webs that are sometimes gigantic, over one hundred square metres wide, as in the case of social spiders; it can make beehives organised like skyscrapers, as in the case of social wasps; it sets up complex nests like weaver birds and equally complex burrows like moles; it styles its body or shows off its skills by building a patio, as gardener birds do. And in adapting its nature to the opportunities of the world, the individual transforms the information it inherited into an evolutionary primordial soup capable of giving rise to new patterns. Phylogenetic memories, then, no longer belong to the past, but become *projects for the future*, material available for the individual's creativity. Research often tends to dwell on species-specific behavioural displays, and we have the impression that non-humans are only repetitive mechanisms that are also highly homologated. But this is not the case at all.

The innate is like a set of hypotheses about the world that have already been subjected to the scrutiny of natural selection—which, as we know, operates as a falsification stress-test. In other words, they are generally good hypotheses that can stand up to adaptive tests, but on condition that they are *available to* the individual, who can use them at any time in new situations or adapt them to make them fit better. An animal's behaviour always presents variations on the theme, which can be defined as individual interpretations or contextual adaptations. This does not mean that there is no species repertoire and that there are no criteria for recognising it: it is rather necessary to know how to attribute them correctly. A behaviour is not an automatism with a uniform-repetitive character, but an expressive range within a perimeter whose area varies according to numerous variables. These include: (i) the specific adaptive value of that behaviour; (ii) the degree of flexibility of the reference species. In other words, if it is true that some species have greater individual expressive variability, it is equally true that, even within the same species, some behaviours have a greater or lesser extent of variability than others. Moreover, it is evident that the individual does not drastically change its phylogenetic heritage, especially if it lives in the same environment that packaged those innate skills and that therefore also contributes to foster expressive uniformity.

In that case, it only makes small adaptive variations to increase the functional adherence of its actions. Creativity does not necessarily have to be striking in order to be recognised as an indispensable factor in an individual's local adaptation process. An animal's creative work is very often imperceptible. A good example is the case of the mason bees of the genus *Osmia*: they use the empty shell of a mollusc as a paedotrophic nest, inserting pollen and nectar into the bottom, then laying eggs and finally obstructing the entrance through a diaphragm of vegetable cement made by chewed leaves. But these operations, apparently uniform and therefore repetitive, actually involve very particular adaptations: first of all because the shells often have different shapes and sizes, and secondly because, between the opening of the shell and the diaphragm, the insect must isolate itself using soil, pebbles and other material, the availability of which varies depending on the location. As if this were not enough, the *Osmia* must hide their paedotrophic nest, observing the characteristics of the environment, that is the opportunities it presents. To do this they move the shell with their front legs, rolling it to the safest places available and then covering it with dry leaves or something else.

Another example of entomological creativity is provided by some solitary wasps of the genus *Sceliphron*, also known as mud-dauber wasps because of their ability to build paedotrophic nests out of congealed mud. In order to build these nests, the wasp has to collect about thirty pellets of wet earth, which it then mixes with its own saliva—a job which is already quite onerous. The females attach their nests to various surfaces, and it is not uncommon to see them on the outside walls of buildings or even inside houses. These surfaces vary considerably in their characteristics, so that the wasps have to adapt their architectural work time after time, which is not easy. In fact, having reached its destination, the wasp lays down the pallet and presses it with its head to make it concave and give it the shape of a tube, but to do this it must carefully follow the orthogonal characteristics of the surface at hand. Next it sticks more mud to the right and left, working to flatten and smooth its work until its appearance to all intents and purposes resembles that of a wineskin. A wasp's instinct certainly involves a series of actions that are repeated in all individuals, and the final appearance of its nest is also characteristic, but if we carefully analyse different nests we will discover peculiarities in each of them.

Animal creativity therefore lies in the inventiveness and adaptation that are necessary precisely to realise the species-specific behavioural repertoire. Of course, every animal learns through experience, however it is not a question of learning in the strict sense of the word, but rather of freedom of action: an animal must be able to rely on room for manoeuvre when it expresses the behaviours of its species repertoire if it wants to actualise them in a context that is always inevitably unique. Some particular behaviours occur, or rather are reported, mainly in captive or confined animals and the reasons for this, in my opinion, could vary. We could hypothesise, for example, that certain behaviours are more difficult to observe, that is that the problem is purely technical, so that in a confined environment one is able to detect them better. We could also suppose that certain behaviours occur in captivity as compensatory responses to conditions of discomfort. But I would like to venture another hypothesis. It could be that in captivity these animals are confronted with environmental characteristics where the expected adaptive action requires a greater dose of creativity—adaptive necessity sharpens the wit, as it were.

One of the most important phenomena of animal protagonism is the adaptation of a behaviour attached to a certain function to a new use. For example, wolves use eth-epimeletic signals, that is requests for parental care, in the communication of submission of one subject to another. Agonistic buffering behaviour, that is the blocking of a subject's aggression through a particular action, is also present in baboons, where the one who wants to contain the aggression of another holds and presents the latter with an infant. Interestingly, this behaviour is also sometimes used in other contexts, for example to be accepted at a meal or to force affiliation. As Shumaker et al write in *Animals Tool Behavior*: "Agonistic buffering [...] is part of a larger set of social behaviors in which primates and probably other animals enlist the aid of others to further their interests in social interactions" (2011, 92).

Birds have a particular behaviour known as anting, in which a bird sprinkles its plumage with ants, taking advantage of the insecticidal, fungicidal and bactericidal action of formic acid. This activity can be passive, when the individual lies down in an area with high numbers of ants, or active, when it picks up the insect and places it on its body. Anting has innate components that are, however, shaped by experience. As K. Simmons notes in the essay *Anting and the Problem of Self-stimulation* (1966, 149), if a bird comes into contact with formic acid for the first time in the presence of objects other than ants, it may acquire a persistent tendency to ant with substitutes. I think it is important to understand this flexibility as the prerequisite for creativity, and not as a simple variability of response. What I mean is that between a random behaviour and a deterministic one there is the reality of animal expression, which involves functional exaptation, reuse, adaptation, accommodation—that is, the innovative use of the resources available. To think that an animal could ever overcome a challenge through random attempts is to disregard the combinatorial boom of possibilities that would make it impossible to succeed. This is why heritage plays an irreplaceable heuristic role, not only in offering the animal ready-to-use solutions for the canonical contexts it encounters, but also as a recipe book of solutions to try out and recalibrate in order to deal with the singularity of every situation.

Works Cited

Dawkins, Richard. 1976. *The Selfish Gene*. Oxford: Oxford University Press.

Dehaene, Stanislas. 2020. *How We Learn, Why Brain Learn Better than Any Machine*. New York: Viking.

Dennett, Daniel. 1995. *Darwin's Dangerous Idea: Evolution and the Meanings of Life*. New York: Simon & Schuster.

Grandi, Guido. 1968. *Gli insetti. Un mondo occulto di dominatori*. Bologna: Calderini Edagricole.

Shumaker, Robert W., Kristina R. Walkup, and Benjamin B. Beck. 2011. *Animal Tool Behavior: The Use and Manufacture of Tools by Animals*. Baltimore: Johns Hopkins University Press.

Simmons, K.E.L. 1966. Anting and the Problem of Self-stimulation. *Journal of Zoology* 149: 145–162.

CHAPTER 4

Being Interested

Premise

The animal condition unfolds through protagonism, which means being interested and participating in events instead of remaining passive, acting in order to achieve certain results, orienting oneself in an elective manner towards external entities and so on. These qualities characterise every animal's expression and differentiate it from any other entity. While inorganic phenomena appear to fall completely within the strict rules of determinism, life, while respecting the normative domains of matter, manifests particular qualities that cannot be explained by physics or chemistry alone. This does not mean that we should envision different rules for biological phenomena, but that life presents emergent characteristics that define one or more layers of reality with peculiar predicates and laws. One of the most evident characteristics of the animal being is its orientation towards the outside world, in a dimension of interest and of defence of such interest, in a logic of Self-expansion. These two expressions of interest also create specificities, however, because each animal is interested in different things and is the bearer of its own interests, in a condition of synthesis that can be defined as *the inherence character*.

All of an animal's features are, in fact, intimately connected to each other, marking out a space of interaction with the world—open to external contribution—which, however, is always directed towards the animal itself, showing a self-centred finalism. Even when this seems to be lacking,

R. Marchesini, *The Creative Animal*,
https://doi.org/10.1007/978-3-031-07414-1_4

as in the case of altruistic behaviour, the result can easily be traced back to the tracks of self-ownership. To be a bearer of interests requires constant attention to the outside world: the two aspects only apparently constitute a dualism. Let us take an example. Predatory interest cannot be separated from the orientation towards movement, whereby being interested implies transforming the world into a field of agency to explore. Having interests necessarily calls for a *eager openness* to the outside world. An animal, in fact, does not merely open its receptive channels, but literally covets what surrounds it and somatises it, so that an event is no longer an external entity but becomes an emotional state, that is a bodily condition. An animal is interested in the world because the realisation of its interests is strewn with opportunities and risks that the subject must readily grasp in external reality.

Animality is one of the many possible expressions of the living; it is a particular form of life, endowed with its own specificities (we could define it as an emergent level of the biological phenomenon), but also a metapredictive dimension marked by a copulative tendency—an outlook on the world through feeling and desire. Emotions and motivations determine the horizon of interests to defend and achieve, and at the same time the animal's eager openness to the world. Again, this dialectic is probably more in keeping with our epistemic structure, which is prone to dualisms, than with reality as it is, but it can help us to understand the development of the interests that guide and characterise animal expression. The animal is open to the world precisely because it is a bearer of interests. This condition induces it to transform the external scenario: (i) into events that define a maximum value with respect to the valorisation of its interests, which can therefore be translated into a state of the body (emotional feeling) or; (ii) into a projection space that establishes the degrees of accessibility for the expression of its behavioural tendencies and therefore the possible objectives towards which it can direct its attention (motivational desiring).

It is not possible to understand an animal outside its condition as an interested entity, which therefore has an internal motive for its actions. Its phylogenetic heritage makes it frantic, as if it had a task to perform and little time at its disposal, but precisely for this reason its context-directed actions do not respond to mechanical-repetitive configurations, but require the ability to extract specific opportunities and criticalities from the singularity of the situation and at the same time to interpret the heritage canon through creativity. The innate endowments, therefore, do not limit creativity but represent its indispensable partner in the adaptive dance. Indeed, it is not possible to understand this being interested outside of a diachronic dimension, where present behaviour always refers to events of the past and turns towards something future. As already well understood by Nikolaas Tinbergen in his *Four Categories of Questions and Explanations*, which redefine Aristotle's explanatory tetrarchy in an evolutionary sense, an animal's action cannot be fully understood if analysed in the present of its unfolding. I believe that the convergence of these four causalities (phylogenetic, adaptive, ontogenetic, physiological) on the subject, through the compass of pleasure, is always central to individual ownership in choosing a course of action.

Experience as Dialogue

The intersection of hereditary tendencies that call on the one hand for an interest in the world to serve one's own interests, and on the other hand for the presence of situational singularities, in terms of opportunities and risks present in the here and now, creates the very conditions for the emergence of a dialogue between the subject and the world. From this process, which involves evaluations and projections as well as questions and extractions, descends the unique and unrepeatable pathway that we call individual experience. We would fail to understand the very character of experience if we were to consider it as mere exposure to a landscape whose specificity is limited to the peculiar characteristics of the sense organs. Experience is a feeling, that is to say, an emotion in relation to what invests our senses or what we grasp between the folds of possibility. Experience is a projection into the potential that this horizon allows us to imagine; it means going beyond the perceived image to give shape to our desires and launch our motivations into flight.

Experience is always connected, albeit with different degrees, to the Aristotelian principle of *thaumazein*, that is, of the participated and somatised. Experience is, in fact, marvel at the world, in the two senses of the term: (i) a Being-there, which finds fulfilment in being able to open oneself enthusiastically to a present-future full of opportunities, novelties and unforeseeable events, that is, being *able* to experience reality as virtuality and exuberance and not only being *induced* to do so; (ii) a Having-already-been-there, which is expressed in finding correspondence with a world that satisfies expectations and supports openness, that is, finding some sort of confirmation in experience. Wonder, an ecstatic condition of eager openness to the world and of resonance of the past, is never simply surprise in the face of the unexpected, but joy in the face of a landscape capable of speaking coherently to the deepest chords of one's being. The presence of the subject lies, therefore, in its being both the memory of confirmed expectations and openness to the protagonism of becoming.

In order to understand this convergence between the exuberance of Being-there, as a sense of energy and potentiality that springs up in the individual, and the excitement of Having-already-been-there as a sense of surprise in finding one's own world, it is useful to observe the behaviour of a newborn animal. After a first moment of confusion and torpor, this energy manifests itself in all its might in terms of sensitivity and desire, of acceptance of the world and somatic vigour. The animal's experience of the world is one of wonder, because it encompasses both *re*discovery and discovery, like two sides of the same coin. What we see is a lively sense of interest: if we place a puppy dog in a room with a hundred objects, it will soon touch them all; similarly, if we place a kitten in any environment, it will explore every smallest recess in it. We will see them get excited, scared, surprised, joyful for every little thing, playing in a blaze of openness towards the world, expressing an energy that overflows from every part of the body and is reflected in the copulative coordinates typical of their species.

In fact, every species translates this explosion of vitality into verbal predicates that characterise and diversify it, so we can say that there is a dimension of wonder for every species. But our focal point would still prove to have poor resolution. Each individual has its own heritage and, to be more precise, each moment of its life establishes specific planes of wonder. This hunger for the world is always produced in singular forms. The encounter with the environmental landscape, in that first approach so rich in suggestions, is significant because it shows how each animal is excited to open its senses to its environment precisely because it is comforted by the familiar cushion of rediscovery. Animality, in its ontogenetic and expressive unfolding, is to all intents and purposes a yearning for events, articulated in a set of calls that refer to an intoxicating and yet soothing horizon. The subject's protagonism stems from this orientation towards everything that surrounds it, an interest in the phenomenal that is never a lack to be compensated for or a blank page to be filled, but a dialogue between what the animal intrinsically owns and what it will acquire in the process of becoming.

It is a diachronic dialogue, a convergence of different times, sustained by the electrifying wonder of being admitted to the great convivium of life and at the same time by the world's responsiveness, whereby the animal finds the table coherently set for its needs. The ancestors' experiences resonate in the individual, like confused images full of unfolding contents that are revealed in the animal's life. And when these memories meet the scenario in which they are set, there is a galvanising effect: the subject thus feels the urge to continue along the path, to explore new opportunities, in a word, to do more. This is a feeling of rediscovery, which is not really a clear memory but rather a sense of coherence, which for this very reason induces one to explore and try new things. Protagonism, therefore, is nurtured through subtle mechanisms of self-efficacy, which the animal receives because of the confirming nods given by the world to its first timid steps supported by the innate.

This awakening in the world only to find what was expected should therefore be interpreted following Lorenz, who views every a-priori as a phylogenetic a-posteriori. But this does not extinguish the individual's protagonism: it rather ignites it, setting fire to the powder of its nature, triggering interest in and openness to the world. Experience, therefore, has the flavour of reminiscence rather than of exposition and vacuity, and has nothing to do with the *tabula rasa* model that still features in the sensist and empiricist conceptions. Experience, moreover, is an evolutionary and generative reminiscence, capable of bringing about the adaptation and compositionality that give rise to the individual's singularity. In other words, it is never simple repetition, but rather means bringing back to life the germ of the past in order to construct the individual's singular novelty.

And here we come back to the interested condition as a founding characteristic of the animal being. There would be no experience without emotional feeling and motivational desire, because the yearning towards the world, and the affectivity that somatises it, arise precisely from these dispositions. Openness, in fact, does not limit itself to welcoming, as if the experienced were simply an outer entity, but translates what is external into a bodily form. Experience is the result of the copulative impulse: it is emotions and motivations that make experiences possible, creating the interest in the world that translates the innate into a throbbing feeling and desire. This is a point of fundamental importance to which I would like to draw attention. Emotional and motivational dispositions carry inherited propensities but, precisely because they are copulative and characterise the animal's momentary condition, they make the subject's sense of the present vivid and conducive to knowledge. Emotions and motivations stimulate creativity, leading the subject to make direct contact with events. What prompts experience is not lack, but copulative exuberance.

It has been said that experience is never a generic opening towards the outside world, but rather a precise intersection that takes place on the basis of the particular emotions or motivations that characterise the condition of the subject at a certain moment. The type of somatisation provoked by the encounter with the world is not always the same, but changes according to the individual's internal condition: it is clear, in fact, that the opening sustained by fear is different from that induced by excitement and joy. Experience, therefore, does not develop through a generic exposure to the world but rather on a precise copulative level, provided by the type of emotions and motivations active at a certain moment in the animal's life. That is where dispositional expectation collides with the specificities of the here and now, giving substance to the questioning of the world that underlies experience. If all this is true, why do we struggle to understand the dialogical and affective meaning of experience? I believe that here, too, the mistake lies in conflating the innate and learned components. On closer inspection, the contents of heritage—the spectral images that establish expectations and operational canons, which we usually define "instincts"—provide the very fabric of our questions, the handbook used to acquire knowledge in Piaget's sense. On the other hand, it is in the affective states of feeling and desire that we find the underlying reason for asking questions in the first place. In conclusion, the innate allows the realisation of the learned, and at the same time it is the encounter with the world that gives meaning to the innate.

The overlapping of Being-there and Having-already-been-there makes the animal feel at home during its experience of the world, but at the same time gives it enthusiasm for adventure, a sense of vigour in its presence, an explosion of fullness. There is an excitatory response that manifests itself through bodily pleasure, when the animal finds what it was waiting for in the world: the sweet taste of a fruit, the sense of energy in physical expression, the adrenalin rush after an unexpected event, a given landscape meeting its gaze. Having-already-been-there in this world means that, in its first relations with the environment, the individual *finds* a reality, rather than discovering it, almost rejoicing in the pleasure of having returned to it. The conjugation to the world offered by phylogenetic heritage is presented as a copulative tendency: it's as if we had many categorical drawers to fill with entities and events that unfold along the way. Being afraid means looking for something to substantiate that emotion, just as the desire to chase probes the horizon in search of possible prey. The animal peripathesis is therefore sustained by an irresistible attraction towards what is external, expressing an interest that is not generic but already preset on precise predicative—emotional and motivational—coordinates.

Precisely by virtue of these innate copulative guidelines that link it to what is happening, the individual has its own title, its own presence, its own leading role in the whole affair. Its perspective is never passive in the face of the world, nor is it predefined or condemned to blind and simple repetition. The animal's emotional or motivational disposition makes it interested in and therefore communicative with external reality; the environment, while confirming its orientation in some respects, in others proposes small variations to manage, particular situations to address, new spaces to colonise. The innate endowments provide the animal with an initial elaborative equipment, providing orientational, evaluative and operational modal instruments to face the challenges of the world, which define its instinctive behaviour. As we shall see, these endowments however also make up a toolbox capable of generating *new* resources precisely through experience. In this chapter I will analyse a second aspect of the innate: the affective states that, as copulas turned to the world, are able to foster and characterise experience.

The Animal as a Desiring Entity

An animal's behaviour is always directed towards achieving a result, such as reaching prey, defending a territory, mating and building a shelter. Ethology is not only concerned with describing the choreograms, attitudes or signs through which a species expresses itself in interaction with external reality, but also seeks to provide explanations, that is causal mechanisms that are able to highlight the motives behind a certain behaviour. Undoubtedly, the descriptive aspect plays an extremely important role in clarifying a species-specific behavioural repertoire; this is why, over time, there have been increasingly refined data acquisition techniques, including the use of video cameras and recorders. Nevertheless, the ethological school developed in the first decades of the twentieth century precisely with a strongly explanatory intent, as demonstrated by Konrad Lorenz's psychohydraulic theory and Nikolaas Tinbergen's Four Questions. Moreover, the proposal of the founding fathers of classical ethology also had a polemical connotation and was openly antithetical to the behaviourist theories that were gaining ground in those years, especially in North America.

If behaviourists saw behaviour as essentially a response to a stimulus, for Central European ethologists, on the contrary, the motive was essentially a drive, that is, it originated from within. In other words, the world was not really a stimulus capable of inducing a certain activity, but rather a field of possibilities that, at most, could recall or facilitate the expression of that behaviour, acting as a latch capable of releasing the drive energy: for this reason it was called a key signal or releaser. Ethologists considered both the drive coordinate (e.g. the predatory motivation) and the key signal (e.g. movement), as well as the instinctive choreograms or behavioural translation patterns in terms of innate expressions, that is prior to the individual's experience, which did not need to be learned and which could, if anything, be completed through social learning or exercise. For ethologists, in short, the main motive behind animal behaviour was always internal and at most could be evoked or allowed by the external environment.

Lorenz himself showed that when motivation is very high, the elicitative action of the key signal is not necessary to induce a certain behaviour, while if the motivational drive is lacking, the presence of the key signal is not sufficient to provoke it. Different motivations are therefore the basis of the various behaviours that can be identified in a species, and the dispositional prevalence of one or the other explains why at one time an animal performs particular activities and at another does other things, or why it is uncertain at times whether to perform one action or a different one. Lorenz referred to a kind of *parliament of instincts*, where whoever shouts loudest has the upper hand. His basic assumption was that the various motivations in an animal do not always have the same expressive cogency, but are stronger at certain times and weaker at others, based on fluctuations related to seasonality, the physiological condition of the subject and phasic variables.

It was soon realised, for example, that every motivation has a phase of high attraction, which makes its expression urgent, and a phase of dormancy where, on the other hand, willingness to express it decreases. Konrad Lorenz tried to explain the increase in motivational intensity using the hydraulic metaphor of a reservoir fed by a tap and connected to an outflow point controlled by a valve, kept closed by a spring. The outflow of the liquid represents the behavioural activity and can be provoked by the pressure of the water in the tank in conjunction with the action of an external stimulus, presented as a plate of a scale connected to the valve. The greater the pressure of the tank, the less weight will be required on the scale plate, that is the magnitude of the eliciting stimulus, to cause the outflow—that is, the behaviour. In some situations, the pressure of the water alone is sufficiently strong to open the valve, thus bringing out an activity that is not triggered by any stimuli.

The ethologists' proposal was revolutionary and still is, if we consider the very common risk of a teleological or finalistic explanation. In 1970, the biologist Jacques Monod introduced the term *teleonomy* to denote predetermined behaviours that show an apparent finalism when, on the contrary, they have simply been shaped by natural selection. Ethology had shown an *ante litteram* teleonomy by explaining behaviour as the expression of an appetitive motivation that simply found the opportunity to express itself in the world. As Irenäus Eibl-Eibesfeldt points out in the essay *Grundriss der vergleichenden Verhaltensforschung. Ethologie*, at first, the animal enters a kind of frenzy or increased restlessness, as if searching for something; this is not a general increase in motility, but "a specific disposition, an impulse to action, proven by the readiness with which the animal reacts to certain triggering situations" (1974, 60). Immediately after giving birth, for example, female rats have such a strong inclination to collect nesting material that they forget all other activities, even grabbing their own tails and laying them in the nest.

The characteristics of appetitive behaviour studied by Wallace Craig (1876–1954), considered one of the precursors of ethology, stand out for the active search for the correlated stimulus capable of inducing consummatory behaviour. As pointed out by Craig (1918), an animal must be flexible and adaptable to the specificity of different situations, sometimes having to overcome obstacles or remember an already-taken route in order to reach the desired goal. Paul Leyhausen (1965) also underlines the importance of appetite in the cat's predatory behaviour: appetite creates an experience with multiple possibilities, so that the moving target can be an object to chase and grasp, but also something to spy on and observe. As Eibl-Eibesfeldt recalls, "a predator with the urge to hunt will react primarily and with great ease to stimuli that trigger hunting actions. If it has not been able to hunt for a long time, it will even make do with surrogates and, in certain cases, after a particularly prolonged deprivation of triggering stimuli, it may even react of its own accord" (1974, 61–62).

Appetence can then be defined as the propensity to perform particular acts—such as chasing, competing and courting. This tendency qualifies the species, characterises the inclinations of the individual and defines the state of the subject at any given moment. Appetence is a projection onto a target that is always also a desired result, not simply a *res* to be reached, so appetence is a *desiring* condition capable of bringing about *desirable* conditions. An appetitive animal is always turned towards the future and projects itself both in a physical sense, as if dangerously overlooking the landscape, and in a temporal sense, as if its present did not contain it. The appetitive experience is therefore a way to go beyond the status quo and experience the dimension of desire, projecting oneself towards something on the basis of exuberance, of a condition of vitality and fullness that demands to be translated into action. Motivational appetence contains the sense of *wanting to do* and *being able to do*, an energy overflowing with resources and ambitions that must be framed in terms of activation, that is, of being charged and overflowing. This is an affective condition that is affected by neurophysiological activation mechanisms.

When we look at the biological mechanisms that govern behavioural orientations and the various activities that an animal performs, we note that all vertebrates share subcortical structures and activation pathways supported by certain—for example, cholinergic, dopaminergic, serotonergic—neuromodulators that are specialised in predisposing the individual to engage in certain behaviours. These are somatopsychic states, which translate into affective experiences, albeit with different levels of consciousness, and can be divided into (i) homeostatic experience, such as feeling hunger, thirst, sleep, arousal, cold-hot, pleasure or pain; (ii) sensory experience, given by sensations, perceptions and aesthetics and subdivided broadly into exteroception and interoception; (iii) dispositional experience, which refers to the emotional system, and therefore to sensing the value of circumstances, and to the motivational system, that is being led to perform certain activities.

These three experiences, which indicate *feeling something*, do not necessarily have reflective or representational contents: an example of this is being in a good mood without knowing why. Affective experience, therefore, precedes cognitive experience proper and can come in different levels of awareness, so that an affective state is present even during sleep. We now know that affective experience arises in a specific part of the central nervous system, namely *the limbic system*, which is in contact with (i.e. receives afferents from) the whole body and in particular the endocrine and immune systems. The gastro-enteric and hepato-pancreatic systems also communicate with the brain. The afferences are mediated by particular chemical molecules that function as *intermediaries of activation and deactivation* produced in different parts of the body, not only by neurons but also by glands and immune cells. The subcortical regions have precise specialisations in predisposing particular affective states, also on the basis of connections and projections—we could call them *affectivity circuits*—where specific receptors make those nuclei sensitive to neuromodulation. The afferent pathways therefore go (i) from the site of production of the neuromodulator (ii) to the areas of the brain where the receptors are found.

Appetitiveness can, therefore, be attributed to internal physiological mechanisms, often in close connection with homeostatic mechanisms related to basic needs. Water-seeking behaviour, for example, is regulated by two factors. The first (i) is osmometric thirst induced by an increase in tonicity, that is the concentration of solutes, in the interstitial fluid detected by particular neurons called osmoreceptors, located in the terminal lamina of the third encephalic ventricle which, by interacting with the hypothalamus, control the production of vasopressin. The second (ii) is volumetric thirst, which occurs when blood volume decreases and is detected in the kidneys, with production of renin catalysing the release of angiotensin which, in turn, acts on the hypothalamic-adrenal axis. At this point, a particular area of the brain, the preoptic nucleus, integrates the two signals by stimulating water intake. As for the sense of hunger, a peptide hormone, ghrelin, acts as a powerful stimulator of food intake. Interestingly, blood levels of this peptide increase during fasting and fall dramatically after a meal. On the other hand, the sense of hunger can be inhibited by cholecystokinin, which is produced by the duodenum; even high levels of testosterone and oestrogen decrease the sense of hunger. These modulators act on the ventrolateral areas of the hypothalamus.

Some neuromodulators like dopamine and vasopressin, which generally have an excitatory effect, underlie the mechanisms of competitive aggression; others, such as oxytocin, increase the sense of security and affiliative behaviour; instead, acetylcholine is responsible for the activation of the arousal system, stimulating alertness, attentiveness, excitement and generally high reactivity of the sympathetic system. We know that one of the main phasic mechanisms is the circadian rhythm, an internal clock also influenced by daily variations in light. This regulatory effect is caused by the suprachiasmatic nucleus of the hypothalamus, which is affected by the action of a particular neuromodulator called melanopsin, produced by the retinal ganglion cells: this is why we speak of a retino-hypothalamic pathway. There is also a seasonal rhythm, for example in the secretion of testosterone in males and in the mechanisms of female receptivity and ovulation. The control of seasonal rhythms involves another structure, the pineal gland, located at the top of the midbrain, which produces melatonin during the night. During long nights in the winter phase, the animal's epiphysis produces large amounts of melatonin, stimulating sleep and inducing lethargy. In birds, seasonality has a considerable influence on behaviour, for example nest-building or migration. For example, great tits (*Parus major*) gather in large flocks in winter, thus demonstrating a gregarious tendency, but this disappears in spring when the flock disperses and each pair settles in an actively defended territory.

The appetitive predisposition to perform a certain behaviour over another or to increase the frequency of its expression is often the convergent result of several cofactors. In 1953, J. Van Iersel studied sticklebacks (*Gasterosteus aculeatus*), and found that their tendency to ventilate eggs is regulated by a combination of internal and external factors. The male, after fulfilling the sexual urge to fertilise the eggs, immediately triggers the ventilation behaviour by waving its pectoral fins. Van Iersel measured the intensity of this behaviour over time and found that the increase in activity is caused by the eggs' consumption of oxygen, so that if the carbon dioxide content of the water is artificially increased, the stickleback will ventilate more frequently. Daniel Lehrman (1953) found that the hatching behaviour of the ringneck dove (*Streptopelia risoria*) is influenced by the presence of both members of the pair, so that if they are separated they stop hatching. The reproductive behaviour of the canary (*Serinus canaria*) was studied by Robert Hinde (1956), who pointed out that the growth of the gonads in the female was stimulated by the duration of the light-dark relationship through the hypothalamic production of the gonadotropin-releasing factor, and also by male courtship. In the female, a series of cascading and recursive processes thus occurred, such as nest preparation, copulation, oviposition and hatching.

When the appetitive phase is high, animals often engage in empty behaviour. This was found by Konrad Lorenz (1937) in the predatory behaviour of starlings (*Sturnus vulgaris*), by H. Kluyver (1947) in Bohemian waxwings (*Bombycilla garrulus*), by Niko Tinbergen (1952) in sticklebacks and by Paul Leyhausen (1979) in cats. Exploratory behaviour, referred to a circumscribed area, or patrolling behaviour, like wandering around the territory, is fundamental in foraging and learning the characteristics of a given environment. Here too we can note the important function of a neuromodulator, dopamine, which increases restlessness, interactivity and exuberance in scouring the territory for opportunities. Whenever there is an increase in dopamine release, animals are more excited and curious: they show a heightened interest in and an active search for opportunities. The lateral hypothalamus is activated in the brain when animals are in this state. Another structure involved in exploratory appetite is the nucleus accumbens (basal ganglion) and the medial frontal cortex, which Jaak Panksepp and Lucy Biven describe as a seeking system (2012, 103–157). These regions are linked together by the medial forebrain bundle (Mfb), one of the most important pathways of the dopaminergic system. The Mfb system is, therefore, crucial for understanding the appetitive condition related to foraging motivations.

Electrical stimulation of these areas also produces excitatory and seeking behaviour, whereas when this system is deactivated animals appear disinterested. In 1954, James Olds and Peter Milner discovered that animals would do anything to obtain stimulation (e.g. press levers to self-stimulate) in the Mfb area. We can therefore deduce that, even prior to consumption or reinforcement, the activation of interest and seeking is rewarding. The gratifying effect *precedes* the arrival of the food or the expression of the exploratory behaviour, that is it comes before a homeostatic need is satisfied. This is why it is not correct to speak of the dopaminergic system as the reward—the latter, indeed, is appetence itself. The Mfb system, in fact, is most active when animals are in a condition of homeostatic imbalance, that is when they enter a phase of exploration and expectation, rather than following gratification. Activation of the Mfb runs from the ventral tegmental area and the lateral hypothalamus: it elicits seeking behaviour and affective states such as expectation, interested anticipation, desire, euphoria and eroticism. When stimulated in this area, animals begin to search minutely and with great interest even the least promising environments. Researchers have found that in such cases rats display intense curiosity, seeking behaviour, and attitudes of excitement and pleasure. Instead, humans report craving and a heightened sense of self, as if they felt capable of anything.

It has been shown that this system is also connected with choice, that is, with desire-interest directed towards something, with the tendency to appropriate a resource and to keep possession of it. Animals stimulated in dopaminergic areas tend not only to explore and interact with the environment, but also to become hoarders, gathering objects of all kinds. The affective states shown by animals and reported by people are strength-security, excitement, euphoria and proactivity-will. The choice of target, in animals, is regulated by the relationship between the activation of the nucleus accumbens, which creates interest and appetitiveness, and the activation of the insula, which, conversely, leads to disgust and thus drives interest away. The seeking system induces a condition of expectation, through the cortical activation of experiences, but also produces a general attitude of exploration and excitement towards novelty. The behaviour of autogrooming and parental care is also inhibited by alterations involving these subcortical areas, or by a dopamine deficit. The seeking system is therefore also referred to by some as the General Motivation System, in that it causes animals to become active agents in their natural environment, displaying proactive behaviour.

Proactivity manifests itself through various actions, such as (i) endogenous behavioural activation; (ii) incentivising salience (how much something is able to activate attention) of the orientation and desiring condition (wanting); (iii) persistence of behaviour or ability to withstand effort; (iv) articulation of behaviour or phase transitions. Some also talk of the dependency effect of the dopaminergic state, for example in terms of overactivity, curiosity, need for novelty. In animals, excessive dopaminergic activation induces a tendency to engage in repetitive and stereotyped behaviours, such as wing flapping in pigeons, walking back and forth in humans, self-cleaning in cats and biting their sides in dogs. These behaviours, which are reduced by dopamine antagonists, tend to emerge more frequently when too much time elapses between appetitive activation and consumption, or when the latter is prevented altogether. Appetitive activation and learning are related because dopaminergic action stimulates animals to go to new places, to explore, to interact, to modify their environment, and all this facilitates experiences.

Caregiving and helping behaviours, also termed epimeletic, are pleasurable for those who enact them, and much research shows that caring activities inhibit the condition of affective distress. Rosenblatt (1990) found that the transfusion of blood from rats that have given birth to females that have not yet mated caused the latter to adopt caring behaviours, such as building the den or taking care of the young. This discovery paved the way for research on the molecules responsible for epimeletic behaviour. Oxytocin has been shown to play a primary role in this, and it is thought that there is a link between epimeletism and female sexuality. In reptiles, for example, vasotocin and mesotocin generate sex drives, regulate parturition reflexes and in many species preside over nest construction. For example, it induces sea turtles (*Caretta caretta*) to head for the beach and dig their nest, and while they do so their pituitary glands secrete increasing amounts of vasotocin. Then, when the turtles have finished laying and have buried their eggs, vasotocin drops to insignificant levels (Panksepp and Biven 2012).

Areas of the brain that promote caring behaviours include the anterior hypothalamus, the paraventricular nucleus and the dorsal preoptic area. Injury to these areas can reduce or erase epimeletic behaviours. In mammals, nursing behaviours are related to lactation behaviours, although they are distinct from the standpoint of physiological mechanisms. Lactation is presided over by the lateral area of the midbrain and the spinal cord that innervates the nipples, whereas nursing behaviours can be traced to the action of oxytocin from the ventral tegmental area to the hypothalamus. Through pheromonal induction, foetal membranes also affect oxytocin release. In this regard, an important role is played by norepinephrine in the olfactory-hypothalamic activation of the mother's attachment process. Norepinephrine, in fact, facilitates the creation of olfactory memories: when it is released at the level of the olfactory bulb, it reduces the inhibitory action of gamma-aminobutyric acid (GABA), so that the neurons relevant to the encoding of odours discharge more vigorously. This experience lasts over time, giving rise to a memory at the hippocampal level, which produces attachment to the olfactory signature of the offspring.

The appetitive phase is therefore supported by the activation of subcortical nuclei through the action of specific neuromodulators. (i) If this occurs in concomitance with a releaser's stimulation, this produces a finalised behaviour. Instead, (ii) if this is not present and yet the motivation is high, the appetitive phase yields redirected or targetless behaviours. In other situations the deficit of the motivation's expressive feasibility causes substitution behaviours, such as self-grooming or stereotypies. Consummation, that is the transformation of the disposition into an action, determines a lowering of the motivational charge, so that the subject is less willing to enact that same behaviour later. But beware. It is not necessarily the achievement of the *result* that lowers the appetitive charge, but rather the expression of that particular behaviour. If for any reason the outcome is within reach and does not exhaust the dopamine charge—for example, hunting in the henhouse for a fox or in a flock for a wolf, where catching a prey is just too easy—we notice the phenomenon of overkilling or playing with prey, as in the case of killer whales with seals or cats with mice. Nor can we speak of simple exhaustion or physical fatigue. W. Heiligenberg (1964) observed that in males of *Pelmatochromis subocellatus* the disposition to fight decreases after agonistic expression although they are not at all exhausted.

We have seen that the motivational condition is induced by precise physiological mechanisms, so research in neurobiology makes it possible to answer one of Tinbergen's Four Questions, that is the one about the mechanisms underlying a certain motivational condition. As for the two remote causes considered by the author, it is clear that the presence of a certain motivation in a species depends on its taxonomic relations, that is from which progenitors it originated, and on the specific selective pressures that have shaped its different characteristics. At this point of Tinbergen's explanatory reflection, only the fourth causality is still open: that is the one I wish to discuss in view of an overall reassessment of the meaning of motivation. One of the proximate causes introduced by the Nobel Prize winner is the ontogenetic factor, that is the development process that is based on learning and brings out the identity of the individual. We have seen that motivation fosters such experiences. Therefore, we can consider motivation no longer as a simple predetermined automatism that simply discharges itself through expression, but as a copulative coordinate that leads the subject to learn and to interpret expression in a certain modality. A cat that engages in predatory behaviour with a mouse, a sparrow, a lizard or a ball is not only discharging an impulse, but is also learning to readily detect these targets, to know where to look for them and in which situations, and to perfect techniques to catch them.

Considering motivation as a copula means attributing to it a value that is not only expressive but also experiential, that is, able to foster the development of individual singularity and the implementation of innovations and immersions in creativity. We can therefore see the motivational condition in terms of psychological disposition—basically an affective state—that predisposes the subject to intersect the experiential space on a very precise plane of conjugation, to elect particular entities-events with respect to the multiplicity of presences, and to perform particular actions. A motivation can be translated into a verb-act that defines the subject's expression, but in doing so it must necessarily include modal content that is acquired through experience. It is thanks to predatory motivation that a cat learns about the habits of mice, their responses to its attempts to catch them, the places where they normally hide, the hours of the day when it is easier to surprise them. The ontogeny of the individual is accomplished on the basis of the experiences granted by the different motivations it possesses.

On closer inspection, therefore, motivations are reminiscent of *verbal predicates*, dispositions that qualify the individual on the basis of the actions it is led to perform: as water flows, so a cat chases and a human being collects. Chasing is one of the most frequent verbal predicates in cats, so that many actions of the small feline, both in survival and in play, are expressed within the verbal coordinate of chasing. Following this explanatory model, the cat only *apparently* wants the ball, because its expressive motive lies in chasing: this is its desiring expression. The ball is only what elicits or enables that particular behaviour, the occasional product of desiring, which sets up a process of learning and modal expression of the motivation. In the end, it is the world that defines the expressive space for the propensity to chase, but without this propensity there would be no stimulus capable of arousing it. In fact, if the ball is stationary, the cat pushes it to set it in motion and only then leaps to catch it, so what the cat really wants is to *chase* rather than to *reach* the target. This sheds light on the concept of desire: the object of desire is always the *consequence* and never the *cause* of desiring.

It is certainly wrong to completely disregard the pleasure of achieving a result, because in any case this can give self-efficacy and often produces gratification. However, a moving ball is not able to recall pursuit in a rabbit, because this species has little or no predatory motivation. Moreover, if it is true that predatory motivation has been established in some species because it is useful for food supply, what extinguishes the motivation is not nutritional satisfaction—that is, a full bowl is not enough to turn off our feline's desire to chase. What does extinguish it is the fact of having performed that particular behaviour until expressive fulfilment is reached. We have seen this when discussing agonistic behaviour in cichlids. Motivations bring out desirable results and objects, but these are the consequence and not the cause of desire, which has its inductive source in appetence. We could say, then, that because cats have a predatory motivation, everything that moves becomes desirable to them.

The motivational theory of ethology tells us about an orientation towards the world that unfolds in phases. At the beginning we have the appetitive phase, in which the subject has a very high expressive disposition and therefore tends to look for elements or situations suitable for action. Then there is the consumption phase, where the animal gives free rein to its behaviour, engaging in an activity aimed at a result. Finally, in the quiescent phase, the individual has achieved expressive satisfaction and is thus in a state of fulfilment that extinguishes its desire to continue its action. It is interesting to note that pleasure is distributed throughout the motivational phases and produces involvement in the subject. Indeed, the appetitive phase comes with interest and a sense of expressive fullness thanks to the action of dopamine, the consumption phase is marked by the adrenaline drive, and the quiescent phase involves a state of fulfilment attributable to endorphins and serotonin.

If an animal is placed in a situation that does not allow motivational expression, a state of discomfort occurs, which has very serious effects on its psychological condition. These include depressive states, which in the long run can give rise to lethargy or alterations in emotional sensitivity, with the onset of phobias and anxiety states. More frequently, the animal tries to redirect its expressive attentions, putting in place compensatory or dislocating behaviours, such as licking itself repeatedly, pacing back and forth in front of a gate, grabbing its tail, exposing itself to the onset of behavioural vices, stereotypies or episodes of self-mutilation. It is not uncommon to observe these behavioural alterations in confined or inactive animals. Even in humans, desire is nothing more than our imagination translating the motivational coordinates present in our species, for example the tendency to collect or adopt. This shows that desire is never caused by the lack of something, but that it is the fullness of expression that punctuates our world with desirable things.

The Sentimental Condition of Animality

As we have seen, the complexity of cortical structures in the so-called higher vertebrates, the neocortex in mammals and the striatum in birds, is not directly responsible for affective states that, on the contrary, rest on very ancient structures, such as the limbic system and the hypothalamic-pituitary axis. This leads one to think that emotional affectivity is not the prerogative of animals with very complex and voluminous brains, but is in fact a basic condition that is widely present in the animal world. In the essay *The Archaeology of Mind* (2012), Jaak Panksepp and Lucy Biven emphasise the importance of affective experience in the condition of subjectivity, starting from the fact that affective systems are genetically homologous, that is evolutionarily related, in vertebrates; they took shape because they were exceptionally useful to be able to immediately deal with the world and learn about its potentialities. As the authors write: "If one accepts affective feelings as a fundamental form of consciousness, there are many ways to distinguish those states of mind from the kind of information processing that constitutes cognitive consciousness" (2012, 14).

Emotions should not, therefore, be considered simple physiological responses devoid of affective experience, because they are able to convey intrinsic evaluations that influence the psychological state of the animal, that is what it is feeling at a given moment. These are raw affects, which then activate the associative areas whereby the individual becomes better at comparing what it is feeling with the knowledge accumulated through previous experiences. For this reason, cortical areas can be considered an enormous database capable of providing the affective state with a representational and elaborative type of comparison, referring to specific accumulated knowledge. Moreover, it has been shown that in mammals the neocortex plays a role of containment and mitigation, and not of incentive, of the affective condition: this is demonstrated by the fact that in teratological forms of lissencephaly the emotional state is emphasised. A similar condition of emotional excess is found in newborns that, in the first months of life, are kept in a condition of low sensory stimulation (sensory deprivation) or a deficit of environmental enrichment (experiential deprivation). As adults, these animals that, due to environmental deprivation, have had fewer opportunities for early learning show increased impulsivity and emotional hypersensitivity in the face of events.

John Bowlby's (1907–1990) studies have shown that in mammals the relationship with the mother is fundamental to achieve the complete and balanced development of the young's behavioural profile. For the newborn, the mother is a sort of referential centre (secure base) whose welcoming, constant and tutoring presence allows the young to successfully embark on its journey into the world. In dogs, it has been shown that the more reassuring and encouraging the mother is in the pup's early experiences, the more the latter will grow to be capable of dealing with different situations with adherence adaptive properties and a low level of impulsiveness. On the contrary, the early separation of the puppy from its mother—that is before the third month of life—causes a particularly problematic syndrome called maternal deprivation, which is characterised, among other things, by the prevalence of states of alertness, high activation of arousal and aversive behaviour. As we shall see when discussing social learning, the mother plays a central role in fostering the young's experiences (learning facilitation) and in passing on information or behavioural models. Here, too, we can see that knowledge deficit does not cause a fall in affectivity, but rather emotional emphasis.

Emotions perform a plurality of functions in animals and cannot, therefore, be exclusively regarded as pre-adapted responses to basic categorical situations, such as the presence of a risk for fear, a constraint or impediment for irritation, a dreadful event for sadness or an unexpected opportunity for joy. Of course, if we consider that one of the main characteristics of subjectivity is *to bear intrinsic interests*, it is evident that every emotion is related to a condition of the latter (the intrinsic interests) with respect to the events being foreshadowed or recorded. In this sense, fear is about a threat to one's interests, anger an obstacle or a frustrating factor, sadness a compromise, to name but a few examples. The study of the adaptive significance of emotional responses, as factors that have been established in animals from very early times, reveals the widespread presence of these affective states in the animal world. Only an anthropocentric bias leads us to attribute them exclusively to our species, but even this should not come as a surprise.

Emotional experience is the furthest thing from Cartesian mechanism: it brings out an ability to feel that cannot fit with the theory of automatisms, because it inevitably implies a perceptive Self. Also, a simple examination of comparative anatomy, neurophysiology and evolutionism shows that conceiving affectivity as a human prerogative makes no sense. According to Carroll Izard (1923–2017), father of the Differential Emotions Scale, the different emotional states reside in subcortical programmes specific to each emotion. "One does not learn to be afraid, or to cry, or to startle, any more than one learns to feel pain or to gasp for air. Of course, one does learn to fear specific things, to be angry or ashamed, or excited under certain conditions" (Izard and Tomkins 1966, 91). These considerations show the innate nature of the affective state and likewise the copulative character that enables us to somatise certain entities or events by transforming them into bodily states.

Another important scholar of emotional affectivity was Walter B. Cannon (1871–1945). Cannon (1871–1945), in his book *Bodily Changes in Pain, Hunger, Fear and Rage* (1915), challenged William James' (1842–1910) theory of the posteriority of emotional sensation, pointing out that activation of the subcortical system produces both affective experience and physiological changes at the same time. Cannon's proposed prevalence of limbic efference over visceral response does not, in fact, take into account the complexity of the bidirectional responses that link the body to the brain. Stephen Porges' *The Polyvagal Theory* notes that "peripheral organs do not float in a visceral sea. Rather, they are anchored to central structures by means of efferent pathways and are continuously signalling central regulatory structures along their abundant afferent pathways. Thus, the bidirectional connections between autonomic and central brain structures are becoming apparent" (2011, 21). Antonio Damasio, therefore, refers to emotional states as somatic markers, capable not only of producing trains of physiological and behavioural responses, but also of representing the body's condition at a mental level. Indeed, we know that the bodily state—for example, metabolic status, fatigue, hormonal fluctuation, chronic inflammation—is able to act on emotional experience, particularly mood, through somatopsychic action.

For a long time, studies on emotions in non-human animals made little progress because it was not possible to detect affective experiences in them. Thus, wishing to avoid the sword of Damocles of anthropomorphisation, it was preferred to refer only to detectable physiological and behavioural changes. As pointed out in the *Dizionario di etologia* edited by Danilo Mainardi (1933–2017) "When we observe a bird engaged in nest-building or foraging, we do not attribute these behaviours to a particular emotional state. If, however, the animal interrupts these activities to fight with a conspecific or to seek refuge, then we usually do refer to emotional states" (1992, 277). However, if it is true that the emotional response, as an explicit attitude that punctuates particular situations, occurs only in certain circumstances, Damasio is right when he states that an animal is *always* experiencing a basic emotional state.

One of the first scholars to study emotions in non-human animals was Jane Goodall, who investigated the behaviour of chimpanzees at Lake Tanganyika: her work revealed the presence of highly complex emotions such as jealousy, resentment, sadness and depression in these primates. Marc Bekoff, in his essay *The Smile of a Dolphin: Remarkable Accounts of Animal Emotions* (2000), offers a fascinating and informative look at the complexity of emotional life in the animal world. Also worth mentioning is the work of Frans De Waal and in particular his book *Mama's Last Hug: Animal Emotions and What They Tell Us about Ourselves*, which explores the theme of the emotional life of animals, emphasising how certain social emotions that are normally thought to be exclusive to humans—such as shame, resentment and hatred—can be recognised in primates. Keith Oately, Barbara King and George Wittemyer are some of the scholars who, also thanks to fundamental research in primatology, studied emotional experiences in the animal world. Today the subject of affectivity in non-humans is no longer taboo.

Charles Darwin, in *The Expression of Emotions in Man and Animals* (1872), already hypothesised that, like intellectual and mnestic functions, emotional states also had a long evolutionary history and could be identified at different phylogenetic levels. Darwin was particularly interested in the *expression* of emotions, such as is found in grimaces, postures and gestures, rather than in emotional experiences *per se*. In his 1872 essay he illustrated three behavioural principles related to emotional states that are widely present in the animal world: (i) the principle of utility, whereby certain responses serve to alleviate or satisfy particular needs; (ii) the principle of antithesis, whereby the expressions of an animal preparing to attack appear opposite to those linked to a submissive attitude; (iii) the psychoactive principle, which states that a strong activation of the nervous system is transmitted to various systems of the body, causing physiological changes. As Robert Plutchik reports in *The Psychology and Biology of Emotion*, "The evidence [Darwin] presented was designed to illustrate the basic continuity of emotional expressions from lower animals to humans. He suggested that the baring of the fangs of the dog or wolf is related to the sneer of the human adult" (1994, 23). After all, it is clear that some visceral reactions associated with fear, such as defecation and urination, are equally found in mice, cats, dogs, humans and other primates.

Many defensive reactions related to fear or anger involve a magnification of the bodily volume: for example, the swelling of the mane in lions, horripilation in most mammals, the ruffling of feathers in birds and the expansion of sacs and fringes in amphibians and reptiles. The universality of certain zoosemiotic expressions, such as wrinkling the nose or curling the lip to show the canines in order to indicate a threat, means that such menacing interactions can happen among different species, as can be seen in carcass disputes between hyenas and lions. The work of the English naturalist opened up a further field of research, in addition to that of the phylogenetic and adaptive bases of emotional responses—and consequently of comparative research on emotions: namely, the study of the communicative meaning of emotional expressions. In the last chapter of his essay, Darwin hypothesised that some emotional expressions, such as those referring to fear or anger, were primitive and therefore present in taxonomic units much larger than the species or genus, an argument that would be illustrated by Konrad Lorenz in *On Aggression* (1966). The study of the relationship between emotional states and communication was also taken up by Detlev Ploog in his "Biological Foundations of the Vocal Expression of Emotions" (1986), which showed how emotional signalling, in particular vocal signalling, is very important in processes of social regulation.

Ploog's approach, termed *neuroethology* precisely because it closely links neurophysiology and ethological findings, compares the vocal behaviour of different types of animals in an attempt to link differences in both lifestyle and brain systems. Using this methodology, Ploog and colleagues were able to diversify about fifteen emotional signals developed in social interaction, such as warning, protest, challenge, engagement and pleasure. Ploog concluded: "Since social signals are expressions of emotion, we must concede that vertebrates—whether at a high or low evolutionary level—express emotions when they communicate via signals". Similar research, conducted on dogs by Turid Rugaas in her essay *On Talking Terms with Dogs: Calming Signals* (2005), shows that interaction between different individuals always expresses the emotional state of the subject and aims to quiet the alertness and aversiveness-agonism system by trying to reduce the tension of the encounter. That is why the Norwegian scholar speaks of "calming signals".

In his essay *Grundriss der vergleichenden Verhaltensforschung Ethologie*, Irenäus Eibl-Eibesfeldt points out that many animals exhibit alarm signals that warn their fellow creatures about predators: "spermophili (*Citellus citellus*) and marmots (*Marmota marmota*) cry out before fleeing, as do many birds (…); chicks still in their eggs fall silent and stop scratching their shells when they hear a chicken's alarm call" (1974, 180). In some animals, certain alarm signals are specific to the type of danger present, so that they indicate not only their affective state, but also the content that provoked it. In an article published in 1980 in the journal *Animal Behaviour*, the ethologist Robert Seyfarth and colleagues showed that the grivet (*Chlorocebus aethiops*) makes specific signals for various types of predators; these signals provoke different reactions in its conspecifics, consistent with the best way of escaping the given danger. Nevertheless, alarm signals are not necessarily expressed through the sound and hearing channel. We know, for example, that many mammals, such as dogs, cats, rabbits and rodents, release specific alarm pheromones that trigger emotional states in their fellow creatures. Aquatic animals sound the alarm by emitting particular chemicals; this has been detected in molluscs, for example in the snail *Helisoma nigricans*, in fish and in amphibians.

The state of fear very often increases the cohesion of the social group: primates can be seen hugging each other, buffaloes huddling in rows, birds shrinking into flocks or fish in schools. As Irenäus Eibl-Eibesfeldt points out in *Liebe und Hass* (1970), fear-induced cohesion has its roots in ancient times and plays an important role in all social animals. For animals living in large groups, there is also the phenomenon of *emotional osmosis*, which means that the fear or excitement experienced by one or more members of a group soon spreads like wildfire throughout the community. The reason for this is clear: emotional responses have emerged in different species because they are advantageous both in responding promptly to an event and in anticipating something that is about to happen. The time it takes to notice a favourable or unfavourable event and the readiness to provide a coherent response are, therefore, the two most important variables in emotional appraisal. Flocks of birds roosting to peck at a resource can be easily seen taking a sudden concerted flight, allowing each individual to respond promptly even without having directly observed the danger. As is well known, living in large communities has the advantage, among other things, of multiplying the sensors on the surrounding environment, and this effect, associated with the phenomenon of emotional osmosis, is extremely beneficial to everyone.

But let us return to the primordial character of affective states, which makes them widely shared among species. Some emotions, such as fear, are processed in the limbic system in vertebrates, the set of predominantly subcortical nuclei that we find in these animals. Fear, for example, is an affective experience linked to some deep areas of the brain such as the periaqueductal grey (PAG), the lateral area of the amygdala, the ventromedial area of the hypothalamus and part of the midbrain. In mammals, fear tends to increase expressions of alertness, immobility (freezing), a decrease in some self-directed behaviours, such as feeding and grooming, many social activities, such as allogroomig and playing, and sexual activity. The emotional stimuli of fear can differ from species to species, as regards olfactory-pheromonal aspects (e.g. the smell of ferrets or cats for rodents) or visual feedback (e.g. open and well-lit spaces for rats, dark and closed environments for diurnal animals). Most mammals, for example, are afraid of loud noises, and pain can also stimulate fear. Yet, the reverse is not true: intense fear can indeed *inhibit* the experience of pain, since with strong fear the brain secretes analgesic substances, such as endogenous opioids.

Neurophysiologists have pointed out that fear conditioning can be produced by the classic Pavlovian mechanism: if a light is consistently switched on before an electric shock, both rats and cats learn to avoid everything that is present in the phobogenic situation. The process of fear conditioning can give rise to a state called "heightened alarm reaction" whereby, if the subject is exposed to the conditioned stimulus or contour elements and then to a given sound, it will have a much stronger reaction and will easily become sensitised to noise. Some drugs such as benzodiazepines (BDZ) stimulate the most important inhibitory system, the GABAergic system, producing sedative and hypnotic and, in some cases, anxiolytic effects. Benzodiazepines therefore act by enhancing the effect of gamma-aminobutyric acid, a neurotransmitter that inhibits the activity of neurons by reducing their discharge frequency, so these substances have the side effect of reducing learning ability, as well as causing addiction. BDZs affect the entire fear system from the amygdala to the PAG to the pons, which is responsible for the alarm reflex.

Anxiety is often a consequence of repeated experiences of fear or stress, but this is not the case for separation anxiety: this condition, typical of baby mammals left alone for too long, is linked to another kind of affective experience, sadness and suffering, so that benzodiazepines have no effect in eliminating crying and separation-related behaviour. Fear and suffering have different circuits and are handled differently by the autonomic nervous system: (i) fear is expressed by the sympathetic system, which increases the heart and breathing rate, also dilating the pupils to increase vigilance; (ii) sadness is mediated by the parasympathetic system, which slows the heart rate and promotes tearing. Since the fear system is promoted by the sympathetic nervous system, treatment with drugs that reduce the action of endogenous catecholamines, for example, beta blockers, has been used in behavioural medicine. In addition to catecholamines and acetylcholine, the corticotropin-releasing factor and corticotropin itself also produce anxiety conditions, as do cholecystokinin, neuropeptide Y and glutamate, which act on the amygdala. We must not forget that the affective condition of fear develops from the regions of the hypothalamus and the PAG, but has its most important centre in the amygdala. Antonio Damasio's studies, reported in the book *Descartes' Error* (1994), show how lesions or degeneration involving this subcortical nucleus have serious consequences on the subject's general evaluation process.

Once again, emotional experience has a strong copulatory significance, acting on learning processes and stimulating readiness but also creativity of response. In the essay *How We Learn* (2020), Stanislas Dehaene stresses the importance of emotional activation in learning processes, especially through emotions such as curiosity and surprise. A similar consideration of the relationship between emotion and mnestic systems is reported by Joseph LeDoux in his book *The Emotional Brain. The Mysterious Underpinnings of Emotional Life* (1996). In particular, neurophysiological research has shown how the activation of certain subcortical nuclei, such as the amygdala, is able to activate both the brain areas responsible for perception and short-term memory and the attention systems through the action of acetylcholine and the hippocampus, responsible for spatial and referential memory. Fear experiences therefore generate very specific memories that influence the subject's orientation.

A different condition relates to the affective condition of sadness, which is easily seen in animals marked by parental care or strong social relationships. The evolution of attachment and social bonding has paved the way for psychological distress and separation anxiety mechanisms in both the young and the parent. Pups separated from their mother emit separation-evoked distress vocalisations. Research has shown that when parents hear their offspring cry, regions of separation anxiety are activated in their brains. These activation areas, which can be observed using neural imaging tools such as positron emission tomography (PET), are common in all mammals. These are: the anterior cingulate cortex, the dorsomedial thalamus, the periaqueductal grey, the ventral septal area, the preoptic area, the striatum and some areas of the cerebellum. The same structures and neurochemicals are also at play in birds.

There is good reason to believe that the psychological distress system evolved from primitive pain circuits in the brainstem, for example in the periaqueductal grey. When this system is active, animals have an affective experience described as *psychological pain* and are intensely motivated to seek reunion, which extinguishes suffering and produces a wave of pleasure mediated mainly by endogenous opioids. Separation, on the other hand, produces agitation at first and then, in the long term, depression. The physiochemical mechanisms of this process can be traced to a massive release of adrenal stress hormones. The sensitivity of the distress system gradually decreases as puberty approaches, suggesting that increased levels of sex hormones, especially testosterone, play a crucial inhibitory role. Since the pleasure of reunion and the pleasure of opioids are linked to MU receptors, one can also speak of *affective addiction*. On the other hand, it is well known that self-directed somaesthetic activities and mutual grooming play a central role in social relationships and have anxiolytic and rewarding effects, precisely because of the action of endorphins.

We know that episodic memories and those relating to spatial orientation are based on the action of the hippocampus. Unfortunately, it has been shown that distress, through the continuous production of cortisol, seriously damages the hippocampus and therefore compromises learning. For this reason, early and prolonged social deprivation can impair the maturation of the mind in the young and cause degenerative effects on memory. Distress also causes a depletion of biogenic amines—norepinephrine, serotonin, dopamine—as well as a drop in the *neuronal growth factor* and thus in neurogenesis processes. These effects are at the root of depression, a condition that can be seen in animals confined to zoos or continuously harassed within their social group. In *Through a Window* (1990), Jane Goodall tells the unfortunate story of Gilka, the female chimpanzee whose community partner Passion killed her infants on several occasions: her state of sadness compromised her life, leading to an early death. There is, therefore, a difference between fear and sadness and consequently between performance anxiety and separation anxiety. This is demonstrated by the fact that opioids and oxytocin are much more effective for sadness, while benzodiazepines are more effective for fear.

It is very difficult to separate a condition of joyfulness from other forms of emotional pleasure or openness which develop in the face of extremely pleasant consequences, such as receiving care or during play. Following prolonged isolation, many primates become despondent; instead, after reunion they emit cries of joy and are prone to play. Rats exhibit ultrasonic squeaks in the face of pleasant events and during play. It appears that the full development of the frontal and septal regions—those that mature more slowly—inhibits joyful tendencies in adults, which may be why the young are more prone to this emotional response. There is a relationship between dopamine, feelings of joy and a tendency to play. Average levels of dopamine promote joyfulness and playfulness, while excessive doses of dopamine lead to restlessness, increasing seeking and goal-oriented behaviour. Excess testosterone also causes a decrease in playful and joyful behaviour, reinforcing more competitive-conflicting attitudes.

The thalamic regions of the midline, such as the dorsomedial thalamic nucleus, are responsible for joyful responses, but lesions in the ventromedial hypothalamus also diminish playful tendencies and make animals particularly aggressive. Finally, it has been found that endocannabinoids play a major role in playful behaviour, which may be why marijuana causes a tendency to laugh in humans. Laughter circuits have been mapped in rats: they involve the activation of the mesolimbic system and the joint action of dopamine, endogenous opioids and endocannabinoids. Laughing thus appears to be a strong activation of the pleasure and appetite systems. The insulin-like growth factor also promotes playfulness behaviour. As we know, play is fundamental to the processes of creativity and it has been found—even in humans—that dopamine activation associated with endocannabinoid action stimulates creative behaviour precisely because it frees up expression.

The affective condition of anger presents various overlaps with aggression, although not all forms of aggression present affective states of anger. There is also a relationship between anger and frustration and between anger and insecurity or fear, so in some circumstances, when the animal is unable to flee, it may switch from a flight reaction to a fight reaction. Certain homeostatic imbalances, such as hunger, heat or thirst, can evoke anger. Competitive behaviour can also give rise to anger-aggression reactions. Anger leads to increased pressure, and increased pressure makes an animal more irritable, so drugs that lower pressure, such as beta blockers—for example, propranolol—that deactivate adrenergic and noradrenergic receptors also lower irritability. Stress and overstimulation, or unpleasant and annoying stimuli, also trigger anger behaviour in the long run. The main stress axis comprises the pituitary (hypothalamic-pituitary) system, the sympathetic system and the adrenal glands. If these systems are activated, individuals tend to interpret their environment as hostile-problematic.

One of the easiest ways to provoke angry aggression in most adult male mammals living with females is to introduce another male into their environment. The neuroanatomy of angry aggression has been explained by Alan Siegel: it involves the lateral hypothalamus, the periaqueductal grey, the medial areas of the amygdala and the striatum. The most important structure for this condition is the periaqueductal grey. The ***periaqueductal grey matter*** (PAG) is that portion of grey matter in the midbrain that surrounds the cerebral aqueduct. The PAG is also responsible for modulating pain, while the interaction between the PAG and the hypothalamus determines the primary condition of anger. The hypothalamic structure is very important because within it the various nuclei mediate different but interacting responses. For example, (i) the ventrolateral areas are related to anger, (ii) the ventromedial areas to fear, (iii) the dorso-lateral areas to seeking.

Walter Hess was awarded the Nobel Prize in 1949, when he discovered hypothalamic-stimulation-induced anger in cats. The PAG, ventrolateral hypothalamus and medial amygdala determine the affective experience of anger: when stimulated in these areas, animals will immediately and furiously attack everything in front of them. Stimulation in humans produces similar responses, such as clenching of fists, muscle contraction throughout the body and clenching of the jaws. The chemicals that promote anger are testosterone, corticotropin, catecholamines, acetylcholine, glutamate and substance P. Substance P is a short-chain neuropeptide with neuromodulating functions in mammals, involved in pain, vomiting and vasodilatory hypotension; it also increases salivary secretion and diuresis. There are several encephalic receptors for substance P: the NK1 receptor, for example, specifically promotes aggression. The lack or physiological deficiency of substance P—as happens in the naked mole rat *(Heterocephalus glaber)*—leads to reduced sensitivity to physical pain and reduced social space. The naked mole rat is a rodent that lives in East Africa and forms societies that are unique in the world of mammals because they consist of a queen and other worker individuals. This strange rodent has adapted to life underground and oddly enough, despite being a mammal, is ectothermic. But its special feature is its eusocial status: the mole rat lives in colonies of 70–80 individuals, similar to those of social hymenoptera.

Mammals, for example deer and wolves, generally display conflict between males, which can be expressed through rituals but also in the form of far from peaceful confrontations. Dominance aggression is mainly linked to testosterone levels, which can (but doesn't necessarily) result in angry attitudes. In fact, damage to certain regions such as the preoptic area of the anterior hypothalamus, the septum, the nucleus accumbens and the raphe decreases dominance aggression, but encourages anger. Dominance is mainly related to testosterone, but it appears that the neuropeptide responsible specifically for dominance aggression is vasopressin, the synthesis of which is promoted by testosterone. It is necessary to distinguish between aggression and anger, even in the behavioural translation of aggression. For example, predatory aggression is linked to the dopaminergic-appetitive system, not to anger, so that the centres of this system are not activated during predation. Angry irritation can be curbed or diminished by septal stimulation, since this region produces the modulation of this affective state through cortical influence. Therefore, damage to septal areas can increase choleric tendencies.

In conclusion, one can say that the emotional life of animals is very rich and produces affective experiences that are fundamental in learning processes as well as in the construction of personal identity. Emotions, in fact, urge individuals to find solutions that are immediate but above all adherent to the specific nature of the situation in which they find themselves. It is therefore not difficulties that sharpen our wits, but the emotions we feel when faced with the kaleidoscope of the world, where difficulties and surprises activate our inventiveness.

Works Cited

Bekoff, Marc. 2000. *The Smile of a Dolphin: Remarkable Accounts of Animals Emotions*. Toledo, OH: Discovery Books.

Cannon, B. Walter. 1915. *Bodily Changes in Pain, Hunger, Fear, and Rage*. New York: D. Appleton and Company.

Craig, Wallace. 1918. Appetites and Aversions as Constituents of Instincts. *Biological Bulletin* 34 (2): 91–107.

Damasio, Antonio. 1994. *Descartes' Error: Emotion, Reason, and the Human Brain*. New York: HarperCollins.

Darwin, Charles. 1872. *The Expression of the Emotions in Man and Animals*. London: John Murray Albemarle Street.

Dehaene, Stanislas. 2020. *How We Learn, Why Brain Learn Better than Any Machine*. New York: Viking.

Eibl-Eibesfeldt, Irenäus. 1970. *Liebe und Hass*. München: Piper.

———. 1974. *Grundriss der vergleichenden Verhaltensforschung-Ethologie*. München: Piper.

Goodall, Jane. 1990. *Through a Window*. Boston: Houghton Mifflin Harcourt.

Heiligenberg, Walter. 1964. Ein Versuch zur ganzheitsbezogenen Analyse des Instinktverhaltens eines Fisches (Pelmatochromis subocellatus kribensis Boul Cichlidae). *Ethology* 21 (1): 52.

Hinde, Robert. 1956. The Biological Significance of the Territories of Birds. *Ibis* 98 (3): 340–369.

Izard, Carrol E., and Silvan S. Tomkins. 1966. Affect and Behavior: Anxiety as a Negative Affect. In *Anxiety and Behavior*, ed. Charles Spielberger, 81–125. New York: Academic Press.

Kluyver, H.N. 1947. Over het gedrag van een jonge Grauwe Vliegenvanger en van een troep Pestvogels in de winter. *Ardea* 35: 131–135.

LeDoux, Joseph. 1996. *The Emotional Brain: The Mysterious Underpinnings of Emotional Life*. New York: Simon and Schuster.

Lehrman, Daniel. 1953. A Critique of Konrad Lorenz's Theory of Instinctive Behavior. *Quarterly Review Biology* 28: 337–363.

Leyhausen, Paul. 1965. Über die Funktion der relativen Stimmungshierarchie (dargestellt am Beispiel der phylogenetischen und ontogenetischen Entwicklung des Beutefangs von Raubtieren). *Zeitschrift für Tierpsychologie* 22 (4): 412–494.

———. 1979. *Cat Behavior: The Predatory and Social Behavior of Domestic and Wild Cats*. New York: Garland STPM Press.

Lorenz, Konrad. 1937. Über die Bildung des Instinktbegriffes. *Die Naturwissenschaften* 25 (19): 289–300.

———. 1966. *On Aggression*. London: Methuen.

Mainardi, Danilo. 1992. *Dizionario di Etologia*. Torino: Einaudi.

Olds, James, and Peter Milner. 1954. Positive Reinforcement Produced by Electrical Stimulation of the Septal Area and Other Regions of Rat Brain. *Journal of Comparative and Physiological Psychology* 47: 419–427.
Panksepp, Jaak, and Lucy Biven. 2012. *The Archaeology of Mind: Neuroevolutionary Origins of Human Emotions.* New York: W.W. Norton & Company.
Ploog, Detlev. 1986. Biological Foundations of the Vocal Expression of Emotions. In Emotion. In *Theory, Research, and Experience*, ed. Robert Plutchik and Henry Kellerman, 3, Biological Foundations of Emotions. Orlando: Academic Press.
Plutchik, Robert. 1994. *The Psychology and Biology of Emotion.* New York: Harper Collins.
Porges, Stephen W. 2011. *The Polyvagal Theory. Neuro-physiological Foundations of Emotions, Attachment, Communication, Self-Regulation.* New York: W. W. Norton & Company.
Rosenblatt, Jay S. 1990. Landmarks in the Physiological Study of Maternal Behaviour with Special Reference to the Rat. In *Mammalian Parenting*, ed. Norman A. Krasnegor and Robert S. Bridges. New York: Oxford University Press.
Rugaas, Turid. 2005. *On Talking Terms with Dogs: Calming Signals.* Wenatchee: Dogwise Publishing.
Seyfarth, Robert M., Dorothy L. Cheney, and Peter Marler. 1980. Vervet Monkey Alarm Calls: Semantic Communication in a Free-Ranging Primate. *Animal Behaviour* 28 (4): 1070–1094.
Tinbergen, Nikolaas. 1952. The Curious Behavior of the Stickleback. *Scientific American* 187 (6): 96.
Van Iersel, Jan Jozef Arnold. 1953. An Analysis of the Parental Behaviour of the Male Three-Spined Stickleback. *Behaviour Supplement* 3: 1–159.

CHAPTER 5

The Ownership Principle

Premise

The affective experience shows a particular level of subjectivity in animals, which is expressed in the interested condition that leads them to search the outside world with yearning. It could be defined as *feeling something* when looking at one's landscape. In their dialogue with the context, animals are enveloped, as it were, within a fluctuating succession of feelings resulting from the transitory nature of the various emotions. At the same time, they are shaken by the tide of passions, when the appetites of the various innate motivations transform the environment into a space of possibility. Affective systems translate experience into bodily states, because the hedonic characterisation of feeling something in the encounter with a phenomenon inevitably produces somatisation. Experience, therefore, is never something neutral or external to the subject, but takes place in terms of internalisation of experience and personalisation. Affectivity therefore produces a state of self-ownership.

We have seen that a primordial form of *Self-referred* subjectivity can already be found in the basic mechanisms of organised locomotion, since, as Giorgio Vallortigara points out, the difference between a self and a non-self becomes fundamental "when the organism moves actively with some form of locomotion mediated by cells that are separate from those in charge of receiving stimuli". Indeed, to move it is crucial to be able to distinguish "the sensory signals that come from the outside world and

R. Marchesini, *The Creative Animal*,
https://doi.org/10.1007/978-3-031-07414-1_5

those that result from the organism's own movements in the world" (2021, 107). We can say, then, that the affective condition produces a second level of subjectivity, which can be defined as *for-Self* or referred to the condition of pleasure.

The affective subjectivity of animality speaks not only of the dispositional character of the individual's *coming into the world*, but also of two other things. First, (i) the aesthetic translation of sensory reports, whereby windows onto the world provide not only a landscape but also more or less pleasant sensations. Second, (ii) the hedonic consequences of homeostatic mechanisms, whereby needs become sources of pleasure (if fulfilled) or of suffering (if denied). At the heart of this level of subjectivity, which I have defined as for-Self, there is undoubtedly the condition of pleasure, which in some ways subsumes the Benthamian principle of "Can they suffer?" or Jacques Derrida's inverted version of exposure to pain—as well as the Freudian principle of the Es. Affective subjectivity leads animals to constantly gauge their actions against the pleasure they may derive from them, so the individual's fundamental compass is the search for pleasure, which can originate from various sources and is never predetermined. This brings us back to the principle of animal peripateticism, the act of "strolling" through the world seeking small sprinkles of pleasure deriving either from mitigating situations of need or from achieving gratifying objectives.

One of the most important principles of learning, although overemphasised by the behaviourist school, is the law of effect developed by Edward L. Thorndike (1874–1949) in the first decades of the twentieth century. According to this law, learning is affirmed by a mechanism analogous to Darwinian selection, whereby a behaviour becomes more and more likely in the individual—just like a character within a population—the more its consequences are able to bring pleasure (positive reinforcement) or alleviate suffering (negative reinforcement). The hedonic system, constituted by the dialectic between pleasure and suffering, undoubtedly induces a given behaviour, for example when an animal wishes to escape from an unpleasant situation or to alleviate a particular yearning or restlessness. Talking about motivational dynamics, we have seen how (i) the appetitive phase gives rise to a condition of exuberance which, however, is also unrest that seeks expressive possibilities. We have also seen how, (ii) in consummation, the subject reaches a sense of fulfilment that extinguishes the restlessness and sometimes adds a hedonic supplement through the gratification resulting from achieving an objective.

However, while being central in inducing or confirming behaviour, pleasure has so many sources that it is more of an orientation tool, a sort of rough indicator, than a predictive factor of behaviour. Pleasure is certainly the final goal of any action for-Self, but affective subjectivity presents a very wide range of potential hedonic returns. In other words, except for situations of expressive cogency, when for example there is a peak of motivational appetence or an urgent need, in most situations the animal finds itself at a table set with many courses, where some require a certain amount of effort while others are easily accessible snacks. This means that the plurality of options, even if marked by the hedonic meaning, leaves room for the ownership of choice. Thus, attributing a "for-Self" dimension to animals means no longer considering them as puppets passively driven by pleasure, but rather as subjects who, precisely by virtue of the affective condition, have reason and power of choice. Hence an aspect of subjectivity that is marked by complexity, indeterminacy, state singularity and above all creativity. Affectivity lays the foundations for choosing one's behaviour at any given point in time, and this always involves a creative component.

The factors of affective influence are multiple, the situations encountered in the world are variable and indeterminate, and the mixture of emotions and motivations experienced by the individual in the here and now is always unique: these are the reasons why the animal is able to choose. The complexity of affective influences leaves open a space of freedom that the animal fills through its choices, whether conscious or not. We could say, then, that the more plural the interests and influences in the subject's present, the greater the space of existential ownership, because it is precisely in ambivalence that the subject needs to be able to make decisions. I shall return to this topic when speaking of the instruments of knowledge—the third level of subjectivity. For now, let's just say that the greater the number of determining and influencing factors, the wider the scope of subjectivity. We must not, therefore, imagine the subject as an entity managing the different indications that come from the body, the mind or the environment, but rather as a space of superintendence that emerges from the plurality of factors at play. For this reason, the affective condition does not produce a cascading directionality of behaviour, but enables the ownership of choice in arranging a multiplicity of pleasant conditions.

Indeterminacy as a Space of Subjectivity

Supporting the idea that there are precise hereditary contents in behavioural expression, as opposed to the concept of *tabula rasa*, ethology has highlighted how individuals of a particular species show fixed behavioural modules, acting in a standardised manner that is scarcely influenced by experience. Consider, for example, the way in which the greylag goose (*Anser anser*) retrieves eggs that have rolled out of the nest (Tinbergen 1951), the male guppy's (*Lebistes reticulatus*) courtship through a sigmoid posture (Baerends et al. 1955) and the way in which passerines scratch their heads (Wickler 1961).

Selective reactions to stimuli through stereotypical attitudes also point in this direction. The male robin (*Erithacus rubecula*) responds aggressively not only to the sight of a rival, but also to reddish feathers (Lack 1939), just as the male stickleback reacts in a confrontational manner to a model with a red belly (Tinbergen 1951). Likewise, in estrildid finches feeding is stimulated by the presence of particular markings in the chicks' beaks: the study of these effective stimuli was therefore carried out through the use of decoy-birds. In certain shapes, the latter can even become hypersignificant, precisely to confirm the innate dialectic between releaser and behavioural pattern.

A particularly interesting case concerns the use of a model with an asymmetrical structure that can be interpreted, depending on the directionality of movement, as either a goose or a hawk. As Robert Hinde reports in his essay *Animal Behaviour. A Synthesis of Ethology and Comparative Psychology* "Lorenz and Tinbergen found that this response could be elicited by a relatively simple moving model (…) the model elicited a fleeing response if moved with the short end in front ('hawk model') but not if pulled the other way ('goose model')" (1970, 66). This research is still extremely useful today in showing that animals are not passive to external events. In fact, there is no doubt that, from an innate point of view, different species present fixed behavioural patterns and selective reactivity to particular stimulus configurations. Both emotions and motivations express a way of being interested in the world, as they have a copulative and predicative value. However, precisely for this reason, they rely on modal orientations, such as the Gestalts capable of stimulating an emotion or a motivation, as the response patterns to such stimuli. These orientations are innate expressive contents, which represent the behavioural spellbook of each species, namely the fundamentals that allow the subject to navigate in the world.

It is wrong, however, to deduce from these unquestionable considerations that animal behaviour is based on simple automatisms. The detection of species-specific expression through ethograms cannot go into the detail and finesse of individual behaviours. As Hinde himself points out, "a complex stream of behaviour must be described and broken down into units suitable for study, which then must be classified into groups according to common properties" (1970, 10). If what I am interested in is not individual variability but the definition of a species-specific style, I will adopt a level of generalisation such that it can outline common qualities. In fact, it is evident that expressive standardisation depends on the scale of evaluation adopted, so that individual variations can always be reduced to noise by adopting a certain metric. On closer inspection, the classical ethological approach has a very precise epistemological structure that necessarily produces a certain type of feedback, just like a particular lens—a wide-angle or a telephoto—deforms reality to obtain a specific perspective. This does not mean modifying the picture, but it does encourage a certain angle on it.

Moreover, it should be noted that most of the standardised choreograms highlighted refer to behaviours of high adaptive value and strong environmental influence, such as the response to a threat, courtship, parental care and competitive behaviour. As Irenaus Eibl-Eibesfeldt points out in his monumental *Human Ethology* (2007), standardised responses are also easy to detect in human beings, with little sensitivity to cultural or experiential superstructures, when it comes to behaviours such as fear reactions, courtship cues, and competitive or challenging expressions—an aspect already highlighted by Charles Darwin (1872). Yet we would never say that such evidence suggests that courtship and agonism are two behavioural automatisms for humans. Innate responses do not take away from the individual's psychological dimension or limit one's expressive range. If one observes mammals during exploration or play, one will see that the innate configuration presents itself more as a range of possibilities than as the mere repetition of a fully predetermined pattern.

In other words, there is an epistemological problem that should be addressed when one claims to shift from categorical evaluation to predicative considerations about the subject. If one adopts given measurement scales and boundary factors that favour certain findings over others, it is not correct to say that this evidence is proof of expressive mechanism. This is not to say that such evidence is an artefact of observational methodology, but the type of research undertaken undoubtedly highlights certain aspects of reality and not others. It is no coincidence that when, in the last decades of the twentieth century, we concentrated on aspects of individuality, we began to detect animal personality phenomena—see the essay by Claudio Carere and Dario Maestripieri, *Animal Personalities* (2013). There are peculiar characterisations even among different ant-hills of the same species, in terms of a tendency to reuse expressive endowments in new contexts. Even the cells of a *Polistes* or *Vespa crabro* hive may look perfectly identical from a distance, but if you look closely they are always slightly different. This phenomenon can be defined as a scale gradient, a distorting effect that shows how innate endowments certainly define expressive perimeters, but these must be interpreted as a range.

As we have seen, the phylogenetic heritage does not only provide the animal with representational tools—for example: what meaning to attribute to a phenomenon or what use value to an entity, how to deal with a certain challenge or adaptive function, what stimuli to extract from the context or recompose in a *Gestalt*. It also generates expressive drives, which support the openness-connection between the subject and the external world and initiate the expressive experience as a process in becoming. The emotional and motivational expressive drive is never self-referential, despite being aimed at ensuring the individual's interests—in line with the for-Self—but always presents a copulative dynamic, linked to circumstantial elements. The variability of the application context and the singularity of the subject's psychological state determine the expressive need and therefore the particularity (indeterminacy) of the result. In order to understand the subject's ownership in interpreting its heritage, it is not necessary to appeal to a presumed insufficiency of the innate, but rather to the latter's copulative exuberance. It is the heritage that produces singularity.

If we want to fully understand this process, which through heredity produces individuality, we must regard affectivity not as a repetitive series of automatisms, but as a pull towards the world: an experience that favours the emergence of singularity. As already anticipated by the drive theory of Konrad Lorenz's classical ethology, the world can recall or arouse a certain behaviour but not provoke it, because the motive that sustains it is always internal, resulting from the presence of a certain disposition. Affective components lead the subject to be interested in the world and thus set the individual in motion. They do not only trigger animal expressivity, but also place it on a well-defined plane of intersection: the space of choice is indeed referred to the characteristics of the world, but also organised by the copulative mode that the experiential subject puts into action at a particular moment. To be interested means to selectively extract something from the sensorial landscape, experiencing it in the performance of the conjugative action itself: when a cat plays at chasing a ball it learns about itself and its possible choices through its predatory character.

Emotions (e.g. feeling fear or joy about something) and motivations (e.g. being led to perform particular activities such as chasing or gathering) have a copulative function because they are always directed outwards; at the same time, their predicative nature transforms experience into a space open to creativity and occasional choice. Let us take an example. Predatory motivation, as an innate tendency to chase, leads the animal to search for a moving target and for circumstantial complements, that is the contexts, situations, modes and times to do the chasing. The predatory motivation causes the subject to experience different moving targets, different ways of reaching them, different contexts in which to find them; it defines a certain interest and a certain experience. The predatory copulation is different from, say, the syllegic one, that is the propensity to collect and hoard objects: in the latter case, the focus is on iconic structures, what emerges from the context, such as a fruit, a mushroom, a flower, a shell.

An animal that encounters the world through a certain affective state makes a selection with respect to what it can experience, but the field of expressive opportunities, precisely because the motive is internal, remains undefined, that is open to possibilities. Moreover, an animal's affective state is never stable, but tends to change on the basis of the feedback it receives in the course of the experience itself, so even the copulative aspect is in becoming. And even if there were a stable motivational prevalence—for example, the predatory one in a lurking cat—there is an evident modal difference in the resulting experience, depending on the species being hunted and sometimes even on the circumstantial characteristics of the hunt or the qualities or skills of the prey. Therefore, a subject could not implement any behaviour without ownership: it would be condemned to perpetual failure. In short, I am not denying the characteristics of specific endowment, but rather the traditional way of regarding them as automatisms driving the individual. On the contrary, they underlie a complex subject-world dialogue which transforms experience into a process that is unique and unrepeatable each and every time.

The affective component places the subject on a copulative plane of intersection, giving rise to a search field: freedom of movement among the eligible opportunities endowed with hedonic content. The copulative level—offered not only by the emotions and motivations prevailing in the subject's *hic et nunc*, but also by latent (i.e. not compelling) needs and aesthetics—defines an experiential virtuality containing choices, experiences and proposals that are characterised by contingency and singularity. This field is also open to invention, attempts, probing and creativity. As we have seen, experience is always a dialogue which, precisely because of its intrinsic constitution, has an open and indeterminate structure. Experience is therefore not predefined in its results, but can only be assessed as a field function: the affective (dispositional, homeostatic and sensory) components play the role of *strange attractors*, capable of influencing the space of action but not the specific trajectory or the deterministic outcome of the process. This definition may seem to involve a weak conception of subjectivity, since it goes beyond free will or consciousness, but I think it is a good starting point to get away from the deterministic view based on automatisms.

The affective components influence the copulation space and therefore also the perimeter of experiential virtuality that the subject can perform. For this reason, the experience of a child (prevalently syllabic) cannot be equated to that of a kitten (prevalently predatory) when both are placed in the same situation, for example a meadow—nor can it be completely predetermined. There are too many variants, and the individual's circumstantial subjectivity dwells precisely in this space. The experiential result is guided, though not predetermined, not only by the sense organs, but also by the affective dispositions prevailing in *that* subject at *that* moment. A given emotional sensitivity, for example, can determine salience, that is what the individual places in the foreground in its relationship with external reality, while a certain appetite can push for more desirable results. However, the influence of affective systems draws a space of possibilities within which the subject's ownership is realised. Just as the sense organs define a landscape of access, but not exactly *what* we perceive, in the same way the affective state places us in a certain copulative space, but does not decide what experiences we are going to have: it simply outlines the field of probable peripateticism.

It has been said that emotions and motivations in behavioural propositions are predicative functions that, as such, rely on complements drawn from the world: towards what to direct affectivity, how to express it, with what timing and in what situations. The copulas link the individual's disposition to external elements, giving rise to hybrid propositional structures: for example, in the case of a cat, chasing the ball around the house, performing an expressive dynamic open to all possible variations on the theme. In this case, therefore, it is not simply a question of linking motivation to an object and experiencing it, but of transforming the ball and the house into entities that free the cat's creativity and allow it to experiment with new motor choreographies. Watching a cat play with a ball, we recognise the prevailing motivational structure (predatory), the emotional condition and the state of arousal, just as we know what our house looks like and what the properties of the target are. However, we cannot go further than a space of expressive possibilities: we will never know exactly which playful choreographies the cat will enact. But, I would like to add, the cat most probably does not know either. Its game is made up on the spot and that is all there is to it. Affectivity opens up a space of expressive ownership.

The affective state in the intersection with the world creates a field of possibilities and thus allows the subject that freedom of interpretation that makes it the owner of its own behaviour. In this case, paradoxically, to be a subject is to let oneself go to a creativity that escapes even decision-making coordinates. Experience is the inclusion of this principle of behavioural indeterminacy. If behaviour is comparable to a propositional expression, a sentence governed by affective predicates, its conjugative structure (the modal and circumstantial complements it envisages) is a field function—that is to say, it is only conceivable as a space of probability, not as a linear and predictive sequence. In fact, the expressive choreogram does not precede the action, but develops in the course of experience itself, and therefore the configurative endowments instruct it only in part. Precisely in virtue of the copulative action, behaviour is an entity in progress, an open construction site: it arises from a subject-world dialogue that amounts more to probing the space of possibilities than following a predefined path: an animal interacting with the world takes one road or another because the condition it is experiencing gives it optional freedom.

The copulative coordinate, in fact, refers to modal indicators that are only hinted at in general terms, so that experience becomes the circumstantial and occasional translation of this general textbook. If affectivity can be defined as the evolution of a for-Self that transforms the world into a table set for pleasure, animals are always spoilt for choice, unless there are pressing needs to be met. Documentaries have accustomed us to seeing animals only in moments of expressive urgency, such as the confrontation between a prey and a predator, the fights between males during the mating season, courtship or parental care for the young. But those who observe animals in the field are well aware that they spend most of their time doing almost nothing, exploring and testing, sometimes searching the environment for new opportunities, colonising new spaces and acquiring additional information about their world. To be interested means to widen one's gaze whenever there is a break from urgency—that is, most of the time—to stroll around the space of interest produced by affectivity.

In this sense, each animal receives a drive from its heritage, which leads it to experience external reality a certain way: this flywheel on the one hand is a conjugation to the here and now, and on the other is a revisitation of the phylogenetic past. But these innate propensities, precisely because they increase the animal's interest in the world, open the subject to the circumstantial condition of experience and to interactive nomadism. In a way, the individual makes its appearance on the stage of reality as a tourist who has already received quite a lot of information on the country they want to visit, the sites not to be missed and the recommended routes. The tourist, however, has extra time and, having reached their destination, embraces a degree of serendipity or customises their route according to indefinable variables. Precisely because of their wide interest in the destination, they do not blindly follow the guidebook, but stay open to possibilities. This interested openness, that is being fascinated by the places they can visit, increases the tourist's ability to adhere to situational particularities and random occurrences, thus laying the foundations for singularity and mirroring, summed up in their personal travel experience.

In many ways, an animal moves just like a tourist in its existential story: while it is true that it has important needs to fulfil, it is equally undeniable that its life is not limited to these. The essential property of somatic movement also has a metaphorical significance and points to a peripatetic existential condition that subsumes the ontological characteristics of animality. Being an animal means being able to explore one's own space of possibilities, sometimes favouring particular events that are only hinted at on the stage of reality, and sometimes ignoring calls that, on the contrary, should draw one onto pre-established tracks. Affectivity primarily defines a condition of psychological motility within the spaces of the possible, because yearning ignites the imagination. The condition of being interested is what leads an animal to follow new paths and to sound out the terrain in search of opportunities, giving free rein to the creativity that gives it expressive protagonism. Each individual participates in making the history of the world through their personal contribution—no matter how little. Affectivity has to do with the dimension of *thaumazein*, which is first and foremost excited openness, shakenness, creativity in power.

And this is where the unpredictable path of the individual's story begins: having copulas means being drawn towards the accidentality of worldly events. Innate tendencies are conjugative flywheels and therefore *facilitate* experiences and do not reduce them, contrary to what is often claimed. Experiences involve the subject in the form of *states of non-equilibrium*—such as yearning, appetite, restlessness, desire—which seek solace in dialogue with the world and express themselves through the pleasure of fulfilment. The need to feed oneself does not lock the individual into a condition of food-focused captivation, as Martin Heidegger (1889–1976) suggested, but allows for the experience of exploration, research and foraging as moments of openness to external reality. We must not, therefore, consider the innate as a mere repetition of a resonance of the past, but as an opening to the world that, while making use of precise coordinates of interaction with reality, makes the singular experience possible. An animal is always in a dispositional state that can never be ascribed solely to internal or external conditions, but is due to the dialogical dimension that takes shape in the psychological here and now of the subject. Precisely for this reason, experience escapes all forms of determinism, whether hereditary or environmental, giving rise to events that belong to the individual; but beware: they are such insofar as they are dialogical and not self-referred. The indefinable is the strength and lightness of the subject, where every traced path always has the flavour of a singular invention, which would most probably not be repeated if one could go back in time and relive that particular moment.

The "Need" Character

As mentioned above, the animal is a heterotrophic organism, a condition that inevitably makes it dependent on the world for the intake of energy resources that allow it to maintain its negentropic structure. The need to constantly consume food translates into a series of activities that the animal must perform, which in ethology are called foraging behaviours. It is not only a matter of having skills in acquiring food, but also of presenting a *yearning condition*, whereby evolution selects the motivational tendencies that increase interest, research, knowledge of one's environment and the propensity to test one's abilities. In its relationship with one's habitat, the need to find food resources always produces challenges that an animal must solve with a dose of creativity. Obviously, finding food is easier in some circumstances, while it is more difficult in others, but this is not a question of assessing the degree of creativity—rather, the point is that an animal cannot resort to mere repetition.

Of course, creativity appears to us in full force when an animal uses a tool or invents a new strategy, but these situations of creative brilliance only become apparent on certain occasions, for example when a species is placed in a different environment from its canonical ecosystem, in species that tend to colonise different environments, in situations of captivity or confined spaces. Many authors—such as Adriaan Kortlandt (1918–2009), Hugo van Lawick (1937–2002) and Wilfrid Le Gros Clark (1895–1971)—have noted that the use of tools is correlated with environmental poverty. In particular, Kortlandt (1964, 327–332), observing the behaviour of forest chimpanzees compared to savannah chimpanzees, noted that the latter had developed a more complex and articulated use of tools. In his essay *L'animale culturale*, Danilo Mainardi states: "it is not so much what animals do in their environment, but what they *can* do under changed and adverse conditions, that ultimately determines how they will adapt their behaviour" (1974, 33). In general, overtly creative behaviour is possible because there is always a dose of creativity, however small, as a basic condition of behavioural performance.

Eating is a physiological necessity that translates into a psychological condition, one that has to do with research, exploration, expectation, yearning, projection towards a goal, the construction of strategies and tactics and often the development of tools. A need does not determine an expressive insufficiency but, on the contrary, amplifies subjective conjugations and the realisation of singularity processes related with the indeterminacy of outcome. The need opens the subject to the world, so that the necessity of nourishment inaugurates the pleasure of eating, the curiosity of seeking and trying, the satisfaction given by satiety, the restlessness of waiting, the effort to invent new solutions, the accurate knowledge of one's environment and the desire to colonise new areas. To go back to Mainardi's analysis of new competences, we read: "The use of tools can represent an opening towards new possibilities of adaptation, towards new evolutionary directions" (1974, 34). This is a very interesting topic that I will return to. For now, I wish to note how a necessity is often what paves the way for research and creativity through the exploration of possibilities and the construction of new adaptive niches. We can say, then, that the need for food is not simply a need for nourishment that does not produce a constraint with respect to indeterminacy.

The same can be said of reproductive needs, which we know to be the litmus test of the transgenerational transmission of traits and thus the true basis of fitness in the processes of natural selection. It is often said that animals mate to reproduce, but this is an error in the causal configuration of the phenomenon: we should say, in fact, that because they mate, they reproduce. Indeed, reproduction is the consequence, not the cause of the process. But if we look beyond coitus as the final moment of an expressive dimension, we will understand that sexual reproduction opens up spaces of experiential indetermination through particular affective conditions. Sexuality creates the charge of agonism, the pangs of the mating season, the frenzy of courtship. In certain cases we have the creation of a couple, antiphonal duets, wedding gifts, the honeymoon and so on, in short, an experiential dimension that each time confronts the subject with an ocean of possibilities. Sexuality fills an animal's life, making it passionate, restless, inclined to magnify its presence by giving full expression to its peculiarities, in order to be chosen. When we observe how some deer adorn their antlers or how finches practise singing to improve their performance and be even more convincing, we understand how a need can open up space for subjectivity, just as a problem sharpens the wit.

The need character indicates something already inscribed in the animal's physiology which, however, should not be represented as a gap to be filled, but rather as a point of hybridisation with the world. We could define it as a window on the indefinite, capable of multiplying the possibilities and therefore the indeterminacy of the outcome. The body defines needs which, by their very constitution, can also be interpreted through feedback mechanisms, such as the need to maintain a certain tonicity and blood volume. Some thresholds trigger specific receptors capable of inducing the behaviour needed to restore balance: for example, osmoreceptors can trigger the intake of water. However, in animals this need gives rise to a psychological condition—in the case of thirst we speak of homeostatic affectivity—which is always a *feeling* of something, mediated by the hedonic system that causes suffering or annoyance until the need is satisfied and pleasure when the animal fulfils it. Affectivity means experiencing the world through yearning, but it also means opening up to possibilities.

A similar argument can be made for all the innate behaviours that refer to the subject's adaptive needs: they define expressive guidelines, but then leave it up to the individual to interpret them. It is not enough, therefore, to say that every animal is the bearer of a heritage provided by phylogeny; it is necessary to understand what the consequences of this inherited endowment are: whether they limit the subject's room for movement or, on the contrary, implement interpretation, precisely because they are capable of expanding the subject-world relationship. The innate characteristics of a species are those that have magnified its reproductive potential, but correlation also increases the individual's dependence on environmental availability and thus the incidence of experiential indeterminacy. Some species, for example, have offspring in need of parental care and attitudes that predispose parents to care will be rewarded in their innate heritage. On the other hand, caring means building a nest or den, looking for useful material, carefully monitoring the surrounding space and finding the best site. The epimeletic dimension creates maternal love, attachment, the construction of interactive archetypes and social learning.

Therefore, neither foraging nor epimeletic behaviour can be treated as mere bodily needs, but are existential dimensions that give rise to articulated relationships between the individual and its surroundings. Ultimately, the body's state of openness, that is having specific needs, produces the copulative condition—being a bearer of interests—which, in turn, becomes a flywheel of experience and therefore of singularity. It is the non-autonomy of the body that paradoxically makes it a protagonist in its intersection with external reality. If we observe the way in which bumblebees (*Vespa crabro*) build their nests, adapting them to the various characteristics of the host environment—which can be a hollow tree but also a hollow recess in a house—we discover how the species-specific architectural canon is modified to suit particular opportunities. Similarly, in the Sphecidae of the genus *Mellinus*, pedothrophic cells can be made from a large number of materials, depending on environmental availability. In the insect world too, therefore, there is no oppositional or even complementary-juxtapositional relationship between innate and learned, but rather one of mutual implementation.

When we speak of a proactive disposition in animals, we are referring to the tendency to take the initiative, that is not to be at the mercy of external events, while at the same time not being detached from them. Having needs means being pushed and involved in interaction with the world and, consequently, being open to the specificities of any situation. In other words, the copulative character of animality creates the very conditions for the individual's protagonism and for its singular experience. Insofar as it is projected into the world by its multiple interests, an animal is never passive and at the same time constantly produces novelties, sometimes very limited but at times of great impact. Having physiological needs is a condition that runs through all the levels of the body, spreading horizontally over the various apparatuses and then translating into psychological states, so we can speak of an intrinsically open existential dimension. If physical needs, innate tendencies and environmental calls converge in the individual to support its entrance into the world, it is the affective translation mediated by pleasure that organises the experiential dialogue.

Being a bearer of needs, that is implicitly relying on the world, is the basis not only for the copulative process (turning to the context to receive what is constitutionally presupposed), but also for experiential singularity, because the subject relates to a world that is constantly changing. If the animal were an autonomous entity, it would be completely introverted and static, fixed in its essential qualities, but not conducive to singularity. Animals need the world in order to unfold, and yet the world is in continuous transformation: this shows how reference to the outside is the basis of subjectivity. It produces: (i) the protagonism of action, which leads the animal to pursue its own interests; (ii) the singularity of state, both biographical and expressive, since in a fluctuating environment it is not possible to repeat oneself; (iii) self-ownership, because being a bearer of interests means having needs and a for-Self. The implicit heteronomy of the body is, therefore, fundamental to understanding how animality dwells in the world, bearing an agenda of copulations in fieri, which defines a consequent orientational compass.

Being a body undoubtedly means having needs of a physiological order, such as eating and drinking, resting and maintaining the correct body temperature, performing the most important metabolic functions, breathing, digesting, assimilating nutrients and expelling waste, preventing damage to physical integrity, mating and reproducing—to name but a few. Through the body, the animal is therefore "demanding"—constitutionally open to input from the world and therefore exposed to the availability and opportunities that the environment provides. Openness is thus a condition of implicit uncertainty, because the animal never knows if and how it will be able to fulfil its needs: it lives in a continuous state of wandering through its habitat in search of opportunities to exploit and risks to avoid. We might see the animal as complete in its adaptive equipment and fully ensconced in its own niche, in a condition of perfect overall harmony, but this is an epistemic illusion. Having needs means having to work in a world that is always one step ahead of one's expectations and that in certain cases can be extremely uncomfortable. This is why, in the animal world, need is never mere consumption, that is a condition of being captivated in fruition, but brings out mental states that, while respecting the need at play, search the environment for opportunities.

It has been said that the phylogenetic memory of each animal creates very precise expressive appointments, which translate into ways of relating to the context and therefore of gaining experience: for example, the condition of warmth in mammals establishes precise availabilities towards conspecifics and modes of engagement. However, if we examine the interactive dynamics present in different chimpanzee communities, we will see that the relational configurations have to do with the historical development of that particular social group, as Jane Goodall's experience has shown. The physiological rhythm of dispositions generates affective states which, rather than expressing standard responses, generate opening functions, depending on occasions and occurrences: what the individual finds in the environment in an unpredictable and singular way. Phylogenetic memory, therefore, is not a finished script that the individual simply performs, but gives life to experiential dimensions, creating a link between the subject's experience in the here and now and all that has preceded it. Ethographic appointments are always marked by the indeterminacy of their outcome.

In conclusion, we can say that the body presents intrinsic needs and the innate defines styles of connection with the context, but these are not algorithmic functions, led to express themselves through predetermined cascades, but rather openings to the possible. Innate endowments do not only concern the affective dimension, for example the motivational menu available, but also the modal equipment through which the individual realises its needs. It is a set of extractive, evaluative, operational and orientative traits that belong to the repertoire of that species. Each animal, for example, is equipped with perceptual *Gestalten* that allow it to acquire certain stimuli or respond to particular prompts through an initial attribution of value to the phenomenon. Even with regard to proactive behaviours—caring, agonism, affiliation, territoriality, to name but a few examples—phylogeny provides distinctive operational choreograms, recognisable as a species repertoire. These innate endowments, however, together provide a basic toolkit that the individual uses to accrue new knowledge and skills. We must therefore regard phylogenetic content not as a limitation on creativity, but as useful material for individual interpretation. Experience is therefore the result of the adaptation of previous endowments that evolve in the interested interaction between subject and world, as suggested by Jean Piaget (1896–1980).

To be interested in and supported by species-specific modal endowments is to transform the outside world into something that has to do with the individual itself. For a predator, movement is not something generic that is happening in the environment, but something that affects it directly and urges it to experience it. The need thus produces an orientative condition, which is not a generic openness to the world, but is not a deterministic exposure to the stimulus either. Being a body transforms events into somatic conditions that directly concern the individual and activate its experiential wandering. The need becomes a psychological state and establishes a dimension of feeling that transcends the presumed objectivity of the phenomenon itself, influencing the type of research and the evaluation of opportunities. In fact, the same event, if it occurs at different times, receives different responses from the individual. However, except in cases of extreme gravity, it is not a simple reaction, but rather an elective orientation.

Facing the world with needs means testing what is available and finding a strategy that fits that particular circumstance in order to achieve a result. But in order to do so, it is necessary to come up with something, undoubtedly drawing on one's own inventory—whether innate or learned—but adapting it to the specific situation, that is relying on a Self that not only demands but also inevitably evaluates. Evaluation is an important operation in the process of adapting one's action, but it clearly depends on possessing interests and, therefore, on being able to judge the degree of coherence and the hidden possibilities. Recognising the animal's condition of need means admitting a convergence between needs and predispositions on the one hand and opportunities and risks on the other: a confluence of influencing factors that define a singular state condition. Nothing can be repeated and the elements involved follow a chaotic course, so expression cannot be explained through the causal linearity of determinism.

The condition of openness lays the foundations for a historical dimension of the individual, whose passage through the world always has the connotation of events that can be analysed without disregarding the ownership of the circumstantial choice. Being open to the world implies a good dose of serendipity of outcome, linked not to chance but to the type of orientation produced. If, in fact, a carnivore has the copula of chasing, this factor will increase the likelihood that the animal will pay attention to moving targets. However, if we observe a hundred hunts on the part of the same individual, even in the same context and perhaps with similar prey, we will find that the hunting configuration always comes with a character of indeterminacy. The internal and external variables can be traced back to influences, almost as if they were hidden suggestions on what to do, which the subject listens to but which ultimately do not take away the creative task of behavioural preparation. The condition of need ignites the fuel of creativity, because it puts the individual in the condition of inventing something to make ends meet, as they say. Expressive ownership is not a frill, but the simplest way to respond to need.

Similarly, when we talk about the predatory or territorial motivation of an animal or analyse courtship behaviour, male-male competitions or parental care, we must consider these attitudes as real needs that demand to be fulfilled. Just as an animal suffers when it is unable to find food, so it experiences great discomfort when it is unable to exercise the behaviours that are part of its repertoire. Motivations push the subject to deal with external reality, transforming the entities-events that appear on the horizon into fields of expressive possibility or opportunities that allow the individual to meet the world through its own proactive coordinates. For this reason, motivations should not be considered simple drives that find an outlet by being expressed on pre-established targets: they are rather expressive tendencies that allow the subject to gain experience, and therefore to construct situations of singularity, by acting on pre-established levels. Admitting dispositional components does not mean, therefore, lowering the terms of the animal's subjectivity or protagonism. Owning propensities that demand expression lays the groundwork for opening up to the occasionality of the situation. And it is precisely because the animal presents different propensities and encounters different conditions at every moment that the expressive and experiential singularity is realised, since mere repetition would make no sense.

Dispositions open the individual to the occurrences of the world, though not in an amorphous and objective way—if this were the case, the outside would dictate the subject's expressive and experiential results—but through specific coordinates, so that the final effect resembles a creative dialogue between the subject's orientations and the opportunities present in the environment. On the other hand, being susceptible to movement and inclined to chase after objects in motion defines a psychological need in any predator which is impossible to escape. A need, therefore, never amounts to skimming the world, but means being *involved* in it, both in the here and now of expression and in the experience's effects on the ontogenetic construction of the subject. Subjectivity is often confused with individuality, the character by which the individual is endowed with features that make it indivisible in certain respects and unique in others, differentiating it from all the other beings of its own species. And yet, on closer inspection, there are innumerable entities that present such characteristics, even though they lack subjectivity. On the contrary, an animal's individuality is the product of its subjectivity, of the experiential protagonism that allows it to build a biographical structure based on its leading role, given by the succession of relationships established.

Hedonism and Somatisation of the World

Having interests means not being indifferent to what happens to us. As opposed to a machine that, however skilful its performance, is always imperturbable in the face of circumstances, animal susceptibility is unmistakably one of the essential qualities of subjectivity. One cannot understand the animal dimension without starting from this *participation* in the events that take place—an involvement that, however, is a selective sensitivity and not a simple open door to events. Susceptibility indicates, in fact, an opening that is accomplished by sifting through the events of the world, viewing some as relevant and others worthless or even letting them fade into the background. For this reason, not being indifferent means pondering events and inserting them into the warp of one's somatic fabric, that is translating them into physiological beats. If one has interests to defend and pursue and is endowed with an affective menu, capable of detecting the coherence between the world and such interests, there are good reasons not to be indifferent to what is going on. A mentally elective subject—since physical needs produce an oriented affectivity—is driven by the hedonistic quest, that is by the urge to escape conditions of discomfort or anxiety and to seek pleasure, whatever form it may take.

This involvement is not limited to filtering out events, as concentrated attention would entail, because non-indifference leads to action and creativity: opening up to the world is more like throwing oneself into the fray of experience rather than enduring it. The orientation drive is so strong that it induces animals to modify the environmental parameters they encounter, devising a table set for themselves. The animal's *Dasein* is never simple dwelling or existing in a pre-established context, but means acting on the circumstances to increase their coherence with one's own needs. I believe, therefore, that the animal should be considered a *creator of worlds*, as we shall see when discussing the niche theory. But, beyond this aspect, which calls into question the simple adaptation to, or inhabitation of, a given environment, each individual inevitably builds an individual niche, endowed with singularity. The body's needs, starting from its constitutive morpho-functionality—the reason why gills require a liquid world—on the one hand determine pre-established appointments with the ecosystem to which it belongs, on the other hand open the individual to the unknown of the ocean and to a matrix of events, where the only valid reference is given by the hedonic system.

Non-indifference is therefore the basis of the individual's historical dimension. Phylogeny undoubtedly lays the foundations for adaptation through the definition of a perimeter of encounter; however, the instability of the environment and the affective fluctuation of the individual call for a diachronic development, an *ad hoc* fulfilment of behaviour guided by the pleasure of feedback during the action itself. The animal's protagonism is thus expressed in various ways, from sifting and searching the surrounding space to find what may be hidden beneath the surface of appearance, to moving to colonise new territories or extend the area of relevance, from modifying the characteristics of the habitat to literally building its own dimension of life. An animal therefore takes an active role in its encounter with the world, modifying it in the image of its needs. Furthermore, this process is recursive, so the modified environment in turn gives rise to new demands: we must always avoid the assumption of linearity. I said earlier that the animal is "demanding". This term, however, indicates a character of request and expectation, an effort—akin to Spinoza's concept of *conatus*—that supports the individual's proactivity, the force through which demands are to be fulfilled.

These demands manifest themselves in different ways, as: (i) impulses that ask to be satisfied within a certain time; (ii) restlessness that aspires to find consummation or else produce states of unease; (iii) orientations that establish sensitivities or preferences. The animal moves through the world guided by these interests and uses its somatic adaptation resources to adapt to them. It learns easily, through the non-associative processes of habituation and sensitisation or the associative processes of prediction, attributing precise hedonic meanings to events. The emotional colouring with which the subject faces different situations—such as fear if interests are threatened, anger if thwarted, sadness if compromised—offers another example of how non-indifference is always the centre of an individual's life, influencing interaction with its surroundings. In the same way, the motivational horizon transforms the world according to precise parameters of operability, which again range from the suffering of restlessness and frustration to the pleasure of consumption. In a nutshell, the animal searches for its own dimension in the environment, referring to what it likes to do or is inclined to do, passing from the unpleasantness of yearning to the pleasure of satisfaction.

Being a bearer of interests, the animal is not neutral towards reality, but always demands something from it. The individual asks questions of the world with regard to (i) the intersection between external events and the subject's interests and (ii) the type of operability that the context offers to its behavioural tendencies. Emotions, for example, respond to type “” questions, indicating coherence between the here and now and the subject's interest; motivations respond to type “ii” questions, probing the usability of a certain context. But these are never shy requests: an animal never treads the proscenium of the world in a passive whisper, but makes every effort to shout out its interests, sometimes drawing on the aggressive resource. It is this strong claim on the animal's part that leads me to use the term “demand”. The animal, in fact, bends external reality in its action on the world, in such a way as to increase the context's coherence with its own interests. In this sense, subjectivity manifests itself in will and effort: it is never a passive entity that simply accepts the state of affairs. As far as it is able, the animal does anything and everything to get what it wants and to increase its state of pleasure.

The hedonic system translates the subject's experiences, in whatever part of the body they occur and through whatever receptive apparatus, into affective memories of the world, in a process that we can define as the somatisation of feedback. Every element encountered that has been relevant to the dialectic of pleasure is somatised, becoming an integral part of animal physiology and producing affective effects through mnestic functions, just as when a phenomenon stimulates a homeostatic or dispositional state. Pleasure, therefore, cannot be simply considered a momentary condition of the subject in interaction with reality, but is rather an introjection process that shows the ownership of the individual in the course of experience, its progressive somatic expansion along the path of its nomadism. This process of hedonic translation of experience into physiological beats reveals how the indefiniteness that accompanies an animal's life becomes a constitutive character of its being.

Emotional experience, for example, produces an affective condition that is indescribable in its complexity, and can only be outlined as *feeling the world*, through the attribution of values and mnestic introjections. The individual's path is therefore written in the deep threads of its physiology. This exemplifies the hedonistic translation of demands, establishing a somatic link between the individual and the world that also implies an inner state and a subjective bodily predisposition to respond to what it encounters on the peripatetic journey. One of the most fascinating aspects of animality is the emotional sensitivity to events, the condition of being interested in being present, which means not only being there, but being involved. Emotional engagement entails the anticipation of events and, at the same time, the arbitrary assignment of colouring to different situations, transforming them into an experience. This produces an immediate recognisability, which differentiates the animal from the machine: despite being endowed with performative capacities, the latter is totally detached from events and unmoved by the world—it does not palpitate in feeling. Feeling is clearly about giving value and this involves acting in relation to the world, making choices, orienting oneself in a certain way, taking the initiative.

Motivational affectivity also produces somatisation events through the pleasure mechanism. Having motivations means, in fact, being involved in an act as part of the expressive eros—the desiring condition that launches the individual to the conquest of the surrounding space. The erotic charge of the motivational experience is freighted with *furori*, to take up Giordano Bruno's thought, and passion, charging the projection towards the contents of reality with hedonic contents. The animal transforms the world into its playground, where every performative endowment becomes an instrument to carry out a vitality that springs from within and is translated into actions. The animal acts for the sake of acting, so that appetitive activation can, but doesn't necessarily have to, be aimed at an adaptive result. The desired result is never objective, but is a product of the animal, of its way of assigning a particular adaptability to the circumstance. The tension lies in the need-question: what is possible in this situation? Action can be translated into a game or a pastime, a way of distracting oneself, because what matters is being able to act, to actualise one's contents. In the motivational demand, the world is never a counterpart, but represents the connection required to express oneself, so that one can experience pleasure in the presence of operability or, in its absence, experience boredom and frustration.

Sensitivity, which of course is not merely the ability to feel pain, but means not being indifferent to events, reveals the animal as constantly searching for something, scouring reality and bending it according to inherent principles. The body is scattered with orientative sensory receptors, which help define the psychological affective condition in its constant dialogue with the hedonic system. Pleasure travels on the rails of sensory aesthetics, along the tonicity of muscles, in visceral functions, on the roads of appetite as well as satisfaction, in curiosity as well as in wonder and even in the adrenalinic thrill of the unexpected, of the overwhelming, of fear. It is a memory, an expectation, an omen, a possibility that the individual only imagines or barely glimpses in the patterns of appearance, but which charges the dopamine impulse to dominate the course of events. Sensitivity opens the relational pores of the individual, so it is not just a matter of finding the correct answer, but of converting the events of the world into a subjective memory. This condition produces, as a consequence, the need to turn towards the world to write one's own unique and unrepeatable presence, to change the circumstances in one's favour, somatising them. Not being indifferent makes the animal peripatetic, that is, intrinsically nomadic in reality, but this transient situation is its home, its existential dimension.

The motor system, since its first reflexogenic expressions, gives us the clearest image of this non-indifference and also constitutes the basic toolkit to transform the affective disposition into a concrete orientation on the theatre of the world. The neuromuscular interface enables us to move towards or away from elements that may have some bearing on the body's needs. With the development of more and more complex nervous structures, orientation, no longer reflexogenic, has given rise to increasingly articulated psychological conditions overseen by a central regulatory system. On the other hand, when I speak of the nomadic character of *Dasein*, I mean a condition that precedes action and that does not always manifest itself in movement or orientation, but is based on the subject's internal non-equilibrium—a self-insufficiency that necessarily makes it lean towards the world, where it fulfils its interests. A machine is a static entity, obviously neither because it is incapable of movement nor because it is unfit to process information from the world or to learn. Its static nature lies in the fact that it does not feel any need towards reality, it does not have any internal motive sustaining its activities—in other words, it is indifferent to the world.

Sensitivity, therefore, is not only a reaction or a responsive predisposition, but the introjection of external entities into one's own biographical structure—the transformation of events into flesh. If an event causes fear, it loses its connotation of emotional neutrality to become a fundamental emotional reference point, and the same can be said for the target of motivational appetite. Saying that an animal is a sentient being means much more than it can feel pain: it indicates the condition of non-indifference to worldly events. Pleasure is a kind of indicator of the fullness of the individual's presence in its here and now, both as libidinal tension towards the world and as full satisfaction of present needs. Thus, pleasure turns the condition of demand into an engine to experience the enjoyment of appetite and the serenity of satisfaction—a biphasic compass in the dialogue with the world that goes beyond the mere fulfilment of needs. An animal also experiences pleasure in strong desire, in the exuberance of vitality and volition, not only in the serotonin calm of satisfaction. Pleasure is also found in excess, playful excitement, display of power, as well as waves of endorphin heat.

Obviously, experiencing pleasure or banishing suffering are in themselves valid reasons for doing or not doing certain things. Often, however, we misunderstand the extent of this system, focusing exclusively on what may appear particularly pleasurable, such as eating, comfort or rest, which indeed have to do with subjective hedonism, but do not exhaust it. The individual may in fact follow a seemingly antithetical path of pleasure. For example, many behaviours that produce fatigue and sometimes physical pain, but are sustained by strong motivational appetites, produce pleasure, whereas inactivity may turn into a Procrustean bed. During the mating season, if certain ungulates are kept in a state of full comfort, without having the opportunity to perform in tournaments and courtship activities, they enter a state of great suffering. This shows that there is no full overlap between pleasure and welfare, as even muscular fatigue, through the action of endorphins, is capable of producing pleasure, just as the expression of basic behaviour or sensory activity through endocannabinoids produces euphoria. Thus, everything that has to do with the expression of introjected subjective characteristics, whether attributable to phylogenetic heritage or experiential background, is pleasurable.

In general, the urgencies of the body are measured by the pleasure system, so that not only eating, but all the basic organic functions give feedback in terms of physiological adequacy by means of various pleasure notches, placing red marks when conditions of suffering begin to appear. This system, therefore, is a meta-component, that is to say, a quality that oversees all the expressions and conditions of the individual. Pleasure or its opposite—that is states of suffering, discomfort, restlessness or annoyance—therefore makes up an alarm system that testifies to the coherence of the present experienced by the subject with respect to its interests. Pleasure can manifest itself in various ways, such as a sense of lightness, euphoria, invigoration, enjoyment, relaxation, attraction and aesthetics. On the contrary, one can feel discomfort, restlessness, heaviness, difficulty, pain, irritation and so forth. An animal's behaviour is strongly influenced by the hedonic state, which, however, can also present situations where the various interests and the related pleasures appear in an oppositional form, such as when interest in physical safety clashes with that of foraging or mating. In these cases, there are situations of oppositional orientation that must necessarily find a synthesis—Lorenz already regarded this aspect as a sort of parliament of instincts.

From the physiology of the body to the psychological condition of the individual that emerges from it, pleasure is always the most important reference in animal orientation. When I am asked why an animal performs a certain behaviour, my answer is always the same: because, for some reason, which is not always easy to identify, doing so gives it pleasure. Just as fitness, that is reproductive capacity, is the ultimate judge of a certain character's compatibility with the evolution of the species, so to understand individual behaviour we must turn to the hedonic magistrate. On the other hand, we are mistaken if we refer to it simply in terms of enjoyment or pain, interpreting it exclusively according to the laws of welfare. Pleasure is a constant presence in the individual's life; just like memory, it has a predominantly operational meaning, overseeing the tasks to be carried out and the priorities in place with respect to the agenda of needs. This is why I call it a meta-component. The hedonic dimension is the most explicit form of *being the bearer of interests*: it acts like a thermometer, passing from the condition of lightness to that of discomfort, manifesting itself in physical pain or in enjoyment, in explicit suffering or in mood alteration. The hedonic dimension indicates whether and to what extent the interests of the body and mind have been fulfilled.

In reality, the hedonic dimension is not limited to assigning positive or negative values to our experience, but enters directly into the orientation processes, not only through the antithesis between getting close and moving away, but by influencing the phasic course of behaviour and the prolongation of each activity. It is pleasure that determines when to start doing a certain thing, how long to carry it out and what other action to follow it with. It is, in fact, this system that establishes whether it is time to stop an operation—whether out of satiety, tiredness or satisfaction—or to continue with it. Therefore, pleasure is not a momentary condition, but rather a progressive transition from one phase to another, with the temporal parameter as a reference. We do not experience pleasure at a certain time or in a certain activity, but we are always surrounded by a hedonic marking of the expressions available at each moment of life. Pleasure dictates the sensitivity to calls and the gradient of need embedded in the animal's interests.

Pleasure is the most explicit reference to what I have defined as being interested in the world and being endowed with sentience. We would be wrong, however, to separate sentience from subjectivity. Indeed, if it is true that there are objective parameters to evaluate harmful elements that produce suffering or impediments that determine discomfort, as far as pleasure is concerned, much is related to the individual's undefinable condition, namely the subject's momentary state and biographical features. An individual has ownership of its definition of pleasure and of the hedonic characterisation of experiences. In the latter case, pleasure can often be traced back to situations of behavioural addiction, so that each individual learns to build its own ways of achieving it. Subjectivity, therefore, is also characterised by hedonic creativity, linked to the fact that each experience produces specific modes of gratification and satisfaction. When, for example, a dog becomes accustomed to the game of tug-of-war, the rope is elected as a source of gratification and the act of grasping with the mouth and pulling becomes a reward, in an inseparable link between pleasure and experience.

On the other hand, pleasure is clearly linked to the subject's specific physiological state, so that what produces enjoyment at one time may even cause discomfort at another. Generally speaking, the hedonic condition develops in three phases. First, there is (i) yearning, which indicates the pleasure of the subject's activation and call to action, a feeling of attraction towards the world that ignites the senses and makes the context worthy of attention and rich in opportunities. Secondly, we have (ii) consummation, where pleasure is expressed in action and takes on the aspect of gratification when the need is fulfilled, the target is reached and can be enjoyed. Finally, there is (iii) fulfilment, when the action has concluded its course, reaching complete satisfaction, so that the subject can finally indulge in rest and satiety, and the feeling of restlessness and unease vanishes in a temporary dulling of the senses. Pleasure, therefore, resembles a wave that shakes and sets the individual in motion and then gently subsides.

To understand the hedonic system in its phases, let us look at two typical situations: the need to feed and the mating drive. The first activates the subject and triggers a search behaviour that does not necessarily, at least at first, lead to unpleasant sensations. On the contrary, the shock of activation is stimulating and energising. Impatience arises, however, if the need does not find fulfilment, while if the individual is able to reach consummation, pleasure increases, confirming the animal's orientation. Once the action has been completed, the subject enters a condition of fulfilment, that is a state that combines the satisfaction of the result, physical exhaustion (gently leading the subject to rest) and a sense of satiety with regard to the completed activity. A deficit of appetite corresponds to a condition of apathy or abulia, that is, a lack of yearning and therefore of activation and invigoration in the individual, as occurs in depressive situations. If, on the other hand, there is a lack of consumption, the individual experiences feelings of suffering linked mainly to frustration, which tend to have different hues, but sometimes risk extending to physical pain. Lastly, the lack of satisfaction creates restlessness and unease which, in the long term, can lead to serious problems such as anxiety, distress or compulsive behaviour.

From what we have said, pleasure is not only about post-consumption satisfaction or the attainment of excellent welfare levels, but it also has to do with physical vigour and vitality, which are expressed in the very condition of being interested in the world. These are qualities we can appreciate in a puppy: a kind of *joie de vivre* that expresses itself precisely in copulatory excess. When we admire a puppy's curiosity, interest in the world, interactive eagerness, joyfulness and wildness of movement, qualities that inevitably show us vitality, we are observing just such copulatory excess: the puppy's need to take in the world. The yearning can take the form of a strongly targeted desire or of a generic appetite, of curiosity towards an event or of exploratory interest, of evocative vulnerability or of proactive drive, and can assume the connotations of inclination or of simple sensitivity.

Pleasure is basically the result of the different demands and their state of fulfilment in the subject's here and now. Then, there is a cause-and-effect relationship between needs and pleasure, whereby pleasure indicates the degree to which the individual's inherent interests are unfulfilled, in terms of being absent or jeopardised. Now, interests are sustained by particular needs, which, however, fluctuate in time and often change in the course of the subject's life. The world, on the other hand, presents risks and opportunities, favourable or unfavourable conditions that change from time to time and are, in fact, unpredictable. For these reasons, pleasure is not only a thermometer to indicate the degree of coherence between the animal's state and its needs, but also acts as a driving force for the subject's creativity. In other words, pleasure must act on orientation, preside over learning processes and store up the most important experiences at a mnemonic level, but also stimulate the imagination and the subject's ability to explore the potential of reality. The hedonic system undoubtedly represents one of the main features of subjectivity, because the fact of experiencing pleasure, in itself, gives a title of self-ownership to the animal, which follows its own north star.

Works Cited

Baerends, G.P., R. Brouwer, and H.T.J. Waterbolk. 1955. Ethological Studies on Lebistes Reticulatus (Peters). *Behaviour* 8 (1): 249–332.

Carere, Claudio, and Dario Maestripieri. 2013. *Animal Personalities. Behavior, Physiology, and Evolution*. Chicago: University of Chicago Press.

Darwin, Charles. 1872. *The Expression of the Emotions in Man and Animals*. London: John Murray Albemarle Street.

Eibl-Eibesfeldt, Irenäus. 2007. *Human Ethology*. New York: Routledge.

Hinde, Robert. 1970. *Animal Behaviour. A Synthesis of Ethology and Comparative Psychology*. New York: McGraw-Hill.

Kortlandt, Adriaan. 1964. *How Do Chimpanzees Use Weapons When Fighting Leopards?* Penrose Fund. Report of Committee on Research; Geological and Biological Sciences.

Lack, David. 1939. The Behaviour of the Robin. *Proceedings of the Zoological Society of London* A109 (2–3): 169–219.

Mainardi, Danilo. 1974. *L'animale culturale*. Milano: Rizzoli.

Tinbergen, Nikolas. 1951. *The Study of Instinct*. Oxford: Clarendon Press.

Vallortigara, Giorgio. 2021. *Pensieri della mosca con la testa storta*. Milano: Adelphi.

Wickler, W. 1961. Über die Stammesgesclfichte undden taxonomische Wert. einiger Verhgltensweisen der Vögel. *Zeits. für Tierpsychologie* 18 (3): 320–342.

CHAPTER 6

Endowments as Tools

Premise

If creativity is the ability to go beyond competence, to not limit oneself to already tested operational resources, but to add something more, one must ask oneself what relationship there is between the two terms. The question, therefore, is whether creativity produces its own results or whether it makes use of the resources of competence. Observing the behaviour of certain species that profoundly modify the context by creating a sort of "niche environment", like beavers, we can see that, while certainly following some general guidelines, these animals undeniably operate each time in different situations, always facing very specific problems. It is not correct, therefore, to reduce their behaviour to a cascade of standardised and closed expressive units simply because the result is preprogrammed by phylogeny. In fact, achieving that pre-established goal requires a great ability to evaluate the particular situation, to put into practice specific actions, to continuously deal with difficulties by identifying effective interventions, to address any setbacks and so forth.

In this, animals display not only great versatility of goal-directed behaviours, but also great flexibility in evaluating the starting and ongoing conditions during their reconfiguring action. Certainly, when an animal performs trials to solve a problem, it cannot act at random, because the possibilities would be so many—the problem of combinatorial explosion—that it would inevitably be faced with a stalemate. As with entropy,

R. Marchesini, *The Creative Animal*,
https://doi.org/10.1007/978-3-031-07414-1_6

disorder, given by unsuccessful attempts, would be immensely more probable than order, that is successful attempts or approximate solutions. In order to work by trial and error, one cannot rely on the mechanism of randomness but narrow down the field by using heuristics (Herbert Simon 1955), that is probably useful solution recipes. The principle underlying the theory of heuristics is that one does not always have the immediate answer to a problem, nor can one guess at it at random, so one uses targeted attempts or solution strategies. In addition to Herbert Simon (1916–2001), other authors who have dealt with heuristic approaches to solutions include Amos Tversky (1937–1996) and Daniel Kahneman, who addressed decision-making aspects and wrote the essay *Judgment under Uncertainty, Heuristic and Biases* (1974).

On the other hand, a heuristic is a resource of the subject, in that it belongs to the latter's pool of competences, but it is such if the subject has it *at its disposal*. The same can be said for invention, which cannot arise *ex nihilo*, but always relies on previous resources. Even the use of a tool or a particular strategy that appears in a certain population within the species, and therefore has the appearance of invention and cannot be traced back to species-specific patterns, can very often be attributed to lateral behaviour or reuse of phylogenetic repertoire activities. An example is the case of blue and great tits piercing the tinfoil cap of milk bottles, as highlighted by Robert Hinde and John Fisher (1951): this behaviour can be traced back to these animals' tendency to debark trees in search of insects. In many primates, the aversive behaviour of shaking branches to ward off an intruder, sometimes causing a branch to fall, may be the basis for the invention of using sticks against opponents. In certain cases, it is sufficient to hybridise two pieces of knowledge or to apply an operant outside its ordinary attribution to have a creativity effect. We must, therefore, conclude that creativity is strongly linked to competence, even if it goes beyond it.

But there is another question which in a certain sense precedes the relationship between competence and creativity: how are we to understand the operational endowments provided to the individual by the innate and then by learning? It is evident, in fact, that if competence were not an available resource *at the individual's disposal*, but only worked as a behavioural automatism, we could not connect creativity and competence, as the former would depend on chance, like mutations in the genetic heritage. As Donald Griffin points out in his essay *Animal Minds* (1992, 112–141), when we are faced with great versatility in project implementation, as in the arbours built by garden birds or the dams erected by beavers, it makes no sense to think that the enacted behaviour is based on expressive automatisms, because if this were the case we would have to hypothesise as many sub-programmes as there are variants. The plausibility of an *ad hoc* explanation is inversely proportional to the number and complexity of the behaviours that are referable to the predisposition.

This is not to deny the presence of projects predefined by phylogeny and of innate endowments characterised by recognisable patterns. Rather, I wish to question two things: (i) that the project is realised as a cascade of predefined actions; (ii) that the innate endowments are *unavailable* to the individual but are simple automatisms. The Cartesian paradigm of the animal as a machine driven by automatisms, even if revisited through all the formulations of the twentieth century, would not allow for an availability—a title of use—of competence. This is most likely the reason why solution processes and inventions have historically been given a separate explanatory domain, such as random trial and error or productive thinking, both of which are elegant ways of avoiding a real explanation. In my view, it is not possible to talk about animal problem-solving skills and inventiveness without questioning the interpretation model of innate resources and learning processes.

The Problem of Endowments

Let us start with the first point, namely inherited endowments. The analysis of how we should think about the phylogenetic equipment of the species, traditionally interpreted as a series of instincts triggered by internal or external variables, is one of the fundamental aspects of the study of animal subjectivity. The reason for this is easily explained. If the innate endowments are considered automatisms acting directly on the animal, they become *fully* responsible for, and exhaustive of, its behaviour. If we think of phylogenetic competence as the expression of many disjointed threads, which analytically drive the individual-puppet, it is evident that we could find no basis for either versatility or out-of-context reuse, let alone creativity. We would have to assume that for every tactic adopted by an animal to achieve a particular strategic goal there is a corresponding automatism, going against the most basic form of explanatory parsimony.

In *Cognitive Maps in Rats and Men* (1948, 189–208), Edward Tolman had already suggested a distinction between behavioural molecularities (strategies) and implementation molecularities (tactics), the latter being more free in their adaptation to the specific context. In the early 1960s, G. Miller, E. Galanter and K. Pribram published *Plans and the Structure of Behaviour*, which can be considered the first complete formulation of the cognitive perspective. Its authors claimed that behaviour was configurational, that is structured in hierarchical planes and not as a sequence of domino-like responses. Instead of a sequential model—the classic chaining of the behaviourist tradition or the reactive chain model described in M. Washburn's *Movement and mental imagery: Outlines of a motor theory of the complex mental process* (1916)—they propose a top-down structure, taking up Edward Tolman's intuitions (molar-molecular). Each molar unit is presided over by several molecular units, so that behaviour is organised.

Consider the behaviour of animals far removed from mammals, such as insects or spiders, in their individual implementation of species-specific projects like the construction of a termite mound, a paedotrophic nest or a spider web: time after time, the individual has to start from a different situation that it must necessarily readapt to its need, in a DIY fashion. Studying the larvae of the trichoptera *Lepidostoma hirtum*, Hansell (1972, 1974) highlighted this insect's extreme versatility in building its case, even repairing it using the material available. Guido Grandi (1968, 13–30), studying the paedotrophic nests of the Scoliidae, Sphecidae and Trichinae, underlines the versatility of the strategies adopted by these species. In the 1980s and 1990s, I myself studied the nests of European bumblebees (*Vespa crabro*), noting that, although the architectural scheme remained unchanged, each community used different cavities (hollow trunks, pigsties, window frames, wood sheds) and inevitably had to adapt the nest to the specific characteristics of the chosen site.

We must therefore assume that an animal's innate endowments define—in addition to the emotional and motivational aspects seen in the previous chapters—specific overall or molar configurations, like the architecture of a bumblebee hive, and species-specific molecular behavioural patterns. However, in both cases these endowments must give the individual a degree of versatility for adaptive translation in line with the singularities of the environment. All inventive behaviour, therefore, does not magically arise from nothing, but relies on innate endowments. In this sense, the latter should be regarded as resources available to be used in a variety of ways, as heuristics needed to deal with unknown situations and not only local specificities. The predicate of *availability* therefore becomes central to understanding how an animal can make use of the innate endowment to cope with the range of situations it has to face. This also avoids the risk of hypothesising *ad hoc* behaviour for each variant, with a great gain in terms of parsimony; at the same time it provides a basis for creativity, rather than considering it as appearing out of nowhere. Ultimately, in order to talk about subjectivity, it is first necessary to question the automatism model.

We also know that experience and the various forms of learning give the individual new behavioural skills, adapting innate endowments or giving rise to new resources, for example through associative mechanisms. Again, if we consider the outcome of learning to be nothing more than conditioning—that is a mechanism that, given stimulus x, produces response y—it is not enough to assert that animals gain experience and modify their innate behaviour in order to deduce that they have subjectivity. The model proposed by the behaviourist school is based, in fact, on an automatism which, as such, is not at the individual's disposal, but regulates its behaviour, imposing it. Conditioning is an explanatory model that can be superimposed on that of instinct: it involves two automatisms, one learned, the other innate, which act in a direct and analytical manner and are therefore directly responsible for behaviour. They are not at the individual's disposal but rather imperative on it. In order to bring the learned back into a condition of availability, it is necessary to shift from the concept of conditioning to that of knowledge.

We have seen that conditioning runs along the lines of automatism: in this case we do not have innate determinism, but learned determinism, yet the explanatory gist is the same. Conditioning drives the individual, on the basis of the stringent relationship between stimulus and response, so it is not a resource available to the animal: the latter, in fact, cannot use a conditioning, but only submit to it. If, on the other hand, we consider learning as the acquisition of one or more information units, aimed at extending or modifying an applicative resource or at generating a new one, we can no longer speak of conditioning but of knowledge, of an endowment in the making which yields new competences and can be used in several ways. It is only then that the learned endowment becomes a tool for the subject to use, adapting it to the situation, reusing it in other contexts or applying it creatively. Here again, it is clear that both versatility and creativity are based on being able to use one's own equipment of pre-existing competences (innate and learned) and cannot arise when subjected to the deterministic regime of automatisms.

And it is not a question of behavioural complexity because, when I speak of creativity, I also mean those small adjustments or deviations that could not be explained without two things: on the one hand, the subject's *ownership* of its endowments, and on the other hand, a condition of *availability* for use on the part of the endowment itself. It is not complexity that makes the automatism model untenable—even the most complex behaviour could be transformed into a sequence of automatisms—but its insufficiency in the adaptation to the specific situation and its unavailability for creative use. An automatism cannot be *used*, otherwise it would no longer be such. On the other hand, a further question arises: how could a mind, lacking available resources, be able to produce anything? If we abandon the idea that creativity is solely attributable to human ingenuity and look not only at the inventions made by other species, but also at the small events of adaptation and reuse of resources in new contexts, we will realise that this *minimal creativity* is constantly at work in the animal world. We thus discover that even this creative undergrowth would be impossible if innate and learned equipment were not fully at the individual's disposal: the endowments' availability is therefore a prerequisite.

Creativity is not based on consciousness, but on the willingness to use the resources of competence, and does not reveal itself in the complexity of its results, but in the diuturnal refinement and adaptation of subjectivity. Sometimes this work reaches surprising peaks, but it can only do so because these anomalous waves rest on a moving sea: the productive mind working on its resources. I also believe that the tendency to establish a sort of dualism between simple behaviours (or those considered such) and complex ones, often simply because we are reminded of some of our own behaviours, is due to a misinterpretation of Conwy Lloyd Morgan's (1852–1936) principle of parsimony. Formulated in 1894, this principle, while not expressly denying the mentalistic explanation, discouraged attributing behaviour to superior faculties when it could be explained by less elaborate ones. On the other hand, on what basis do we interpret a faculty as superior or inferior? This is a question that Donald Griffin asks on several occasions in his essay *The Question of Animal Awareness* (1976).

Morgan's canon can undoubtedly be considered a psychological translation of the *novacula Occami* or Occam's razor, set out in the fourteenth century by the philosopher and Franciscan friar William of Occam: when faced with several hypotheses for solving a problem, results being equal, one should go for the simplest one as a methodological principle. Now, it is not necessarily the case that explaining behaviour by means of individual automatisms, each responsible for a very specific action—which evidently involves as many automatisms as there are different behavioural expressions in an individual—corresponds to the most economical explanatory hypothesis.

In communication, for example, through a limited set of words or signs, it is possible to create an almost infinite number of messages, precisely because there is no overlapping relationship between signifier and signified. Admitting that the resource of competence, in its condition of availability for use, is workable by the mind—through adjustment, reuse and innovation—means not only recognising an emancipation of the resource from the constraint of mere automatic repetition, but also giving a foundation to all those (more or less complex) behaviours that cannot be explained through the trigger model. Any non-repetitive behaviour, and *a fortiori* creativity, can only exist on condition that the endowments are at the subject's disposal, that is that the individual can *use* them instead of being driven by them.

We must therefore overcome an analytic or non-mentalistic explanation for animal behaviour, whereby each individual endowment acts on its own, albeit in a logic of converging combination or in sequence with other endowments. According to that model, the individual's expression is only apparently subjective, because it is actually driven by automatisms. One must therefore foresee as many automatisms as there are micro-actions that an individual carries out in order to achieve a certain result. It is necessary to be very consistent with this argument: an automatism is such precisely because it is detached from the domain of use, because its triggering by a variable produces a well-defined reaction and because its presence and configuration are exhaustive in explaining behaviour. Whenever we observe adaptive behaviour, if we wish to avoid the infinitesimal multiplication of automatisms, we must assume that the system is arranged on several hierarchical planes. We must therefore admit: (i) that each endowment allows for a multiplicity of functions and therefore entails its availability for use; (ii) that the subject exercises superintendence over its behaviour, though not necessarily consciously.

On the other hand, moving from the traditional way of considering endowments, based on the concept of automatism, to another which, on the contrary, interprets them as resources available for various uses, means changing models radically. Too often has this aspect been overlooked, thinking that it was enough to admit creativity or consciousness to suddenly transform automatisms into versatile behaviours, but this argument simply does not hold up. Likewise, an explanatory framework that places conscious behaviour or insight alongside innate and learned endowments, locked in the determinism of trigger expression, is inevitably fragile. This, in my opinion, is the weakness of a kind of cognitive ethology that is only concerned with complex behaviour, which cannot be explained by the associative model of traditional paradigms. This approach relies on properties that seem to emancipate themselves from innate endowments, so as to do without them, proposing such properties as an alternative. I understand that the reason for this choice has something to do with Morgan's principle, but this is a significant epistemological error, for many reasons that I will discuss below.

In order to talk about subjectivity, creativity and, more generally, the use of resources, there must be a basic model capable of explaining animal endowments in a totally different way. This prerequisite is necessary and cannot be bypassed if these behaviours—defined as the result of adaptation, reuse, solution, innovation, invention—are to be included in an explanatory system that actually works. We need a model that is not divorced from the subject's innate and learned endowments and that does not ask consciousness to explain in a tautological way something that otherwise would have no chance of sitting at the explanatory table. If we refer to what I have defined as minimal creativity, we are talking about common behaviours that are consistent with the specific species dimension and far from anthropomorphic, but which in ordinary application reveal processing by a superintending systemic entity. But in order to be processed, they must be available for use. In other words, the resource must be explained through a modelling different from that of automatisms.

We have to think of resources as ingredients available for reconfiguration every time they become part of a particular psychological milieu: in other words, they are able to relate to other resources within the mental dimension. This model is therefore *systemic-productive* in the sense that it envisages a dimension, the mind, capable of putting resources together and adapting them to the particular situation. There can be no creativity without the possibility of building on previous endowments, using them in a specific way according to the needs that gradually arise in the individual. In this model, innate and learned endowments are seen as tools at the subject's disposal, translation resources that the individual can use to realise its affective drives. The individual can apply them precisely because they are available resources, which the subject has full ownership of. They can thus be used (i) in a canonical sense, making corrections to adapt them to the specific situation; (ii) in a co-optative sense, that is by reusing them in new contexts, with a sort of ontogenetic exaptation; (iii) in a totally innovative way, by modifying them for use, or in an inventive way, by generating new resources through the recomposition of those already present.

Innate and learned competences are resources that allow one to translate an affective condition, that is a motive—such as being afraid, hungry, eager to chase—into a finite behaviour, through their modal contents. The endowments of competence indicate, therefore, the way in which the subject realises its motives and therefore refer to the above-mentioned drives. First, there is (i) the ability to recognise the reference target of the affective state, that is what is frightening, what is food, what is to be chased. Second, we have (ii) the way to exercise affective propositionality, that is how to avoid, forage, catch. Finally, there is (iii) the identification of the actuating circumstances of the motive, that is where, when and under what circumstances to express that behaviour. Therefore, we need an explanatory model that refers to all the orientative, predictive, operative and solution resources through which the subject can translate affective motives into behaviour, attempting to realise their contents. Shifting the interpretation of modal resources from the automatism model to the tool model explains how an animal adopts adherent responses and freely uses its endowments.

Of course, it is not enough to state that endowments are tools and not automatisms in order to change an explanatory model. First of all, it is necessary to explain what the differences are between an automatism and a tool (i.e. what is meant by a tool, or what differentiates it from an automatism) so that the latter can never be reduced to the former. Secondly, one must clarify why this allows for the ownership of use that underlies any non-repetitive expression of the individual. Furthermore, it is essential to show how the proposed model is consistent with the neurophysiological and ethological knowledge of animal behaviour, while never forgetting the importance of the aforementioned canon of parsimony. In this proposal I will try to show how the tool model is actually much more parsimonious than the model based on automatisms. In addition, but no less importantly, a new model must also be able to explain inconsistencies that the previous model could not account for, having to rely on explanatory epicycles to make ends meet—a bit like the Ptolemaic paradigm.

A final aspect, fundamental in the epistemological field, concerns the need to be subsumed by several explanatory models. Today, in fact, we are faced with three ways of explaining animal behaviour, resulting from what I view as the messy amalgamation of three approaches: (i) the classical ethological approach based on drive psychoenergetics; (ii) the associative approach that can be traced back to the behaviourist school; (iii) the elaborative approach developed from the insights of Wolfgang Kohler

(1887–1967) and Edward Tolman (1886–1959), and then introduced into cognitive ethology through the work of Donald Griffin (1915–2003). These three models present a large number of inconsistencies in their mutual relationship, bearing in mind that, as Joseph LeDoux points out in *Synaptic Self: How Our Brains Becomes Who We Are* (2002, 90–134), all that is expressed in behaviour is also inscribed in the same neurobiological matrix. We therefore need to find a common model capable of explaining the various aspects of behaviour, overcoming the classic dichotomies of innate versus learned or simple versus complex and above all not leaving creativity unexplained.

The Cognitive Level of Subjectivity

There is no contradiction between having endowments of phylogenetic inheritance—that is being the bearer of a specific project and having skills that precede the individual's experience—and being endowed with subjectivity, but only on condition that we recognise the instrumental value, that is the availability, of these resources. The real compass guiding animal behaviour is the hedonic system, which induces certain behaviours in order to satisfy cravings or reduce restlessness, or by simply producing pleasure. The whole body and every behaviour can thus be used by the individual to procure that momentary release of endorphins that make it feel better. Animals, therefore, should not be regarded as automata caught up in their functions, mechanically locked into fixed goals, because they are in fact inveterate hedonists who seek a share of pleasure in every action. There is no doubt that their constitution is the work of the selective dialectic, whereby pleasure also predisposes the subject to perform activities with a fitness function—the most striking example being the erotic tendency. Nevertheless, once the hedonic colouring is in the hands of the individual, the latter uses it with its own discretion, because it owns it. Otherwise, we could not explain certain behaviours that are very common in the animal world, such as masturbation.

We have seen that, in order to understand the foundations of animal subjectivity, it is necessary to start from two metapredicative levels concerning animal being in the generic sense of the term. These can be traced back to the evolution of (i) a *Self* distinct from the non-Self, capable of differentiating the changes produced by its own action from those that occur outside; (ii) a *for-Self* referring to the subject's interest in the world, given by affective states, that is psychological conditions that translate into real demands, which in turn are regulated by hedonic feedback. The resulting subjectivity appears as a three-dimensional emergence of the individual from the background world: the individual thus rises up, detaching itself from external reality and referring any external presence or event to itself. From these two levels we also perceive the subject's sense of self-belonging, whereby on the one hand it owns its actions in the surrounding landscape and, on the other, it is the bearer of interests demanding to be met. The subject is, therefore, one who is able to transform the environment into a space of usability conformed to its own dimension, that is, interpreted according to the specificities of its nature. For any baby mammal, the environment is almost always a playground, but the game is always a little different depending on the species: in other words, without prejudice to the objective constraints and opportunities of the environment, accessibility is a function of subjectivity.

The animal condition is based on openness to the world. In a certain sense, it is correct to say that heterotrophy characterises the ontological condition, so that non-self-sufficiency lies in that basic peripatetic dimension, which can be expressed in multiple ways—even standing still and waiting for food to reach the animal through the current, as is the case with many marine organisms. We must not, however, think that this results in some vague nomadism in the world, because this openness to external input involves a kind of *life project*, that is, the construction (and not just inhabitation) of a niche. The aforementioned cases of beavers and social wasps exemplify this concept, which I will return to when discussing phylogenetic protagonism: being born a beaver means, essentially, having a dam in mind, just as being born a hornet means designing a cardboard hive. An animal is therefore open to the world to the extent that it bends what it encounters, bringing it back into line with its own design model—this is what is meant by attribution of operability. But for this very reason, an animal cannot do without assessing the starting conditions and the environment in which it has to carry out its project, which obviously always has its own particularities, posing specific problems that have to be solved artfully, that is, through a multiplicity of actions.

There is no doubt, then, that each species is innately equipped with modal competences that enable it to tackle problems, understand their characteristics and deal with situations through effective operational solutions. There is a species-specific typicality in how to achieve the desired result, so that we notice pre-established and species-specific behavioural modules, useful to carry out all the necessary operations within the given life project. But although there may be a specificity in the way animals deal with everyday tasks, so that we recognise different species of gardening birds by the way they build their arbours—not only in terms of the end result but also in terms of the operations they carry out—it is clear that these behavioural patterns must be adapted to the particular situation at hand. In other words, there cannot be just one way of carrying out an action, but rather a style of execution that the subject always adapts to adapt its action to the specificities encountered. Any observer can see that the expressive pattern, rather than an automatism that the individual would be forced to repeat, is a style, a tool offering a multiplicity of different uses. While having an undoubted functional domain—a hammer evidently lends itself only to a certain range of functions—nevertheless, within that range, a tool allows for several uses.

Here, then, I would like to dwell on a third aspect of subjectivity, which concerns the principle of ownership of one's endowments. This aspect could be simplified as follows: *the subject uses its endowments instead of being driven by them*. Moving from a vision based on automatisms to one based on tools means, in fact, bringing out the ownership principle. This does not deny the fact that each species is, so to speak, contained within a life dimension that envisages a niche to be built and useful tools to do so. Let us now return to the concept of *umwelt*, as presented by Jakob von Uexküll in his essay *Streifzüge durch Umwelten von Tieren und Menschen* (1956), in particular in the chapter dedicated to purpose and design. The Estonian naturalist speaks of an overall natural design that characterises each species and that in a certain sense encloses the animal within its own immersive dimension. On the other hand, this does not necessarily mean turning the animal into a simple effector of mechanically determined functional cycles, nor does it mean considering its *umwelt* as a world that is already totally structured and horizontal in its predicative characteristics.

An individual's life, although played out within a dimension that characterises it, accesses a plurality of possible variations, carried out in spaces of belonging positioned on several levels. The taxonomic differentiation of the animal world has given rise, in fact, to the fractal repetition of common spaces: for example, we recognise a shared dimension (a particular existential style) within the world of birds or mammals, which means that being a mammal implies and enables a certain way of being in the world. This condition of belonging is supplemented by other, progressively more specific conditions, attributable to the sub-levels of orders, families, genera, species and even individual populations. Each space of belonging has a range of interpretation that allows each animal to create its own style, acting on different levels: the *umwelt* is thus a dimension to be personalised. An animal does this not only on the basis of its own experiences, but also in relation to its own interpretation, like an actor reciting a text. This ownership cannot be taken away from it, because it is inherent in the very principle of a life project. Recognising the presence of a phylogenetic pre-definition means admitting, therefore, this field of interpretation.

At this point, a third level of subjectivity becomes evident: that which refers to an animal's life project and modal endowments, which the individual owns insofar as it can adapt them to the situation while being free to interpret them. These resources make it possible to translate the previous levels—the sensorial subjectivity of the Self and the affectivity of the for-Self—into completed actions, defining how to use one's life project and competences to fulfil one's hedonic needs. This is a cognitive or modal plane of behavioural expression, referring precisely to the endowments that enable the animal's customary operations. These include finding useful information in perception, assigning a certain meaning to an event, being able to make predictions about what might happen, knowing how to deal with a certain risk or opportunity, having solution recipes to guide one's attempts, and so forth. There is subjectivity if and only if the individual has a certain ownership of its cognitive endowments, because otherwise we would fall back into the traditional image of the animal driven by automatisms and therefore lacking a decisional Self.

These are the endowments of competence mentioned in the premise and which, as we know, are partly inherited by the individual and partly learned in the course of its life. On the other hand, this plane of subjectivity—which we could define as an *in-Self*, an internal theatre that allows one to correctly process the input data and construct the best representations of the various extractive, evaluative and operational aspects—can exist if, and only if, the endowments do not act autonomously and directly on behaviour. The in-Self condition can exist as long as the endowment package is not made up of automatisms that, in an autonomous, imperative and analytical way, condition the animal through a trigger mechanism. If this were the case, in fact, the individual would lose any form of ownership over its own behaviour. When I speak of ownership over the resources of competence, I mean the individual's ability to use its endowments with a certain margin of freedom, thanks to their availability. Mind you, I am not talking about consciousness or free will, which I will address later, but about behavioural expression as the result of adaptation, reconfiguration and synthesis of resources carried out by the mind. Automatism, in fact, acts on its own, even if its action can be enhanced or mitigated by other automatisms or organised in sequences through the chaining mechanism. In order to speak of cognitive subjectivity it is clearly necessary to go beyond the automatism model.

So, in order to talk about cognitive subjectivity, and therefore creativity, it is necessary to propose a different interpretation of the modal endowment, based on the concept of tool, something that lends itself to use and offers modal competences to the animal's action. Such a tool (i) does not act autonomously, but lends itself to use; (ii) cannot be considered exhaustive in its explication, because it provides for a multiplicity of different uses; (iii) does not drive the animal through a triggering action, but is called upon by the subject; (iv) cannot explain behaviour through an analytical approach, because it always refers to a systemic condition. I believe that overcoming the automatism model is fundamental in order to speak of an elaborative-representational subjectivity. Consciousness is not sufficient to achieve this—indeed, invoking it sounds like a *petitio principii* or the action of a sort of *deus ex machina*. We can see consciousness as a particular function of the mind that is able to make explicit—we could say, illuminate—certain mental states, while keeping others in the shadows; however, it is evident that thought, not necessarily conscious, must be explained by going beyond automatism.

At this moment, therefore, I will suspend the discourse on consciousness and its different levels of intentionality—in Daniel Dennett's use of this term in his book *Kinds of Minds: Towards an Understanding of Consciousness* (1996), recalling Franz Brentano (1838–1917). Rather, I wish to focus on cognitivity, that is all those elaborative-representational activities that translate the individual's affective tendencies in a modal sense. To speak of consciousness would only confuse the presentation of this model. Moreover, the proposal of a new explanatory paradigm must meet certain basic requirements. I would like to suggest five: (i) *clarity and simplicity*, that is the ability to show, by means of comprehensible examples, how that resource contributes to behaviour; (ii) *productivity*, that is the ability to explain phenomena that the previous model cannot; (iii) *parsimony*, that is the ability to explain a variety of behaviours with a lesser use of endowments; (iv) *adherence*, that is coherence with the characteristics of the neurophysiological system; (v) *subsumption*, that is the ability to unify in a coherent manner the tripartition of the models, which today are used together despite being incompatible.

Let us start from a very simple assumption that has accompanied this entire discussion: the singularity of reality. If the world constantly proposes small deviations in the situations to be faced, behaviour cannot fully adhere to the given specificities without the adaptive endowment being available to the individual. Every competence, when faced with events, requires adaptation and can never be applied by mere repetition. The individual must be able to use the endowment and not be driven by the latter. Being in tune with the situation means knowing how to manage the margin of novelty, whatever challenge it presents, even if it is extremely small. In other words, it is a question of being able to use one's own resources. Usually one has to make simple adjustments to adapt one's action to the particularities of the situation, but one is also often faced with a problem that is somewhat *new*. In these cases, one has to find a solution using one's skills, so that the endowment is used beyond its operational domain. Once again, it would not make sense to make blind attempts, since in order to get a result, it is necessary to carry out tests targeted at the type of problem at hand: the field of possibilities must be narrowed down.

What I am saying, in practice, is that although selection in phylogeny and learning in ontogeny predispose endowments that make the subject adaptive, the latter must nevertheless readapt them or use them experimentally, because situations never occur unchanged, whether in terms of risk or opportunity. Therefore, the subject must be a protagonist in the implementation of behaviour, since everyday life requires a solution effort, however minimal, to achieve adherence to events. It is not enough to mechanically respond to a stimulus or to discharge an urge in some activity, because the world always presents challenges and an animal must necessarily be a problem solver. As such, it is required to play a full part in ordinary life and not just in exceptional situations. Versatility is a prerequisite for adaptation, whether you are a wasp or a beaver, so without protagonism in the use of endowments, phylogenetic skills would not make sense. The ability to use an endowment *ad hoc* applies in every situation the individual encounters, not only in those that jump out at the researcher as surprising. Protagonism is not a high-profile endowment to be used in special and extraordinary circumstances, but is the common, normal and ordinary condition of animal existence.

That said, one must establish the necessary and sufficient requisites for the freedom of action that allows for "cognitive protagonism". These are (i) using one's own skills with a certain margin of freedom of adaptation and creativity; (ii) being productive in the face of the tide of events that affect the individual, instead of simply seizing what is offered by the environment. I have already discussed the first point; as for the second, I mean the individual's ability to take the initiative and not be at the mercy of inductive factors. The individual cannot simply solve the problems that the world directly presents to it but must pressure the environment to achieve things that are not blatantly there. An animal does not merely respond to various stimuli but investigates the potential of a situation and turns it into a problem requiring a solution by actively intervening in the world. Most opportunities arise as a result of some action to change the existing order of things, so the individual must always go beyond appearances. It must be able to make hypotheses, investigate and probe reality, and try out methods that are not consolidated or only partly accredited by previous experience: it must therefore have its tools available.

In the same way, risks are hardly evident: most of the time they have to be extracted by applying anticipatory modules, looking for similarities but also for differences—in other words, by using one's own appraisal equipment very freely. Appraisal is based on internal representations—consider the hidden unit model proposed by Jones and Hoskins (1987)—that arise on the basis of input connections. Nothing ever presents itself for what it is: life is a never-ending masquerade ball where you have to pick up the clues; it's like an incomplete jigsaw puzzle, so that you cannot help but venture an answer based on slim assumptions. And there are no common denominators to justify behaviour by minimal triggers, because every attribution of meaning stems from a family set and recurring clues are most often misleading. As John Benjafield points out in the essay *Cognition* (1992), knowledge is distributed in the global configuration. All this shows that a merely reactive view of stimuli does not correspond to the attribution processes that an animal normally carries out when it has to make a prediction. Situations change and its forecasting equipment must be able to get an idea of risks or opportunities based on the sophisticated use of its tools.

A fundamental author on this subject is Urlich Neisser (1928–2012): for him, one cannot claim that the stimulus acts directly because there is a process of elaboration starting from the very first perceptive moment. Perception involves a process of configuration-recognition, so according to Neisser there is a close relationship between perception and memory. In *Cognitive Psychology* (1967), the German psychologist states that there are different stages of information processing before arriving at a behaviour, and that recognition is central to the attribution of meaning. According to

James Gibson's (1904–1979) ecological approach, experience always has to be reconstructed from partial information about the world, and in the course of processing it is necessary to integrate information coming from the stimulus with information derived from the environment. In the essay *Outline of a Theory of Direct Visual Perception* (1969), Gibson emphasises the subjective nature of information which, therefore, has to be collected. Hence the importance of considering the individual as an active entity, capable of selecting the structural invariants to which to pay attention and to place them in an overall framework by taking advantage of other references scattered throughout the environment.

Perception as an active faculty is emphasised by Donald Griffin in *The Question of Animal Awareness* (1976, 18–22): in it, Griffin reports how, through echolocation, animals like bats and porpoises literally interrogate their environment through the use of ultrasound, adapting it to the specific situation each time. In *Listening in the Dark* (1958), Griffin shows us how the orientation process not only relies on the echolocation mechanism, but also on spatial memory, especially in familiar environments, so placing new obstacles may make it more difficult for them to avoid them. The cognitive versatility of an animal, therefore, does not necessarily reveal itself in successful or high-performance behaviour; on the contrary, a mind in action is most easily revealed when difficulties arise. Very often, in fact, we see animals struggle when faced with unforeseen situations or circumstances that diverge from the ordinary. It is then that their capacity for improvisation becomes evident, most often through an innovative use of their skills.

Ultimately, how should we understand endowments? This is the paradigmatic question we must ask ourselves before climbing to the heights of consciousness, if we do not want to base this faculty on nothing and use it tautologically. Subjective freedom does not lie in awareness of a certain function, but in the fact that the function cannot be directly deduced from the endowment. If the affective components place the subject on a precise orientational plane towards the world, cognitive endowments do no more than make themselves available, as tools, for the translation of the affective tension into a propositional attitude, enriched with modal components. It is up to them to specify how to express a given tension: for example, by recognising the appropriate target, the most effective action, the level of approximation achieved, the location or circumstance of execution. If, as I said, affectivity defines the subject's verbal predicates, endowments complete the behavioural sentence by adding circumstantial, modal, evaluative and anticipatory complements.

In short, the mind resembles a factory of thoughts about the world, which mixes the affective and cognitive components in the vast cauldron of the here and now, giving rise to a flow of propositions that sometimes, but not always, translate into behaviour, and sometimes give rise to conscious experiences, but most often recall an immediate excitement that runs through the body and generates actions before one realises. The cognitive level lies in this expressive freedom of the individual, who thinks even when totally submerged in the unconscious; it's a magmatic nature that allows thoughts to pop up like mushrooms after rain because, in the great ecosystem of endowments, the possible interactions are almost infinite. Endowments are always interpreted by animals, both because adaptation to the specificity of occurrences requires it, and because they do not act autonomously but must be subject to a systemic configuration. If the mind is a thought factory, this is possible thanks to the conjugational plurality of endowments, which transform competence into a resource that can be adapted to singular needs. This is the only way in which each product can be said to be different from all the others and adhere to the needs of the given life project.

Cognitive protagonism is characterised by these two prerogatives, namely: (i) not being passively subject to the influence of external factors, but acting in a proactive manner, while taking into account the elements present, as if they were route references; (ii) having knowledge and skills at one's disposal, as useful tools to be adapted to the situation. It is therefore a question of placing the responsibility for action back on the subject, instead of assigning behaviour to individual mechanisms whose function is disjointedly caused by triggers. The mind thus becomes the place of convergence of influential variables, both environmental and affective, and is therefore capable of combining the different demands that press on the individual. At the same time, the mind is a productive entity, capable of standing up to the situation at hand, to face it and to chart a viable course time after time. Considering endowments not as automatisms acting autonomously, but as utilities available to the individual, makes it possible to meet both pre-requisites. This is for two reasons. First, (i) the expressive motive is brought back to the subject, understood as a unitary entity capable of implementing a "propositional synthesis" between the various demands, avoiding the analytical and peripheral conception of individual mechanisms. Secondly, (ii) endowments are no longer conceived as devoid of functional availability (an automatism cannot be used), but as entities that present a range of possible applications and are potentially capable of modifying their usage title.

In order to admit the animal's protagonism, it is necessary first of all to put the subject back at the centre. The subject, however, must be understood as a system capable of bringing into play two things. First, (i) a sense of unity, through the convergence of influential factors, but also through the productive property of the mind. Secondly, (ii) freedom of choice, releasing the subject from the impositional view of endowments whereby the latter are supposed to atomically and mechanistically determine behaviour. It must be clear, however, that this apparently intuitive operation—how else could an animal adapt its behaviour to the circumstances?—actually determines a sort of Copernican revolution in the explanatory modelling of animal behaviour. In the traditional view, in fact, the individual is driven by triggering factors and by the conformation of its automatisms, so that the deficit of protagonism is due to the fact that there is: (i) no supervising centrality, but rather single mechanisms; (ii) no freedom or ownership of use over the endowments, whereby the subject is rather driven by them.

In order to recognise individual protagonism, it is therefore essential to move from a disjointed—I would say almost scattered—conception of competences, which act autonomously on the basis of external triggers or inputs, to a systemic vision where each endowment makes itself plastically available for the solution of a certain problem-need and does not impose a behavioural standard. After all, it is a question of restoring the concept of mind and admitting the psychological condition of the subject. This then paves the way for the cognitive plane of subjectivity: by regarding endowments as tools that the individual uses, the subject can emerge as a cognitive superintendent of its own resources. This way, the individual is put back at the centre of expression, recognising its capacity to make innovative proposals and depriving endowments of the autonomy and deterministic character that made the individual a slave to its own innate or learned mechanisms. Clearly, when I speak of a propositional synthesis of the individual I do not mean its total emancipation from endowments: the latter remain the instruments with which the subject is called upon to face the challenges posed by the world.

This is not just a change of perspective, but the recognition of a systemic entity that is responsible for its various processes, because it emerges with respect to individual modal endowments and is not contained in the simple summation of influences. On closer inspection, protagonism lies precisely in the functional outcome of the mind: this innovative product is capable of receiving and bringing together the influential factors, but at the same time can go beyond their simple addition, putting forward a singular proposal. This paradigmatic shift places the behavioural initiative in the hands of the subject: (i) what was previously a trigger is now downgraded to a reference, an ingredient of the synthesis; (ii) what used to drive the individual or impose a certain behaviour now becomes an exploitable and available tool; (iii) what used to rigidly define an expressive mechanism, that is a standard function that could simply be repeated, now appears as a map that allows for multiple routes, multiple functional adjustments, multiple innovations and possible inventions; (iv) what used to be deterministic about the subject is now influential in probabilistic terms, but not prescriptive, so that behaviour cannot be deduced from the variables present.

On closer inspection, therefore, admitting a cognitive level in subjectivity means changing the image we have of animals, granting them productive psychological states and an instrumental use of endowments, which do not drive the individual, but put themselves at its service. It is the subject that uses a competence, especially when it has to make it work (adaptation) in a particular circumstance or apply it as a targeted attempt (co-optation) to the specifics of a problem. This is far from infrequent because: (i) even the most responsive behaviour requires adjustment from time to time; (ii) when faced with an unusual situation, one cannot afford to make blind attempts. The adaptation of endowments to manage the margin of situational novelty and the co-optive use of skills to make targeted attempts are, if you like, the simplest forms of cognitive protagonism. If an animal is to be able to adapt a pattern to the specificity of the situation or use it as a solution recipe to face a challenge, the competence must be "available" to the subject, it must be a tool and not an automatism, it must be a utility that the individual is entitled to use.

In order to admit animal protagonism it is therefore essential to rethink the model used to represent innate and acquired competences, moving from (i) a vision based on triggers driving the individual, over which the latter has no power but is fully passive to one (ii) where the scheme is an available tool which the animal uses, sometimes in a canonical way, sometimes with a certain margin of freedom in adaptation or co-optation. It is clear that this shift cannot simply be stated, but must be accompanied by a corresponding new model. When I speak of a tool as opposed to an automatism, two problems immediately arise. First, one must (i) find a different model, capable of explaining the predicate of availability underlying the ownership of use. Second, (ii) one must characterise this superintendence, so as not to fall back to another version of the *homunculus* and regress to infinity. I will develop this argument in the next chapter; for now I will try to explain how to model the concept of tool.

From Automatism to Tool

When I speak of the animal's cognitive subjectivity I mean the availability of innate and learned endowments—what we may call the individual's competences—to be used as resources. Endowments, therefore, (i) present multiple functional possibilities or performance outcomes, like a map that contains a considerable number of possible routes to the same destination. Secondly, (ii) they allow for different scopes of application: sticking to the map example, this means that several destinations can be reached with the same tool. The first point shows the flexibility of the tool that, unlike an automatism, has a range of functions that allow for adaptation to the existing situation. If a route in the itinerary is obstructed, thanks to the map it is possible to find alternative paths. At the same time, a tool is available for reuse in other contexts, that is, it does not only provide *one* performance area. As Martin Lindauer points out in the essay *Schwarmbienem auf Wohnungssuch* (1955, 263–324), bees use the same dances to define targets that are very different from food, such as the presence of water sources, resinous plant substances, cavities suitable for serving as a dwelling and so on.

As for this third level of subjectivity I would like to show that it is wrong to superimpose the subject on its endowments, because if this were the case the individual would be imprisoned within repetitive schemes and could do nothing but follow the same automatisms. A subject is endowed with a modal application of its skills precisely because the latter do not control it, but are available as tools that can be used in a number of different ways, for purposes that are the most disparate and, above all, open to any occurrence. This does not mean that the subject is not bound by the tools it possesses: if we only have a hammer at our disposal, this will inevitably narrow down the ways in which we deal with the problems we have to face. Each animal is bound by the tools it possesses and by the capacity of its cognitive hardware to update its endowments through learning processes. There is no doubt that some animals have greater learning capacities and a larger memory database than others and therefore show greater flexibility in their behaviour: mammals and birds are an example of this. But the principle that endowments are used, adapted and reused by an animal, acting as tools and not automatisms, also applies to invertebrates.

Among other things, this has two major consequences. Firstly, it means that (i) there is a directly proportional relationship between innate and learned, because the greater the innate endowments the greater the starting equipment that can be updated (this is a fundamental point which I will return to). Second, it indicates that (ii) learning increases the adherence and expressive flexibility of the subject, a result that is easily verifiable and that is justified if the endowment is viewed as a tool, while it would be a contradiction if it was considered an automatism, which would inevitably increase the animal's captivity. In the proposal of the cognitive level of subjectivity, all endowments are tools available to the individual. This includes innate tendencies, patterns acquired in ontogenetic apprenticeship, learned correlative relations which regulate certain flows of response or prediction, and the operational knowledge acquired. In other words, they are like master keys or maps, which define a plurality of functions and a multiplicity of functional applications, so that they are used by the subject instead of directing it. In short, endowments alone are not sufficient to explain animal behaviour: one must admit a surplus, that is a kind of metaphorical craftsman who uses his own tools to make expressive products. The level of cognitive subjectivity rests, therefore, on the principle of *ownership of endowments.*

Simply put, we could say that when a dog learns to "sit", from that moment on it has acquired a key to the human being, which it will use every time it wants something from us or when we propose a problem to it; it will not have a switch that will make it sit mechanically every time we say "sit!" In other words, the dog now has knowledge that belongs to it and which, from now on, it will be able to use in a canonical way (following a request from us), but also as a heuristic, when it is not sure what we want from it or as a proposal to obtain something from us. In this sense, we can say that the dog owns the "sit" action. An animal appropriates all the knowledge it has, whether it has inherited it through phylogeny or learned it directly through experience. In making it its own, it is entitled to use it, so that any knowledge it acquires increases its freedom of action and its subjective field of operation. If knowledge amounted to automatism, the more an animal learned, the more it would become enslaved or, in any case, limited in its freedom of action. But this is contradicted by reality.

Admitting this superintendence—for the sake of simplicity, I have proposed the image of a craftsman who owns his tools—requires, however, reformulating the conceptual model of the endowment itself. If the endowment were a switch, the activation of which leads to a standard, rigidly contextualised and repetitive performance, there would be no room for this metaphorical craftsman, since each automatism would carry out the work in a completely automated manner. An automatism does not envisage willingness because it is like a spring that always produces the same result when triggered. It is not possible to use an automatism, let alone co-opt it to another function, because it is the latter that determines behaviour: it is imperative on the individual. It is a question of overturning this explanatory model, in two ways. First, (i) by putting the propositional subject back at the centre, considering it as endowed with a behavioural purpose, influenced by affective states (homeostatic, sensory, dispositional) and by the hedonic principle. Secondly, (ii) by considering endowments as modal resources that enable the individual to achieve its aims by acting as instruments, which therefore affect how the action is carried out and the goals to achieve.

When I speak of the intrinsic purpose characterising the individual, I do not wish to return to a teleological explanation of nature and the evolutionary process. Let us take up, then, the words of Henri Bergson in chapter 1 of *Creative Evolution* (1998, 41): "It does not accept, indeed it ridicules, the idea of an external finality, according to which living beings are ordered with regard to each other: to suppose the grass made for the cow, the lamb for the wolf—that is all acknowledged to be absurd. But there is, we are told, an internal finality: each being is made for itself, all its parts conspire for the greatest good of the whole and are intelligently organized in view of that end". More generally, the very fact that there is a hedonic regulator in animal behaviour, and that this is affected by the homeostatic and affective needs of the individual, inevitably gives rise to the animal's propositionality. On the other hand, propositionality itself implies the presence of instruments that make it possible: if the animal's aim has a predicative form, such as eating, drinking, chasing or enjoying, it is obvious that these verbs rely on modal contents, that is on instruments that enable them to be realised.

The intrinsic purpose mentioned by Bergson necessarily requires a superintendence over the endowments, availability of one's modal resources and ability to use them with a certain amount of creativity. Most of the time this creativity is infinitesimal, nothing more than adaptation of the competence to the given situation, but sometimes it implies reusing the resource or transforming it into heuristics. I would also like to stress that I do not regard the craftsman as some immanent or transcendent entity within the machine, nor do I wish to link him to consciousness—since propositionality is not extinguished in unconscious or preconscious conditions. I rather consider him an integrating function of the different psychological states or the emergence of the mental system. But let us return to the theme of this chapter: what it means to consider the equipment as a set of tools. We know that a tool obviously cannot take the form of a switch or a wire, otherwise we would fall back to the image of the puppet-animal and we would not have strayed far from Descartes. Cognitive subjectivity demands, in fact, an expressive protagonism that cannot be given by a set of mechanisms, each with its own autonomy and functional self-sufficiency. On the contrary, the model based on automatisms has performed the epistemological function of avoiding subjectivity.

Claiming that animals are driven by external or internal stimulatory entities that act as simple triggers, followed by mechanisms that, like dominoes or a functional cascade, lead to the expression of fixed behavioural modules, testifies to a precise desire to deny the non-human even the slightest glimmer of a possible protagonism, reducing it a totally passive being. In his book *Purposive Behavior in Animals and Men* (1932), Edward Tolman advanced the idea of an implicit finalism, in which mental processes were treated as intervening variables between external stimuli and internal influences, on the one hand, and observable behaviour on the other. However, claiming that the animal is endowed with propositionality is not sufficient if we continue to explain its behaviour through the paradigm of automatisms. If we do not overcome this foundational aspect, centred on the concept of behavioural automatism, any attempt to speak of subjectivity is clearly null and void, because every animal expression will only have the *appearance* of subjective behaviour. For this reason it is necessary to start from the basic model.

According to Stanislas Dehaene's *How We Learn* (2020, 72–92), the brain behaves like a budding scientist who, when faced with a problem, formulates hypotheses on the basis of a-priori models, representing a core of knowledge with which animals come into the world. The author points out that the concept of plurality is already present in newborns after just a few hours of life, and this also applies to monkeys, pigeons, crows, salamanders, chicks and fish. The Popperian approach to hypotheses to be submitted to the Bayesian probability calculation is an example of the instrumental use of knowledge. Formulating hypotheses on the basis of correlative or causal expectations, which are then subjected to confirmation or refutation, means recognising the principle of conjecturality, and thus the use of forecasting resources. Dehane reiterates this by saying that learning is infinitely more effective if there is a wide range of hypotheses to choose from. In his essay *Intentional Behavior and Motivation* (1971), F.W. Irwin also recognises that animals have expectations, that is anticipatory hypotheses that can give rise to alternative outcomes. But an analogous case can be found in operational models. In *Animal Minds* (1992, 110), Donald Griffin talks about weaver birds, noting that while the general pattern of the nest is constant, each male adapts its construction skills to the particular situation and to the shape and position of the twigs.

The tool allows for protagonism because it is available for use, whereas automatism imposes a function: this is the true keystone of the whole question. Only by investigating the structure of these two models is it possible to understand the paradigmatic shift in terms of the cognitive plane. The mechanisms of memory and modelling were studied by Frederic Charles Bartlett (1886–1969): for him, remembering should be considered an active process of schematic reconstruction of available information. In his essay *Remembering* (1932), the author underlines how knowledge consists of *schemata* that make functions available. Bartlett's studies on mnemonic processes show how the individual reproduces the contents of experience, creating schematic representations that retain some aspects and omit others: a schema is a flexible arrangement that can be extended and modified over time. This concept is of great importance in cognitive modelling, as it links up with that of map, used as an explanatory model by Edward Tolman (1948); the latter applied it to justify the way in which animals orient themselves in a spatial context through the memorisation of references.

A map is a network of connections that can be selectively activated to produce an itinerary, but which also allows for alternative activations, that is other itineraries. A map is certainly a schema, which can be traced back to the previously illustrated connectionist modelling, but compared to automatism, it is much more coherent with the fabric of the neurobiological tissue. This is true: (i) both with reference to the activation of a synaptic network, thanks to the joint action of specific activation caused by glutamate and inhibition of everything else caused by GABA; (ii) and with reference to a selective activation of areas containing receptors for particular neuromodulators, such as dopaminergic or serotonergic pathways. A map is a functional representation of something, a mental model that takes up an experience of reality but reconfigures it on the basis of specific use. Therefore, the concept of tool is based on three explanatory models, which are strongly connected: the map, the schema and the representation. The concept of tool brings us back to an interpretation of a-priori and a-posteriori endowments in terms of knowledge, which is why I speak of a cognitive approach in the interpretation of behaviour.

While in a non-mentalistic approach the animal resembles a machine full of switches, the activation of which triggers fixed behaviours, in the cognitive perspective the subject uses these tools to carry out its functions. Much like Bartlett's schematic representations, the cognitive maps hypothesised by Tolman are capable of spatially specifying the references useful for a particular orientative function, without rigidly dictating a fixed itinerary, thereby freeing the animal from mere repetition. The cognitive approach based on the concepts of schema and map sanctions the decline of the merely associative and trigger-based view of behaviour. This epistemological revolution started in the 1960s, mainly in human psychology, gradually bringing forward the idea that the individual forms mental models or representations of the various phenomenal entities, which are endowed with meaning and therefore with a predictive value (what a certain phenomenon foretells) or an operational value (how to deal with a certain phenomenon). Following this model, knowledge would be nothing more than a representation of reality endowed with meaning, capable of acting as a dictionary for what the individual has to face—a resource that can be used in evaluation and action.

The cognitive approach clearly differs from the traditional one, not in the tautological sense that "animal think"—a phrase that actually explains nothing—but because the basic model called upon to define the behavioural endowment is different. This epistemological shift therefore requires clarifying the differences introduced by the tool model in its various versions, that is in the images of schema, map and representation. The first aspect that immediately jumps to the eye is the difference between structure and function presented by the cognitive model: the structure does not impose a function but makes itself available for functional use. If we look at a map or a schema we get a network of possible functional states where, as opposed to an automatism, there is no identity between structure and function. Just as the map of a city permits—but does not designate—a given route, so a tool envisages several potentially feasible uses, albeit within a range of possibilities; therefore, it is never possible to extract the function from the characteristics of the endowment. Essentially, an instrument provides for a title of use, something implicitly impossible in an automatism. Likewise, a schema lends itself to being corrected, enlarged, enriched in its internal patrimony of functional possibilities; similarly, a representation does not necessarily have to be led back to a common element essential for recognition, but must take place through approximations.

For example, in learning this means passing from the concept of conditioning—given the stimulus comes the response—to a process of "cognition", which evidently cannot be explained through the process of priming. This is demonstrated by Shepard and Metzler's (1971) studies on the mental rotation of objects, a test in which some animals, such as pigeons, perform better than humans. Knowledge, in fact, does not produce automatic responses, but acts as a resource that allows the mental work of orientation, recognition, predictability and coping, obviously within a fairly precise range of expected application possibilities. The conceptual and functional difference is enormous, especially in terms of expressive protagonism. Automatisms can be added together and give rise to long, linear or recursive chains, but they always operate autonomously, despite being connected, and always perform a certain task. An automatism cannot be used, but must simply be obeyed. The explanation based on automatisms also encourages an analytical and decompositional reading of behaviour, because each automatism is responsible for a specific function. An automatism is a transducer of input variables that produces the behavioural output, so there can be no productive activity and no operational freedom on the part of the animal.

However, the difference between the two models is not merely terminological—a sort of explanatory circularity or a simple euphemism. Likewise, the same instrument cannot be broken down into a set of automatisms, because these two models are conceptually and concretely very different from each other—I would even say antithetical. I would therefore like to focus on the functional differences between the two models, noting their intrinsic characteristics, and on the relationship between their structure and what they enable from a functional point of view. A model, in fact, is a way of visualising a phenomenon or an endowment in order to better understand a process: in this case, the behavioural expression of the animal world. Obviously, this is an exemplification; nevertheless, the two proposals testify to two profoundly different ways of considering expressive endowments. I am often asked what difference there is between considering endowments as tools or as automatisms, and the answer to this question is fundamental to understanding the cognitive approach.

An automatism is like a sort of switch, the activation of which always produces the same effect, like turning on a light bulb: it is not open to multiple functions and therefore to some ownership of use. An automatism can only perform one task: the ratio between "what it is" (the structure) and "what it does" (the performance) is 1:1, so that, once triggered, it produces a certain behaviour in a predictable and autonomous way. The classic example of this approach is conditioning as studied in the behaviourist school where, given the stimulus (S), I have the response (R). This happens (i) directly, that is without any network variation; (ii) imperatively, since there is no escape once the conditioning is triggered; (iii) predictively, that is not randomly, so that if S then R and vice versa; (iv) deterministically, so that the performative outcome is always the same; and (v) exhaustively, since no other element is required to explain the functional outcome (how it occurs) and the performative outcome (what it produces). Even if I add intervening or modulating variables to an automatism, the above functional structure does not change: there will only be a summation effect.

On the contrary, a tool, being structured like a Swiss army knife, envisions several uses—the relationship between structure and performance is not 1:1, but 1 structure corresponds to a range of potential uses (F) and of possible performative outcomes (P)—so that the result does not depend directly on the trigger, but rather on the type of use the subject makes of it. In other words, a tool resembles the map of a city, which allows one to extract one of many itineraries: a map lends itself to being used in a singular way by a subject in order to come up with a congenial route. Therefore, considering the endowment as a tool means renouncing: (i) functional exclusivity and explanatory exhaustiveness, because the map in itself can neither produce nor justify the itinerary undertaken; (ii) performative mechanism, that is the parameters of implicit predictability, determinism of outcome and productive imperativeness; (iii) the image of the puppet-animal moved by strings, since the tool belongs to the subject: it is the latter who uses it freely, flexibly and sometimes even creatively. If I consider endowments as automatisms that move the animal, there can be no room for subjectivity, whereas if I consider endowments as tools, subjectivity becomes an indispensable condition.

There are several aspects that make the "tool" model preferable to the automatism-based model: first, (i) it allows the individual to deal with the novelties it is confronted with, given the fluctuations of the context; second, (ii) it can be used not only as an operant, that is to deal with the situations for which it has been configured, but also as a useful recipe (heuristics); and third, (iii) it is much more parsimonious, since several functions can be carried out with a single endowment. This last aspect, which responds to the aforementioned canon of parsimony, is possible because an endowment has no implicit correspondence between structure (what it is) and function (what it does), but provides for a plurality of functions. The "tool" model, as opposed to the automatism-based model, allows for a considerable saving in terms of endowments, which can be seen by comparing the parsimony of a map with a multitude of predefined itineraries.

Undoubtedly, there is one aspect in particular that leads us to consider animals as cognitive entities, able to use their endowments freely and creatively, just like a craftsman with his tools or a tourist with a map of an unfamiliar city: the world never offers two identical situations. Therefore, the individual is always called upon to manage small margins of novelty, and it can do so only on condition it has ownership of use over its endowments. This is true both in terms of the need to continually correct the course of one's own adaptive performativity, and in terms of the onerousness in the number of endowments one would require. As said, it is much more efficient, as well as effective, to have a map of a city than a million possible routes, and nature always presents the most economical solution. Efficiency, in fact, is a requirement of the living: resources available are always limited, so evolution has not only rewarded systems that are more capable of achieving their objectives, but also those that consume fewer resources. Being able to take advantage of a tool that lends itself to multiple uses is certainly more efficient than having an automatism for each individual performance. In my opinion, these two approaches are not even comparable.

Moreover, if endowments are tools with a wide application range, when an animal is faced with a problem it will not be forced to make random attempts—if this were the case it would never arrive at a solution, for simple probabilistic reasons—but it will be able to use a tool whose functional scope is close to the required performance, modifying it if necessary to adapt it to the problem at hand. Having a heuristic approach, based on solutions that are likely to be useful in solving problems, greatly reduces the combinatorial explosion of possibilities that, conversely, would occur in the approach based on random attempts. To think that an animal solves a challenge by trying at random is, in my opinion, an explanatory epicycle aimed at forcibly supporting the idea of animals as totally devoid of reflective capacity. Here too, the tool model beats the automatism-based model not only in effectiveness—we could say in how quickly it leads to a solution—but also in efficiency, because it provides for a consistent reduction of trials and errors.

In order to understand what is defined as "schema-representation" it is necessary to separate the two concepts that form the model itself, namely that of schema and that of representation, with regard to the contents they respectively translate. The term *representation* refers to its "being-in-place-of", that is having an analogical and functional relationship with a particular entity/event. This is exactly like a topographical map of a city that "represents" the latter in that it defines an analogue matrix, though not in an absolute sense, but with respect to its orientative function. By modifying the functions, while retaining the entity/event (in this case the city), the representation will change. The analogue matrix often moves far away from the entity/event it is supposed to represent, as in the case of an underground map, favouring the functional aspect over scale fidelity. This becomes even more significant if we move from map representations to predictive or meaning-referred representations: we could hardly find an overlap between the sight of a bowl and the food expectation that is shortly built up in a dog.

The term *schema*, instead, refers to a model that presupposes several combinatorial possibilities among its elements, where each configuration corresponds to a specific functional mode. Sticking to the example of a map, the schema indicates a multiplicity of vectors, the streets, whose activation in the itinerary requires additional information from the outside. The schema is characterised by a "space of functional virtuality" rather than by operational determination: this makes it open, available and non-deterministic. If we look at the characteristics of the neurobiological system, we immediately realise that we are not dealing with a set of connected but separate devices, but with a network that can give rise to different configurations starting from the same elements. In other words, what is activated is not an element that has a life of its own, but a pattern that can change each time in terms of functional configuration. In other words, we are dealing with a system that works through connections. The character of a schema is given by the possibility of assuming several functional states according to the specific configuration assumed at a given moment by the elements that are part of it.

These are two fundamental aspects of cognitive modelling. One could also define these qualities as: (i) non-self-sufficiency in functional definition; (ii) usability of the schema in a plurality of ways; (iii) limited predictability according to indices of probability; (iv) analogical-functional relationship with the entity it represents or to which it refers; (v) evolutionary capacity in further schemas. The schema-representation model is an important step in the explanation of animal expression: from a trigger-based reading, sensitive only to the start signal, and an associative one, involving a switch linking an input to an output, we pass to a use-based view, involving the endowment's availability as a tool, and an analogical one, able to represent the entity/event to which it refers. In this model, the behavioural endowment is totally revisited and proposed according to completely different epistemological criteria based on (i) the virtual combination of elements and therefore the plurality of functional states; (ii) interpretation, or rather the functional reproposition (or usability) of the entity/event. In conclusion, even if we look at the characteristics of the neurobiological system, we will realise that they could never work by automatism from a functional point of view, because it is organised by connections or schemes with multiple and hierarchical structures—it is more reminiscent of a map than a switch.

Works Cited

Bartlett, Frederic Charles. 1932. *Remembering: A Study in Experimental and Social Psychology*. Cambridge: Cambridge University Press.

Benjafield, John G. 1992. *Cognition*. New Jersey: Prentice Hall.

Bergson, Henri-Louis. 1998. *Creative Evolution*. New York: Dover.

Dehaene, Stanislas. 2020. *How We Learn, Why Brain Learn Better than Any Machine*. New York: Viking.

Dennett, Daniel C. 1996. *Kinds of Minds. Toward an Understanding of Consciousness*. New York: Basic Books.

Grandi, Guido. 1968. *Gli insetti. Mondo Occulto di Dominatori*. Bologna: Edizioni Calderini.

Griffin, Donald. 1958. *Listening in the Dark: The Acoustic Orientation of Bats and Men*. New Haven: Yale University Press.

———. 1976. *The Question of Animal Awareness*. New York: Rockefeller University Press.

———. 1992. *Animal Minds. Beyond Cognition to Consciousness.* Chicago: The University of Chicago Press.

Hansell, Michael H. 1972. Case Building Behaviour of the Caddis Fly Larva, Lepidostoma Hirtum. *Journal of Zoology* 167 (2): 179–192.

———. 1974. Regulation of Building Unit Size in the House Building of the Caddis Larva Lepidostoma Hirtum. *Animal Behaviour* 22 (1): 133–143.

Hinde, Robert, and John Fisher. 1951. Further Observations on the Opening of Milk Bottles by Birds. *British Birds* 44: 393–396.

Irwin, Francis W. 1971. *Intentional Behavior and Motivation.* Philadelphia: Lippincott.

Jones, W.P., and J. Hoskins. 1987. Back Propagation—A Generalized Delta Learning Rule. *Byte* 12: 155–162.

LeDoux, Joseph. 2002. *Synaptic Self: How Our Brains Become Who We Are.* New York: Viking.

Lindauer, Martin. 1955. Schwarmbienen auf Wohnungssuche. *Zeitschrift für vergleichende Physiologie* 37: 263–324.

Miller, George A., Eugene Galanter, and Karl H. Pribram. 1960. *Plans and the Structure of Behavior.* New York: Henry Holt and Company.

Shepard, Roger N., and Jacqueline Metzler. 1971. Mental Rotation of Three-Dimensional Objects. *Science* 171 (3972): 701–703.

Simon, Herbert A. 1955. Behavioral Model of Rational Choice. *The Quarterly Journal of Economics* 69 (1): 99–118.

Tolman, Edward. 1932. *Purposive Behavior in Animals and Men.* New York: Appleton-Century.

———. 1948. Cognitive Maps in Rats and Men. *Psychological Review* 55 (4): 189–208.

Tversky, Amos, and Daniel Kahneman. 1974. Judgement under Uncertainty: Heuristics and Biases. *Science* 185: 1124–1131.

Von Uexküll, Jakob. 1956. *Streifzüge durch die Umwelten von Tieren und Menschen: Ein Bilderbuch unsichtbarer Welten. Bedeutungslehre.* Hamburg: Rowohlt.

Washburn, Margaret F. 1916. *Movement and Mental Imagery: Outlines of a Motor Theory of the Complex Mental Process.* Boston, MA: Houghton Mifflin.

CHAPTER 7

The Territories of the Mind

Premise

Creativity is an expression of animality, almost a consequence of the protagonism that this biological condition requires in order to constantly adapt one's behaviour to the singular situations of the world. It is a matter of simple creativity: sometimes it can be astonishing in its innovative solutions, but most of the time it is minimal, nothing more than an adjustment to an already structured performance. On closer inspection, the founding principle of creativity is protagonism, the ability to give the world attributes of operability—a point of view that is subjective because it interprets possibilities. Animal protagonism can be traced back to the emergence of a Self that, in its basic form, arises from the distinction between sensibility and agency. As mentioned, this is the first level of subjectivity, around which there are (1) affectivity, that is, the *for-Self*, which indicates the intrinsic finality of the subject; (2) cognition, that is, the *in-Self*, which indicates the processing contents and the instruments that qualify the subject's identity.

If we look at the animal world we will discover a multiplicity of conditions, both in terms of their gradual articulation and complexity, and in terms of the range of solutions adopted, as shown by the realm of cephalopods. And let us not forget that we also experience a plurality of *subjective* conditions: these are usually divided into the two dimensions of the unconscious and consciousness, but are actually more articulated than that

R. Marchesini, *The Creative Animal*,
https://doi.org/10.1007/978-3-031-07414-1_7

and can give rise to a myriad of states. For this reason, we should not imagine subjectivity as a single condition constrained by all-or-nothing values, but in terms of various dimensions. Now, the Self, as a plane of differentiation from the background world made possible by the activity of an organism, constitutes (until proven otherwise) the *conditio* sine qua non of subjectivity. That being said, however, this first core of worldly perspective is soon joined by other reinforcing predicates. Even in animals that are normally, albeit wrongly, considered almost automata, such as insects, spiders or crustaceans, we can identify extremely complex levels of affectivity and cognitivity.

On the other hand, the planes of subjectivity should not be interpreted as hierarchical levels, but rather as dimensions that can be accessed but are not necessarily always active, just as is the case with consciousness. Now, among the dimensions of subjectivity, it is useful to consider what we call *mind*: the unitary condition arising from the integration of the various components. The qualities we recognise in a subject include homogeneity of expression, whereby the individual can be identified as a unity of purpose, and persistence over time, in the form of a solid and subsumptive whole capable of maintaining its identity through change. Whether we hypothesise unity as an epiphenomenon resulting from the concurrence of converging factors, such as the synchronous movement of a shoal of fish or a swarm of insects, or whether we invoke some ordering principle, the result does not change: what we observe is a process of integration between the parts, showing an overall quality in dealing with contingent situations. In other words, the animal acts as an entity that is not broken up into its various functional specialisations, but as a whole capable of handling situations.

We know that animal behaviour is manifested through attitudes that are somewhat standardised, recognisable in their general specificities—think of the different predatory strategies in spiders—as well as in the single acts used to carry them out. A species is, therefore, identifiable in two senses. First, (1) in terms of the *molar* operational plan, for example in the courtship choreography employed by birds of paradise or in the architectural structure of hives for social wasps. Secondly, (2) in terms of the individual *molecular* activities it includes: consider, for example, the way wild geese retrieve an egg during hatching or the individual operations used by solitary wasps to build their paedotrophic nest or by weavers ants to build their anthill. The possibility of recognising these two aspects within a species-specific habitus leads one to paedotrophic a marked innate component. The dialectic between molarity and molecularity, already envisaged by Edward Tolman (1948), paves the way for a hierarchical-finalised conception of behaviour, as we saw when discussing the work of Miller, Galanter and Pribram in the early 1960s, and can be considered the beginning of an elaborative approach to animal expression that goes beyond the interpretative model based on the horizontality of S-R associative mechanisms.

The individual stands in this dialectic between species-specific recognisability and the ability to manage contingencies as an integrative and creative entity, capable of flexing innate characteristics by adapting them to the particular situations at hand. As mentioned earlier, Donald Griffin's 1976 work inaugurated a new understanding, based on the fact that, between the *project plan* (molarity) and the *operational schemes* (molecularity), both of which are largely innate, there are all the activities through which the individual readjusts its inherited tools—through small modifications, reuses and inventions—in order to adapt them to the specific environment in which it has to operate. Thus, it is easy to recognise a predefined architectural plan and equally precise activities of cellulose acquisition and layering in a hornet's nest. However, given the multiplicity of host sites, time after time bumblebees have to adapt their structure to the characteristics of the chosen location. Even the nest itself will take on slightly different shapes, while respecting the basic characteristics. Animal protagonism lies precisely in this process of local adaptation that arises in the dialectic between project plan (the architecture of the hive) and operational schemes (the cellulose deposit). The superintendence of the individual is always a necessary quality and is located in the way in which, given the general plan and the species-specific tools, each individual manages (adapts) the innate according to the situation.

People often talk about the standard nature of the final design, and especially about the uniformity of the actions that a species carries out in its ordinary behaviour, in order to dismiss the mentalistic hypothesis of animal expression. The epistemological errors of this approach are numerous. First of all, as I said earlier, standardisation is the result of *perspective bias*, whereby the further away a species is from us, in terms of homology (of taxonomic proximity) or analogy (of adaptive convergence), the more its behaviour risks appearing standardised, especially if it presents different characteristics. Secondly, the very fact that there is a relationship between molarity and molecularity disproves the analytical theory of behaviour. Moreover, when one recognises a project plan and dedicated operational schemes, there is more congruence with the cognitive explication, based on the schema, than with the associative one based on S-R chains. But the crucial element, in my opinion, is that it would never be possible to carry out the project without initial and ongoing adjustments—that is, without an evaluative and supervisory presence.

Ultimately, we must assume a sort of superintendence in the realisation phase that allows an animal to reach its objective by freely using its own operational equipment. And this applies to all the complex expressions of animal productivity that Richard Dawkins called *The Extended Phenotype* (1982), including cobwebs, termite mounds, beehives, paedotrophic nests and beaver dams. On the other hand, I believe that this protagonism concerns not only the free use of innate endowments, but also that of learned endowments. Protagonism should be defined as the ability to interpret accumulated resources in an ad hoc manner. In order to understand the individual's power to relate all of its functions to the necessary singularity in a unitary fashion, we must posit the concept of *subjective experience*—a *Dasein*—combining the integration of all components and conjugation to external conditions.

Experience includes six elements: (1) sensitivity to external parameters and to one's own condition (*for-me-ness*); (2) agency or the tendency not only to manipulate the environment but also to differentiate the effects of one's action from the outside world (*by-me-ness*); (3) operational retention, that is working memory capable of identifying the course of action (diachronic condition); (4) sedimentary memory in its various perspectives, that is spatial, procedural, conceptual and episodic (mnestic condition); (5) the capacity to integrate the different functions taking place in the various parts of the central nervous system in such a way as to bring out a unitary Self; (6) intelligence, or the ability to go into depth with regard to situations, raising problems and trying to solve them. If we consider all these points together we realise that, in order to have a subjective experience, it is necessary to be a mental entity. In *Metazoa. Animal Minds and the Birth of Consciousness*, Peter Godfrey-Smith proposes an emergent conception of the mind, whereby, given certain configurations of non-mental entities, a mind comes into being. The mind would be an evolutionary product of the organisation of naturally occurring non-mental components (2020, 27–32).

The question of the animal mind has a very distant origin, considering that Aristotle already regarded animals as entities endowed with more than a vegetative soul. However, it was Darwinian evolutionism that first formulated the concept of the animal mind in evolutionary terms. The idea that other animals have complex mental functions thus developed mainly with the English naturalist's followers. Of particular note is the work of the physiologist George John Romanes (1848–1894), who dealt with sense-motor systems, especially of echinoderms and coelenterates, and who wrote important books on the animal mind such as *Animal Intelligence* (1912), *Mental Evolution in Animals* (1883) or *Darwin and After Darwin* (1897). In a distinctly gradualist view, Romanes sought to investigate the inner world of animals, their subjective experiences and their lives through an analogical approach to humans. Romanes was convinced of the profound similarities between the intellectual and emotional lives of all animals, but his ideas were soon considered overly generous in terms of mental faculties.

At the end of the nineteenth century, partly because of the continuist excess anticipated by scholars such as Romanes, who assigned high-profile faculties like intelligence and consciousness to organisms with a very simple nervous system, a sort of counter-reformation took place. An example of this is Conwy Lloyd Morgan's (1894) aforementioned principle of parsimony, according to which it is not correct to explain behaviour through higher faculties if it can be explained by simpler ones. Edward Thorndike himself sharply criticised the mentalistic interpretative model, preferring associative mechanisms based on reinforcement and punishment. This same period saw the spread of Pavlovian ideas about acquired reflexes in reactive learning (conditioned reflex) as well as Jacques Loeb's theory of tropisms in orientation, in addition to Thorndike's law of effect. The basic principle that inspired research in the first decades of the twentieth century can be summarised as follows: since it is not possible to see what an animal is thinking and it is not even possible to ask it, it is scientifically useless to speculate on the animal mind and above all it is not scientifically sustainable, because it is impossible to test any hypothesis. So we should consider the mind as a black box that cannot be investigated and must be put between brackets, because only behaviour is objectively verifiable. But the step from a legitimate suspension of judgement to the denial of mental functions is a short one, and that's the attitude that went on to characterise the first half of the twentieth century.

The mentalistic approach appears, somewhat contingently, in the work of Wolfgang Köhler (1887–1967) as part of the *Gestalt* school, which was to have a considerable influence on the development of ethology. In the essay *The mentality of apes* (1925), the German psychologist challenges the non-mentalist reading by observing the behaviour of chimpanzees when confronted with problematic situations. According to Köhler, the animal is not passive in the face of external reality, but imposes a certain order on things through an internally assigned organisation. Reality is not perceived through a summation of stimuli, but by looking for a form and often attributing one: the perceiver reconfigures the perceived. Köhler especially studied solution behaviour in chimpanzees and introduced the concept of *insight*, which literally indicates the ability to *see into* a problem and solve it. The Gestaltists' approach is based on the concept of understanding and solving problems through internal simulation processes, as opposed to the behaviourists' analytical approach based on progressive accumulation through random attempts oriented by the mechanism of reinforcement.

Yet the author who marked the turning point in the mentalistic interpretation of animal behaviour was a behaviourist: Edward Tolman (1886–1959). His works, such as *Cognitive Maps in Rats and Men* (1948) and *Principles of Purposive Behavior* (1959), produced a profound transformation in the understanding of behaviour. Tolman's study of rats' explorativeness revealed that they learn inside the maze regardless of specific reinforcers. This showed that they do have purposeful behaviour, and that there is a difference between molar or strategic aspects, which regulate their actions, and molecular or tactical aspects, used to carry out their strategies. There are three aspects that I would like to highlight in the work of this American psychologist. First, he laid the foundations for the molar-molecular dialectic, the hierarchical structure of the behavioural matrices I identified in the relationship between project plan and operational schemes. Second, he took into account the fact that animal behaviour is purposeful, and not simply mechanical as posited by John Watson (1878–1958). Finally, he introduced the concept of cognitive map, which was to be of fundamental importance in the evolution of cognitive modelling, overcoming the 1:1 ratio between endowment and function.

In this debate we cannot forget the work of Jakob von Uexküll, and in particular his concept of *Umwelt*, which was to influence both Konrad Lorenz and philosophers such as Max Scheler (1874–1928) and Martin Heidegger (1889–1976). Uexküll believed that every species is perceptually and operationally immersed in a certain context-world with specific meanings. In the essay *Umwelt und Innenwelt der Tiere* (1909), Uexküll distinguishes between a perceptual environment (*Merkwelt*), relating to the way in which that particular animal is immersed in the reality around it, and an actual environment (*Wirkwelt*), given by the set of modifications produced on the environment itself by the organism. Uexküll therefore talks about what I have defined as the first level of subjectivity: the ability to distinguish what happens in the world from the changes that the individual generates in it. But there is one specific aspect that deserves our attention: for Uexküll, the relationship between the individual and the external world is not based on mere stimuli, but on attributions of meaning that are directly related to the species, which nevertheless evolve through experience.

The ethological school, which developed mainly from the 1920s onwards, is based not only on innatism but also on a systemic view, in line with *Gestalt* psychology—see the concepts of operational schema and key signal. Konrad Lorenz's ethology presents a less dismissive attitude to mental life. The animal world is filled with a-prioris resulting from phylogenetic heritage, so it is nature itself that acts as a teacher. The Austrian scholar, embracing Popperian deductivism, believed that animals approach the world through a critical-hypothetical attitude; he regarded a-prioris as instruments of knowledge that can be revised. In line with authors such as Carl Popper (1902–1994) and Donald Campbell (1916–1996), in Lorenz's evolutionary epistemology, the animal acquires a representation of reality to be used as a working hypothesis. This interpretation would find confirmation in the work of the neuroscientist Stanislaw Dehaene, particularly in the essay *How We Learn* (2020, 81–94): the author highlights how, when faced with events that contradict common sense, animals increase their attention span and show surprise behaviour. Their responses to unusual situations are the best evidence that their approach to phenomena is based on precise expectations, which can be traced back to the laws of naive physics.

On the other hand, in the first half of the twentieth century, in order to avoid being discredited by reductionists, the ethological school was forced to distance itself from the mentalistic model, preferring the drive interpretation—well represented by Lorenz's psychohydraulics and Niko Tinbergen's Four Whys, even if a hint remains also in Lorenz's concept of the parliament of instincts. The picture changed radically in the last decades of the twentieth century with the birth of cognitive ethology. However, one should not forget the pioneering work of Karl Ritter von Frisch (1886–1982) and his disciple Martin Lindauer (1918–2008) in the study of bee communication, particularly the choreography of the abdominal dance. These authors recognised two values in bee communication:

(1) a semantic one, as a bearer of complex messages referring to absent entities, (2) a pragmatic one, as a plural use apt for several situations. On the other hand, with the birth of cognitive ethology, research on the animal mind and the mentalistic approach to animal behaviour took off and produced significant results, freeing itself from the obsessive comparison with the human being and contemplating the idea of a plurality of minds. This multiplicity is often characterised by small differences, as in eutherian mammals, but at other times—as in the case of the cephalopods described by Peter Godfrey-Smith in his *Other Minds. The Octopus, the Sea and the Deep Origins of Consciousness*—it bears unfathomable distances (2017).

The Evolution of the Mind

It is not easy to give a homogeneous and definitive definition of the mind; there is already a considerable difference between those who believe we should talk about *having* a mind and those who prefer the concept of *being* a mind, or those who consider it an epiphenomenon of the various nervous functions and those who think of it as a different realm altogether. For this reason, even choosing one of the many positions, we should be content with a low level of disambiguation. I believe it is important to consider the mind as the property allowing the individual to have a unitary experience, that is, an integrated convergence of several states and functions: we can imagine it, therefore, as an emergent condition that enables the evolution of a *fourth level of subjectivity*—personal experience. The mind and consciousness are often overlapped, but in my opinion this approach is incorrect. Some mental states do indeed present a certain gradient of consciousness, but most mental activities take place in an unconscious condition, even though they profoundly influence the experiential feeling of the subject. Unconscious suffering, for example, may influence the subject's mood, calling up conscious representations that may even be inadequate, that is, arbitrary a-posteriori justifications.

I will refer to the mind as a condition that does not depend on a precise neurophysiological architecture, thus avoiding the so-called *corticocentrism* according to which only mammals with a neocortex have a mind. It is common, for example, to view insects as automata, that is lacking the personal experience that we recognise in mammals and birds, but this is an error refuted by research in entomological neurophysiology. Research by Geraldine Wright, a British insect neuroethologist, has shown that these animals are endowed with emotions such as curiosity, fear and excitement, which are mediated by neuromodulators similar to dopamine and serotonin: an example being *octopamine*, responsible for mood control. Physiologist Scott Waddell of Oxford University also speaks of emotional and motivational states in the entomological universe, pointing out that most learning and memory in insects relies on mushroom bodies, a neurobiological region that can be compared to the neocortex. If today we tend to consider the striatum in birds as a structure analogous to the neocortex in mammals, which can therefore justify the undoubted intellectual qualities of animals such as corvidae and psittacidae, we often tend to underestimate the mental capacities of ectothermic vertebrates and invertebrates; as we shall see, however, the evidence suggests otherwise.

What are we to think? From my point of view, the evolution of living beings has been faced with constraints, opportunities and risks that it has had to translate into problems to be solved, opportunities to be seized and needs to be met. The vast biological landscape has therefore developed structures and functions that are capable of responding to the needs that gradually arose, and has done so by inaugurating a multitude of solutions. Undoubtedly, the two most important apparatuses responsible for behaviour are the endocrine and the neurobiological systems, which have had a specific evolution in the different animal macro-groups, so that we can recognise *baumplan* similarities between mammals and birds and between vertebrates and molluscs or arthropods. This tells us that we can use the yardstick of homology when referring to mammals—obviously even more so if we compare humans with other anthropoids—and to some extent birds. A certain homological gradient also applies to other vertebrates, but already in reptiles we find slightly different neuromodulators, such as *exotocin* in nest-building and oviposition behaviour. The comparison with arthropods and molluscs, animals that have shown great capacity for personal experience, is more complex.

On the other hand, we must not forget the analogical parameter, which is amply demonstrated in all systems whereby, faced with similar performative needs—such as flying, swimming, crawling and seeing—different morphological structures have overlapping functions. The idea of a certain functional analogy between different neurobiological structures should not, therefore, come as a surprise because of the phenomenon of evolutionary convergence: the same selective factors inevitably produce effects with performative similarity. Bees, for example, have been shown to be able to perform highly complex processing functions, such as recognising faces, making inferences and learning rules. As Arik Kershenbaum reports in his essay *The Zoologist's Guide to the Galaxy* (2020, 127–139), it has even been possible to train bees to play football by teaching them to handle a ball and then send it towards the goal. Moreover, complex activities have been found in bumblebees, wasps and ants, as we learn from Andrew Barron and Colin Klein's *What Insect Can Tell Us About the Origins of Consciousness* (2016). In his essay *Metazoa*, Peter Godfrey-Smith discusses the extraordinary manipulation and recognition skills of the cleaner shrimp (*Stenopus hispidus*), the ability of hermit crabs to choose their shells, and the tendency of decorator crabs to adorn their backs (2020, 93–119).

A prime example of how great intellectual performance can be achieved, despite having a completely different nervous system from vertebrates, is provided by cephalopods and especially octopuses, which are able to manipulate objects to such an extent that they can unscrew a cap and throw objects at an annoying counterpart. The fact that they have similar neuromodulatory responses was reported in a study by Eric Edsinger and Gül Dölen entitled *A Conserved Role for Serotonergic Neurotransmission in Mediating Social Behavior in Octopus*, where it was noted that following MDMA administration the latter became more friendly and gregarious (2018, 19, 3136–42). Fish demonstrated amazing skills in the ability to orient themselves in maze-like conditions. They also proved very good at face recognition, use of active movements to gain information about the environment, recourse to electric fields to locate their prey, counting and recognising different musical harmonies—not to mention the incredible selective imitation skills of archerfish. In the latter case, as Godfrey-Smith reports, observer fish seem to be able to change their point of view, transforming the angles seen in a shooter's performance into the angles they will adopt when it is their turn (2020, 207–209).

We can therefore draw eight conclusions: (1) the mind reveals itself in cognitive activities that involve a level of integration between different processing and affective functions; (2) the mind operates through a superintendence between prior resources (innate and learned) and specific situations that require ad hoc adjustments; (3) the mind allows the animal to have a personal experience, that is a unitary dimension of the self; (4) the mind is an emergent quality that occurs when the system reaches a certain level of elaborative and integrative threshold; (5) the mind is a property and condition that can arise from different neurobiological organisations, so there is no specific conformation or area required for its presence; (6) the mind is an evolutionary product linked to the challenges and opportunities that that species has faced and to the selective pressures that have shaped it; (7) the mind generates psychological states that can take on different degrees of awareness, sometimes so basic as to produce practically unconscious conditions that, however, can also emerge in various forms through phantasmal representations or misinterpretations. On the other hand, it is undeniable that an organism—even the most neurobiologically complex kind, the human being—can have other forms of expressive regulation, such as reflexes or neurovegetative reactions, alongside mental functions.

In other words, not all animal manifestations can be interpreted through the mental realm, just as we should not assume that every mental event has a degree of consciousness that allows the individual to make a deliberate assessment or decision. In order to understand the evolution of the mind we must necessarily start from the bottom and observe how sensibility has developed in living organisms, nourishing different forms of response to their needs. In some ways, creativity is part of the biological dimension and the tendency of living things to colonise new spaces, modifying the environment and not simply adapting to it. We are used to thinking of ecological niches as something pre-existing that organisms merely occupy, as if they were rented flats. This may well be the case in some situations, but it is becoming increasingly clear that the living forms build their own niches, so that the Earth was made habitable, as we know it today, thanks to biological processes.

Life is a complex phenomenon that is based on the construction and maintenance of an internal order and the implementation of self-revolving functions. This is possible thanks to the dissipation of energy packets, as pointed out by Ilya Prigogine in *Introduction to Thermodynamics of Irreversible Processes* (1961). This is a contribution that every biological entity acquires from the external world, avoiding entropic degradation, as mentioned by Erwin Schrödinger in *What is Life?* (1944): therefore we can view biological functions as negentropic processes. Living beings are therefore open systems, necessarily oriented towards sources that can provide them with useful energy. It is thus not surprising that living beings are sensitive to the outside world and oriented towards energy sources. An amoeba, for example, determines its orientation based on the concentration of nutrients, changing direction according to the chemical flows it encounters. As Arik Kershenbaum points out in the aforementioned essay, a simpler creature might just go straight ahead eating whatever it encounters, whereas the amoeba has improved on this criterion through the coupling of perception and orientation (2020, 128–129).

Energy supply can be of different types: there are autotrophic, chemosynthetic or photosynthetic organisms, which do not need to consume preformed organic substances, and heterotrophic organisms which, on the contrary, need to feed themselves. Even within the autotroph world, there are different forms of sensitivity and orientation. The ability to have an initial *imago mundi* is already present in bacterial forms, enabling them to detect the concentration of particular substances and to release signal molecules to indicate the density to the community. The mechanism of quorum sensing is a genuine communication process between bacteria: for example, communities of certain vibrios—*Vibrio fischeri*—become luminescent if they are present in a certain number. Each bacterium produces a certain hormone that activates the luminescence process only if it is found in a certain concentration. The molecular reception capacity in bacteria enables various forms of orientation: from distancing themselves from dangerous counterparts to the ability to monitor the presence of target cells. Unicellular life is based on forms of autopoiesis, metabolic autonomy and differentiation from the outside world: we could call them proto-selves. In the plant world, too, there are various forms of orientation (e.g. nastic movements), and there is also a particular internal organisation network, albeit not based on a nervous system. Plants have a molecule, *auxin* (indole-3-acetic acid), which is considered the best candidate for the function of a neurotransmitter. Auxin diffusion is influenced by light and gravity; it controls growth as well as intercellular organisation and communication.

When we enter the animal world, sentience is complemented by the property of agency, precisely because of heterotrophy, which requires proactive resource hoarding. Sponges and placozoa do not have nerve cells, but they already show some first signs of agency, notably a controlled water pumping. In some porifera we can see a kind of precursor to nervous behaviour proper: for example, the cells of glass sponges tend to fuse together, so that from an electrical point of view their organism behaves like a huge single cell, whereby the impulse can travel without interruption throughout its body. Then there are cnidarians, that is jellyfish (*Scyphozoa*), hydroids (*Hydrozoa*) and corals (*Anthozoa*), which do not have a true central nervous system, but filaments of nerve cells: that is, a neural network. In cnidarians, agency is even more pronounced thanks to the coordinated movement of organs, which can extend or contract as well as respond through the attack and defence mechanisms of their stinging cells. In particular, jellyfish show oriented swimming through the synchronised beating of muscles: it is linked to spatial organisation through statocyst cells containing small crystals. As Mat Fournier points out in *Quand la nature inspire la science. Biomimétisme* (2016, 13), jellyfish use a stratagem to move quickly in the liquid medium, namely the exploitation of the vortex created by the displacement of mass. The jellyfish contract their muscles at the frequency necessary to be pushed by the water mass displaced by the previous contraction. In other words, jellyfish move by exploiting the very vortex they create.

Can we speak of a mental condition in these animals or in certain forms of protostomes, such as nematodes? Frankly, I have reservations, even if I remain open to further evidence. Perhaps, starting from this condition, we could speak of a first level of Self, the one capable of distinguishing sensibility and agency, and maybe a rudimentary form of for-Self, but I would go no further. On the other hand, when we speak of arthropods, molluscs and vertebrates, I believe they do have the basic requirements for a mental state, obviously with relative degrees of complexity. The mind is a condition rather than a function or a package of functions, and it is expressed in the capacity to take a propositional stand in situations. Let me clarify what I mean. I favour the conception of *being a mind* rather than that of owning the mental faculty: this means that a mental animal can never abdicate from this *ontological dimension*, even when performing the simplest behaviours. In other words, if you are a mind, you can be in a conscious or unconscious state, but you can never relinquish your mental condition. Thus, it is essential to distinguish between being a mind and the complexity of the processing function performed. A rat can move easily through a maze, not because it has a mapping memory, but because it has marked the paths or follows chemical concentrations, just as a bee is able to convince other workers of the path to take through a mechanism similar to quorum sensing.

I believe, therefore, that it is wrong to forcibly attempt to reduce complex behaviour to simple functions in order to escape the mentalistic explanation. I also believe that it is a mistake to reduce all animal expression to complex explanatory models in order to prove that animals have a mind. Having a mind does not necessarily imply that all behaviour is realised through complex elaborative and integrative processes. Even in human behaviour we find very simple expressions such as spinal or conditioned reflexes, as well as neurovegetative reactions like horripilation or shivering. The mind is an emergent condition that arises when an organism reaches a certain degree of complexity of specialisation and integration in its neural connectome, leading to a unified condition of the self—a presence in the place of action or a point of view about oneself and the world. As Thomas Nagel (1974) put it, to be a mind is to *feel something*, but it is also to maintain a unified presence or perspective, a place to be in relation to the world, as Susan Hurley suggests in *Consciousness in Action* (1998).

The mind is therefore an emergent condition that produces a unitary presence, persistence and propositionality, such that the subject plans, programmes, evaluates, hypothesises and so forth, assuming predicative attitudes with respect to the world. The mind involves functions or integrates them on levels of increasing conjugation, it compares different factors, and in this respect it resembles the parliament imagined by Konrad Lorenz. Any metaphor we may use to capture the *ontological dimension* of the mind, of course, will hardly reveal the meaning of *being a mind*. On the other hand, it is clear to me that the mental condition is not identified with the processing functions or even with the integrative activities that characterise it. One is a mind even when one is not thinking about anything, when one is not doing anything, when one is sleeping, and the mind does not correspond to what we are fully aware of or to our state of consciousness. The unconscious is a very complex dimension, not only in terms of the functions it involves, but also as an overall state of the subject. Having said that, embracing a mentalistic conception of the individual neither implies nor requires a high-profile explanation to account for all the expressions of an animal: in this I believe that Lloyd Morgan's recommendation is still extremely valid.

We have seen that an animal, while maintaining a finalistic attitude, may follow a track or orientation by gradients rather than having a plan in its head. As Donato Grasso reports in his essay *Il formicaio intelligente* (2018, 170–172), the study of ant paths, particularly in the argentine ant (*Linepithema humile*) and the black ant (*Lasius niger*), revealed that the attractiveness of a path is directly proportional to the ant traffic and therefore to the amount of pheromone released along a certain route. This means that, in choosing between a shorter and a longer route, it is not the formulation of a cognitive map that counts, but the simple fact that the pheromone gradient is higher on the shorter route.

In 1959, Pierre Paul Grassé introduced the concept of *stigmergy* in reference to the behaviour of termites, signifying the incitement to work (*ergon*) through the production of marks (*stigma*) released into the environment. Stigmergy is therefore a way of communicating between individuals through traces left in the environment, like pheromones or other clues. It should be noted that the stigmergic mechanism is also based on feedback principles as well as on *quorum sensing*. In fact, when a termite places a sludge pellet on top of another, it increases the attraction potential (positive feedback) of this accumulation rather than other positions, because the pheromone gradient increases. We thus have an autocatalytic effect: the more the mound grows, the more attractive it becomes to other termites. In other situations, the mechanism works in a homeostatic manner (negative feedback): for example, when termites increase the size of the queen's cell, once it has grown in size, they use the pheromone gradient produced by it as a reference to be brought back to the threshold values.

The concept of threshold or threshold system is also very important to understand some animal behaviours. Here threshold means a cut-off point that marks two different states of the system and a quantum leap in conditions. Below the threshold (1), perturbations are dampened through negative feedback mechanisms, whereby the system is homeostatic and works like a thermostat, keeping gradients within a dynamic but stationary range. Above the threshold (2), every smallest perturbation is emphasised through positive feedback, whereby the system is autocatalytic and evolves to a new state of balance. The result is a non-linear but quantum system, providing for a plurality of possible equilibrium states. Some animal behaviours are reminiscent of threshold systems. As Arik Kershenbaum (2020, 156–159) reports, the way bees recruit worker bees for a particular foraging source or swarming site works much like a threshold system: just as an electron must acquire a packet of energy to jump to another orbital, so in the hive a certain consensus value must be reached for the whole hive to converge on a single option. A similar condition is reported by Grasso (2018, 159): "in ants, recruitment pheromones produce collective behaviour starting from simple individual behaviours. This is how ants searching for food employ an appropriate workforce, without the need for a leader". The mechanism is especially revealed by fire ants (*Solenopsis invicta*): the greater the source of food, the stronger the pheromonal marking, which will grow exponentially from the ants' own traffic on the route. Here, too, we find a threshold effect, whereby we see increasing or decreasing trends on the basis of a gradient.

Being a Mind

The mind, therefore, is not measured by the complexity of its functions, and it is incorrect to avoid simple explanations on grounds of assigning this ontological dimension to a given animal. Even humans have both simple and complex behaviours, which consequently require different explanatory frameworks while remaining within the same paradigm, because even the simplest behaviour can at any time call up more complex elaborative functions if the need arises. The mind is a condition rather than a function: it is the subject's unitary positionality and propositionality. Mind means many different things, so its specification is always elusive. So let us look at the definitions given by some scholars: (1) for Fred Keijzer (2015, 311–31) the mind means shaping the action, a self-image organisation useful for the fine-tuning of action; (2) for Stanislaw Dehaene (2020, 90–95) the mind is a predictive machine based on Bayesian calculations; (3) according to other authors, such as Peter Godfrey-Smith (2020, 100–112), it can be traced back to the ability to organise the coordination of parts in the space of movement; (4) for Michael Tye, particularly in *Tense Bees and Shell Shocked Crabs* (2016), it is the ability to consider many factors before making a decision.

There are therefore countless ways of defining the mental condition: first-person experience or being endowed with one's own perspective, the diachronic capacity to connect past and future (what has taken place with respect to predictive or finalistic exploration), the dimension of *Dasein* (being present in one's own here and now) or the capacity to integrate the different cognitive and affective functions that take place in the endocrine-neurobiological system. Through the mental dimension, the living world enters a completely new condition, one characterised by thought and introspection. As Karsten Brensing reports in his essay *Das Mysterium der Tiere* (2017, 170–176), rats are able to decide whether to carry out a certain action on the basis of risk factors; the author concludes that, without metacognition, they could not compare their abilities with risk to make a decision, which indeed they do. According to Joëlle Proust, in her essay *Das intentionale Tier* (2005, 223–244), thinking can be considered on four levels or stages of complexity, ranging from the acquisition of external elements, to a mental image of the environment, followed by a third stage of categorisation, and then to the free use of one's mental images.

I said that the mind is an emergent condition, but in order to clarify this aspect, it is necessary to go deeper into this concept. The latter calls for a theory based on layers of reality, one that, according to some commentators, can be traced back to Speusippus, Plato's nephew who in 347 BC became a scholar of the Academy. According to the hypothesis of layers of reality—adopted by Lorenz himself to justify the irreducible predicates of life—reality has several dimensions based on the organisation that its components assume in the whole. To understand the concept of layers of reality and emergence in complex systems, it is useful to go back to Aristotle's statement in *Metaphysics* (1961): "the whole is more than the sum of its parts". Basically, for Aristotle, a whole, we might say a system, cannot be mathematised, that is, it cannot be reduced to the sum of its individual parts, because the relations between them would then be missing. According to this principle, which we will find again in *Gestalt* psychology and in Ernst Haeckel's ecological theory, the whole emerges from the interaction of the parts, giving rise to a different predicative plane: a level that presents its own qualities compared to the constituent parts. The epistemology of complexity, developed by authors such as Henri Atlan, Edgar Morin, Stephen J. Gould and Mauro Ceruti, takes into account this emergent dimension characterising complex systems. The interrelation of the parts is defined by the Stagirite as *eidos*, that is form: a quality that defines the irreducible qualities of the system and that in psychology is expressed in the concept of *Gestalt*.

This chapter cannot analyse emergentism and complexity in a comprehensive manner, but it is important to emphasise that when I speak of relation I am referring neither to dialectical confrontation or *polemos* nor to the attribution of meaning to an entity based on its relational contract (it is the son of). Rather, I mean a predicative condition such that the entity's participation in the system characterises the predicates it acquires. On the other hand, to oversimplify, the emergentist phenomenon of complex systems has some typical features: (1) the parts of a complex system are more constrained, so they do not express all their qualities; (2) the parts of a complex system undergo an emphasisation of some qualities as a result of the relation; (3) the parts themselves can act in synergy with each other and exponentially magnify their intrinsic qualities; (4) a complex system always introjects a packet of organisational information that is lost in deconstruction; (5) complex systems have threshold effects, that is they behave differently at different levels; (6) a complex system always has a certain level of virtuality, so that it can be influenced from outside (heteronomy); (7) complex systems exhibit chaotic trends, so that they can only be evaluated as field functions where the influencing factors act as strange attractors; (8) the interaction between the parts of a complex system may give rise to supervening qualities, that is qualities not present in the individual parts and pertaining to different predicative structures.

I ultimately believe that the mental condition is a different level from the homeostatic, aesthetic and dispositional affective tensions as well as from the modal cognitive endowments that enable processing, but I think it is also different from the specialised functions that develop in particular areas of the cerebrum. From the various districts of the neurobiological network or from the neuromodulation activation pathways, the mind receives specific stimuli that may concern pain, fear, search appetitiveness and other coherent or dissonant representations, and then integrates them within propositional attitudes that indicate the subject's condition in the here and now. The mental plane is therefore the possibility-presence of this integration, which even in ambivalent behaviour (i.e. when the subject is torn between two possible behaviours, for example continuing to eat or rushing to escape a predator), nevertheless allows for a unitary condition. In a unitary condition there is only one Self, which, like a tightrope walker, can stay in this ambivalence for a certain period of time and then lean in one direction. There is no mental work going on behind the scenes and no proscenium of consciousness, since everything that takes place in the mind can be more or less illuminated by consciousness depending on the situation and the progression of time.

Animal life involves continuous creativity, and examples abound when we consider how animals deal with various unusual situations, the countless tools they come up with or the ongoing tendency to colonise new spaces and produce further niche construction modalities. Indeed, creativity is the most important driving and productive force in the evolution of species. But to fully understand the issue of animal creativity, it is necessary to better frame this fourth level of subjectivity, which refers to the integrative-emergential characteristics of the animal mind, and to try to understand how the concept of mind has changed in the light of new neurobiological findings. Today, neuroimaging techniques allow us to look inside the animal mind, which becomes a glass box to all intents and purposes, enabling us to shed light even on an animal's dreams. It turns out, for example, that depending on the stage of sleep the subject is in, the temporal characteristics of its imagined movement change radically: in certain phases of sleep, rats dream of moving at an accelerated speed, as Stanislaw Dehaene reports in *Consciousness and the Brain* (2014, 82–118).

I said the individual is subjective insofar as it has a certain amount of arbitrariness in the evaluation and decision-making phase, specificity in the orientation choice and protagonism in adapting its tools to the situation. This condition is based on a number of requisites I set out above when speaking of affective states and cognitive endowments. In fact, there would be no self-ownership without three fundamental aspects. The first (1) is an emergent Self with respect to the context, capable of grasping the relation between *reafference* (the changes caused by the subject's actions) and *exafference* (the changes caused by something that happens outside the subject). The second (2) is a *for-Self* that indicates interests to be achieved and defended based on affective conditions, that is, motives that support the behaviour oriented by the hedonic system. Finally, the third (3) is an *in-Self* providing for the ownership of modal endowments, which are characterised by a certain degree of availability for use, based on a schematic-instrumental configuration of resources. Beyond and because of these considerations, in order to make subjectivity consistent, it is necessary to admit a superintendence, capable of translating affective states into proactive and responsive attitudes and the latter into finite behaviour through the modal component. The mind is what carries out the integrative work between the different components of the Self or the combination of functions that characterise it.

In his book *De l'intelligence*, the French philosopher Hippolyte Taine (1828–1893) suggested that the mind is a kind of theatre capable of staging images and making them compete (1888, vol 1, 138–139), an aspect taken up by the cognitive psychologist Bernard Baars in his concept of workspace (1997). According to Read Montagu, in *Why Choose This Book? How We Make Decisions* (2006), the mind has mainly an evaluative-decisional function, integrating a multiplicity of features aimed at responding to the primary need to cope with the uncertainties of the world. For Montagu, assigning values to the phenomenological landscape of the world and making choices based on those values is the main task of superintendence. According to Montagu, therefore, the mind is not an abstract entity, but rather the result of a multitude of integrative functions involving affective and information processing, the primary purpose of which is to attribute meaning to the world's data. Rather than a mental theatre, these are therefore *mental functions* performed in different areas of the brain and capable of integrating memory, the affective components that define attributions of value and purpose, and the cognitive tools used for information processing.

In *Kinds of Minds* (1996), Daniel Dennett warns against the mental theatre allegory because of the risk of the bystander effect or the infinite regression of an observer self. A similar position can be found in Thomas Metzinger's *The Ego Tunnel* (2009), according to which mental experience is nothing more than the emergence of an internal perspective. This perspective arises from a set of integrative functions, useful for decision-making and preceding the taking charge of consciousness, so there is no overlap between mental activity and awareness or explicitness. The most important mental functions include: (1) *evaluation*, the ability to attribute a certain meaning to the events that happen to the individual; (2) *judgement*, the property of weighing an option or an event against its hedonic content; (3) *programming*, being inclined to identify objectives even when the target is not immediately visible; (4) *planning*, the ability to define a strategy and a series of tactical alternatives to implement it; (5) *foresight*, the ability to formulate predictive outcomes, starting from the simple if-then relation; (6) *simulation*, the ability to mentally imagine the outcome of an action or to compare the performance of two tactics; (7) *conjecture*, the tendency to propose causal hypotheses to be tested; (8) *decision-making*, the ability to choose the appropriate behaviour for the situation.

As mentioned above, these functions or mental activities are distributed in different areas of the brain, for example: decision-making processes are mainly attributed to the prefrontal cortex, multisensory images are constructed in the superior colliculi, body representation occurs in the post-rolandic cortex of the parietal lobe (Penfield's *homunculus*), while judgement seems to belong to the rTPJ area (right tempo-parietal junction). However, as a whole they produce a psychic condition or internal perspective that, according to the Swedish neurobiologist Björn Merker (2007), constitutes a sort of self-image capable of operating an unconscious synthesis between the motivational dimension, the representation of the world and the organisation of actions. I therefore think it is important to differentiate the mental condition from the mental functions, although the mind arises precisely when there is integration between the parts. This, however, can occur in a multiplicity of ways, so the vertebrate configuration is not essential to have such an integrative process. We can safely identify integration in different forms of experiential convergence: as Goddrey-Smith suggests with regard to cephalopods, it could arise from the integration of action itself or through the activation of neuromodulator reception networks.

Moreover, it is crucial to distinguish mental activity from consciousness, the overlapping of which prevents us from understanding the psychology of the unconscious. As Delos D. Wickens points out in his article *Encoding Categories of Words* (1970, 37, 1–15), the world is too complex for us to be aware of all the basic dimensions of encoding. In John Benjafield's essay *Cognition* (1992, 71), we read that if we were aware of everything we do, we would not be able to do it because of the excessive complexity involved. Benjamin Libet's discovery of the readiness potential reveals that the conscious decision to perform a certain action occurs half a second after the preparatory potential has been activated. As described in *Time of Conscious Intention in Act in Relation to Outset of Cerebral Activity* (Libet et al., 1983, vol 106, 640), the brain decides to initiate action at a time when there has not yet been any subjective awareness that the decision has been made. In *Consciousness and the Brain*, Dehaene reports a large number of integrated functions performed in a completely unconscious way. These include: (1) blindsight and the underlying capacity to process complex, referred responses; (2) spatial negligence linked to the lesion of the right cerebral hemisphere which nevertheless allows for evaluations to be made in the left space; (3) subliminal priming, based on showing a subliminal image and then another element, the target, producing a processing facilitation; (4) the configuration of complex systems, such as the order of pieces on a chessboard; (5) the implicit learning of rules, frequencies or game modes; (6) the unconscious attribution of meanings; (7) arithmetic and evaluative processes.

Another aspect characterising this fourth level of subjectivity concerns the mind's tendency to refer to imagined contents—for example, a condition, a goal, a possibility—through propositional attitudes, such as wishing-that, fearing-that, believing-that, wanting-that. The mind is able to produce intentional states, that is, to address contents that are not necessarily present in the subject's landscape. Today we know that both operative and episodic memory play a fundamental role in this rising above the situation, turning backwards or projecting forwards, or even referring to imagined contents. In *Metazoa* (2020, 290–97), Godfrey-Smith speaks of offline processing as a state of disconnection from the usual flow going from the senses to action. Being a mind therefore means being able to position oneself *elsewhere*, which allows for imaginations, planning, simulations and of course insight. Following this line of reasoning, in the philosophical tradition *Dasein* does not refer to a total immersion in the here and now, but to being able to be elsewhere. As the author reports, by observing rats dreaming through fMRI (functional magnetic resonance imaging) and thanks to the discovery of place cells, neurons that discharge when the rat is or dreams of being in a certain place, it has been shown that during sleep these animals engage not only in *replaying* paths previously taken, but also in *preplaying* paths that they have not taken in reality.

Being a mind means spending part of one's life elsewhere and expressing it through processes of imagination or simple digression. Mental life is intentional because it is dotted with believed, feared, hypothesised and imagined elsewheres. *Dasein* is the unitary presence of the subject who, on the one hand, participates in what happens, but at the same time is detached from it and follows its own dreams and imaginations, thanks to which it can modify the factual order and go beyond the already given. We shall see how creativity is nourished precisely by this experience of the elsewhere, by the mental ability to leave the here and now, to suspend the order of factors, to go beyond the surface of appearance, to imagine something that is not there. Picking up on Godfrey-Smith's analysis of offline processing, we'll find that the mind can be depicted as a kind of arena, a place to review options, invent things and manipulate images. But more than a place of refuge, I argue that the mind is a dimension of Being, capable of positioning itself in different framings of reality—if we consider the latter not as a construction of the Self (idealism) but as a compatible, that is experienceable, configuration of reality.

In line with this propositional condition of the mind, we have to consider the intentional aspect: the tendency to refer to contents. As we note in Daniel Dennett's *Brainstorms* (1979), there are different degrees of intentionality, depending on the number of propositional contents that a certain thought or communication envisages/produces. For example, a first order of intentionality might be found when a sentinel guenon, in warning a mate, *wants him to climb the tree*, while a second level might concern the content of the message, that is the guenon *wants him to believe that there is a leopard*. According to this perspective, which can be traced back to Franz Brentano (1838–1917) and the phenomenological approach, the concept of intentionality and intending-to are not the same thing. According to John Searle in the essay of the same name (1983), *Intentionality* is the aspect that leads our mental states to be directed, in the sense of verging on or referring to, towards propositional contents such as convictions, beliefs, desires. As John Z. Young points out in *Philosophy and the Brain* (1987, 60–69), an intentional condition always refers to something that is thought about, for example, a purpose that's not in view or a potential risk. We may ask whether intentionality must necessarily involve conscious thought: in my opinion this is not obvious.

The mental condition certainly grants a presence, so we can distinguish the phenomenological dimension of mental activities in terms of action and agent from physical-chemical reaction events. As noted by Young (1987, 70–71), when an animal wants or believes something, it develops intentional dimensions: these processes are far more complex than the simple tropisms that, according to the author, were the most we could understand fifty years ago and that should now disappear from any serious discussion of such complicated matters. The subject's internal propositionality, which interfaces with circumstances in a perspective and temporally situated way, brings us back to the concept of *participatory attitude*, which can be found in the work of the philosopher Peter Strawson (1919–2006). In fact, having a propositional attitude implies the subject's diachronic dimension: being in the here and now, but also elsewhere. *Dasein* is realised through the recognition of unity in the diachronic unfolding—see Frederic Bartlett's (1886–1969) conception set out in the previous chapter—which defines the persistence of the Self. The subject perceives itself as continuity: what we experience in ourselves as a continuous narrative, maintaining our self-image through change. The mind therefore produces a diachronic dimension of subjectivity, whereby being a subject means participating in the present by relating the past and the future.

The subject's protagonism is expressed in a state function, a "psychological condition situated in the moment", a projective presence in the intersection with the world, which refers to its own contents. We have seen that animality is the expression of a continuous copulation with the world, which is manifested on the basis of multiple needs. Some of these transcend individual experience, in that they are inherited; others refer to the individual's biography, others to momentary experience, but all are co-factors in bringing out a psychological state that translates into a presence. Protagonism therefore lies in this convergence of influential elements—hereditary heritage, individual experiences, the momentary physiological condition, the opportunities present in the context—whereby the subject operates a "state synthesis", but is not limited to this; it is proactive insofar as it innovatively produces a *possible condition* that is its own and singular. Presence is revealed in this capacity to go beyond, to mix the influential ingredients into a unique and unrepeatable product, which above all is the subject's own.

The Propositional Model

In order to understand this fourth, integrative level of subjectivity, which refers to bringing behaviour back to a mental condition, it is necessary to unravel this matter further, formulating a model that is as clear as possible. I said that the mind is an ontological dimension that emerges when certain systemic factors are determined, so that a living being endowed with processing faculties, connective capacities and a good organisational and integrative level is able to face reality with a certain perspective, presence, continuity and unity. What are these specific systemic factors that differentiate a mind from other emergent phenomena?

Two factors are generally mentioned: (1) a threshold value of processing elements, basically a certain number of neuronal entities; (2) a strong gradient of integration among the elements. As far as the value of processing entities is concerned, the importance of which is far from being established, there is a consensus that the number of neurons is more important than the relative size of the nervous system, for example the size or weight of the brain in relation to the volume of the body. On the other hand, as far as integration is concerned, the nervous system of vertebrates, in particular that of mammals and birds, is clearly very different from that of arthropods and molluscs, so we must assume different organisational connection systems. For the sake of simplicity, I shall refer to the mammalian brain, which has an overlapping *baumplan*.

We know from neuroimaging studies that, while each species has its own morphological characteristics for its central nervous system, the mammalian brain is generally and functionally organised in a similar way, both in terms of the physiology of the neuromodulators, the hypothalamic-pituitary area and the limbic system, and in terms of the different specialisations of the neocortex. It is no coincidence that most knowledge about the human brain is derived from observations and experiments carried out on the brain of mice, rats, cats, dogs and non-anthropoid primates. When questioning the cognitive faculties of non-human mammals—fortunately this prejudice is increasingly rare among scientists today, but persists in philosophical circles—one should remember that much of the information on human neurophysiology is drawn from experimental studies on these animals. Any textbook of behavioural physiology will show that dopaminergic or serotonergic pathways are drawn based on the encephalic topographies of rats. Let us assume, then, that eutherian mammals have an almost identical nervous organisation and physiology.

A number of things emerge from these studies. The first (1) is that the brain is a massively interconnected structure, thanks to the synaptic network, to the pathways of elective (e.g. cholinergic, dopaminergic, serotonergic) activation of neuromodulators, and to neurons specialised in connecting distant areas of the brain, such as fusiform or von Economo neurons. The second (2) is that the different areas of the brain present marked functional specialisations, such as the various areas of the neocortex; it is possible to identify a physiological topography with the relative peculiarities of the two hemispheres, different activities in the subcortical nuclei and even in their single areas, up to the characterisation of single neurons, often deputed to a particular performance. In addition, (3) the nervous system also has autonomous peripheral responses and neurovegetative pathways, such as the sympathetic and parasympathetic ones, which are strongly visceral-brain integrative. Finally, (4) in many respects it is difficult to distinguish the neurobiological system proper, made up of neurons and glial cells, from the endocrine and immune apparatus, especially with regard to hypothalamic integration.

A model that wants to account for the morpho-functional organisation of the central nervous system cannot ignore these aspects and continue to explain behaviour through the reflexogenic paradigm based on the simple association of a stimulus and a response. I do not believe one can explain behaviour by transcending physiology and what we know about the organisation of the nervous system. First of all, it must be said that nervous activity does not cease when the subject stops expressing behaviour or when it lowers its windows of perception; on the contrary, in many cases we notice an acceleration of activation phenomena, for example during sleep, meditation and relaxation activities. No behavioural process ever has a simple associative and sequential dimension, but develops organised and integrated configurations of populations of neurons, much more akin to Hebb's concept of neuronal assemblies than to Skinner's models. If we look at brain activity we will discover a kind of persistent background noise followed by activations patterns peculiar to each type of activity, for example when the subject is called upon to recognise faces or move through space. We also know that damage to a certain part of the brain always produces overlapping expressive impairments in different patients, even in different species. For these reasons, an explanatory model that wishes to be faithful to the characteristics of the biological substratum can only involve the mind.

On the other hand, behavioural analysis also forces us to reflect on common phenomena that the reflexive paradigm fails to account for, requiring ad hoc adjustments that somewhat recall the epicycles of the Ptolemaic model. We might ask ourselves, for example: (1) why the same stimulus produces different responses and sometimes fails to produce associative effects, as for example in the phenomenon of overshadowing. Or we might ask (2) why variable reinforcement produces more relevant effects than fixed reinforcement, or why it is almost impossible to activate an operant to avoid an unpleasant consequence. Another question is (3) why, as demonstrated by the psychologist Karl S. Lashley (1890–1958), pre-activation often occurs in a behavioural sequence—an effect that cannot be explained following the chaining model. In the essay *Critical Ethology and Post-Anthropocentric Ethics*, written with Marco Celentano (2021, 105), I consider many inconsistencies of the behaviourist model and how the term *reinforcement* itself is an artifice that produces explanatory circularity, rather than clarifying the law of effect formulated by Edward Thorndike.

The concept of mind becomes useful and, I would say, indispensable when we realise that behaviour is always an integrated expression, resulting from the synchronic or sequential concurrence of several activations of specialised areas in the animal brain. In other words, the mind is what emerges from, and enables, the integrative process. The mental condition is the product of the connection between different points of activation but, at the same time, it can generate new qualities precisely because it supervenes on individual activations. Therefore, we must consider the mind in its dual form, encompassing: (1) *the integrative character*, as the connective outcome of several activation areas, each of which is sensitive to particular trigger conditions; and (2) *the productive character*, because it does not give rise to an assemblage but to a product. Indeed, its final characteristics exceed the simple sum of the single predicates, so the mental condition has its own characteristics. It is difficult to disambiguate this duplicity, whereby on the one hand integration permits the evolution of the mind, and on the other hand the latter contributes to a more refined integrative character. It certainly seems clear that being a mind is only possible if integration allows for a unitary dimension of the Self, even in situations where some functions are disarticulated, as is the case with the resection of the corpus callosum. At the same time, the productivity of the mind in terms of propositionality is implicit, for example in the projection into an elsewhere.

Would it be possible to understand these qualities by merely listing the different neurobiological activations? From my point of view, this is a useful exercise only if it is done in a logic of integrated convergence, not by thinking of these qualities as atomic and autonomous in giving rise to behaviour. If we observe single activation areas, we discover that their minute character is so articulated, chaotic and polyfactorial as to give rise to a singularity of state, ascribable to the laws of complex systems and to non-repeatability. A similar thing can be said for reflectiveness, which is never located in a single position in space-time. If we are fantasising, flying to the occasional elsewhere in a certain moment, our brain keeps an open gaze on the horizon. When we are conversing with someone at a party, so that what others are saying is no more than a subliminal hum, if we hear our name mentioned we immediately become aware of what is being said. The mind produces a large number of possible states as well as a pandemonium of images and thoughts where each can assert itself over the others for an infinite variety of reasons. The integrative and productive character of the mind explains both the specialisation of the different areas (and thus the concursive and organismic, i.e. embodied, factor of the mind), and the emergence effect or unitary condition of the Self.

Behaviour is the expression of the subject's mental condition, so it is useful to always refer to both the integrative and the productive aspect. But what is, ultimately, the product of the mind? This leads me to the explanatory proposal of this level of subjectivity. The product of the mind is a *propositional structure*, what we ordinarily call *thought*, given by predicative and modal contents. Being a mind means having a point of view on the world, unfolding it diachronically, that is, using one's inherited or learned information or tendencies to project forward. I will use the term "propositional" in a slightly different way than usual to define the kind of relationship that the *subject's condition* (1) has with the elaborations (2) obtained from the comparison between past information and the here and now. If a predator wants (1) its prey not to sense its presence (2) we have a propositional structure where the predicate *wants* indicates the subject's attitude or projection on the action, while the consecutive referring to the prey is deduced from the predator's previous experience and its conjugation with the here and now of the situation. The subject's projective condition always produces an elsewhere, even when the individual is glued to what it is doing, because its purpose makes it go beyond the here and now.

The mind produces propositionality in the sense that it presents a vast typology of attitudes—such as believing-that, hoping-that, fearing-that—each indicating a particular way of projecting oneself into the elsewhere, of anticipating or predicting what may happen. Being a mind means, therefore, producing thoughts all the time, spending a lot of time in the elsewhere, as noted by Godfrey-Smith. Thoughts have a propositional structure because they relate: (1) the condition of the subject, that is, its projective attitude towards the future, with (2) the contents resulting from the relationship between previous information and existing conditions. The propositional condition of the mind underlies animal creativity, because it allows the subject to imagine something different from what there is. It is often argued that thought is an exclusive faculty of the human race, since it is endowed with language, and the author of reference for this position is certainly Naom Chomsky. But as pointed out by Karsten Brensing, in *Das Mysterium der Tiere*, the idea that thinking is only possible through language is obsolete and was refuted by a large amount of scientific evidence (2017, 215–225).

I would like to go even further: it is not thought that descends from language, but the opposite. Language is a technique capable of translating the propositional structure of thought. Thought always has a syntactic organisation, because a lion chasing a zebra assumes an attitude by using a predicate and projects itself into action even before performing it by referring to its competence and to the circumstantial conditions. The lion places itself on the scene in a projective manner, plunged in a diachronic dimension that goes beyond the image of the zebra grazing in the savannah. Its attitude is dripping with projections that follow one another, such as: (1) fearing that the zebra might hear it come; (2) believing that its behaviour will be effective; (3) hoping that the zebra won't be too quick. There is a kind of *isomorphism* between language and thought, but the reason why is the opposite of what Chomsky believes: language is the translation, not the construction, of thought. Cases of language learning in anthropoid primates have also shown that, by teaching them a communicative technique, such as ASL for Washoe or lexigrams for Kanzi, these animals became able to make their thoughts explicit. They did this by translating their propositional structure, which differed from ours in some respects: the result was that communication also differed.

Some authors, such as Daniel Dennet, argue that language provided these animals with the material necessary to articulate thinking, but in my opinion this claim is untenable because a technique cannot bring out contents, let alone propositional attitudes, *ex nihilo*. Observing Irene Pepperberg's experience with the parrot Alex, one cannot fail to notice the propositional dimension followed in the training model. In the book *Alex and Me* (2008) Pepperberg identifies three components of her didactic system: (1) the reference, whereby the meaning of a word referred to a precise physical object; (2) the functionality, that is the pragmatic use of the word and therefore the reason to learn it; (3) the social intentionality and therefore the type of relationship to which it referred. After a year of work using propositional, that is enunciative, coordination, Alex was already able to refer to seven objects and was learning the names of colours. The difficulties in getting their work credited, encountered by various authors who have been engaged in teaching language techniques to other species—think of Roger Fouts with the chimpanzee Washoe, Penny Patterson with the gorilla Koko, Sue Savage-Rumbaugh with the bonobo Kanzi, Lyn Miles with the orang-utan Chantek—were precisely due to the prejudiced attitude held by linguists such as Thomas Sebeok.

The repeated recourse to the Clever Hans phenomenon seems more of a smokescreen than a proper scientific precaution, because the work carried out since the 1970s has had a completely different experimental depth. Having said this, I believe that a difficulty also lies in the attempt to force the hand a little too much on the communicative overlap between different species. What these experiments show is that, alongside the syntactic-organisational similarities of the message, there were differences that were not always accepted by the experimenters as expressions of a different propositionality. Thus, while detractors emphasised these aspects, which in their eyes appeared to be incongruent and therefore proof of the linguistic incapacity of non-humans, experimenters did not hesitate to correct, amend or even ignore these aspects which, on the contrary, offered an incredible mirror of the way these animals mastered a technique to communicate their propositional contents to us. Why look for the human in other species when there is a chance to build a communication bridge between two different propositionalities? I do not wish to diminish the great work carried out by these researchers, who have the undisputed merit of having crushed a prejudice that is still hard to dispel today. If anything, my point concerns the importance of considering this technique as a revealer of propositionality, that is as a further way of understanding what a subject of another species thinks.

But let us try to better understand what is meant by propositionality and propositional attitude. As we know, in philosophy the concept of propositional attitude means an enunciative structure of a prevalently epistemic character—such as believing-that and knowing-that. We find this concept in authors like Gottlob Frege (1848–1925), Paul Rudolf Carnap (1891–1925) and Willard van Orman Quine (1908–2000). According to Jerry Fodor (1935–2017), especially in *Psychosemantics. The Problem of Meaning in the Philosophy of Mind* (1987), propositional attitudes are the most effective way of describing behaviour. A propositional attitude is therefore a relation between a subject who believes, hopes, wishes and so on and a content that is thought, so that the double nature of thought becomes evident. On the one hand, there is (1) the predicative aspect, given by the type of verbality (to wish, to fear, etc.), and on the other (2) the content, which refers to something. One could visualise propositionality with a few examples: (a) a buffalo hopes that over the hill there is a source of water; (b) a deer fears that the rustle of leaves is due to a predator; (c) a dog believes that every time the person grabs the bowl it will eat. In this sense, my view of propositionality goes beyond rationality and epistemics.

From my perspective, thought is always propositional, for a number of reasons. First, because (1) it is a structure that has to give meaning to the action and therefore demands a specific predicative characterisation, for wishing is different from fearing. Secondly, (2) it always refers to something that expresses the intention in a modal sense and functions as a complement to the proposition, in an objective or content-related sense and in circumstantial terms. It also (3) presents an implicit diachronic structure, because it is based on suppositions resulting from an innate or experiential background but projected towards a future dimension of purpose. Furthermore, (4) it implies a reference, but also a non-overlap between the internal dimension of thinking and the external situation that surrounds the subject. Finally, (5) it envisages a unitary presence of the thinking subject and therefore a presence in the scenic landscape of the action at all times.

In order to understand this propositional model of thought—and therefore the mental level as a factory of propositionality—it is necessary to refer back to what was said earlier about the affective and cognitive characteristics of subjectivity. Thought is propositional insofar as it is syntactic, namely the result of an organised composition of elements. Its simplest form is given by: (1) the subject, which is characterised by a certain attitude towards the elsewhere; (2) the verb, which specifies the type of action that will bring about the goal; (3) the direct object, which establishes what the action is directed towards. It is propositional because before becoming an action it is desired, feared, conjectured, believed. The predicative structures of the propositional syntax are given by the affective components, especially emotions and motives, while the modal and complementary contents are given by innate or learned representational components. If this model has any plausibility, it clearly makes little sense to seek a dialogue with a non-human by claiming to place it on the same propositional level. The non-human animal, however well it may have understood the linguistic technique, cannot go beyond its own propositional dimension, so its expression will necessarily be different, while respecting the technical parameters of the learned language. Its motivations and representations are different, and the predicative and content aspects of propositionality are related to these variables. The aim of dialogue with another species should not be a total overlap with the human being. When, for example, we play with a cat, we do not set the framework on the same parameters as when we play with a child or a dog: to play with the cat you have to be a bit of a cat yourself.

Now, if thought resembles a proposition, even as far as human communication is concerned, we must assume that language derives from thought—it is capable of translating thought into a communicable form. I do not deny that language has had an enhancing effect on thought in humans, providing them with syntactic-conjunctive structures that amplify intentional levels, but I believe that thought actually *precedes* language. This is demonstrated by the fact that the propositional complexity of thought, especially when it involves multiple parallel references, is not always translatable into language. But if it is true that thought recalls a proposition and that this is why behaviour can be described by means of a proposition, propositionality may bear the integrative or compositional mark of the mental condition. Therefore, thought is intentional even in its unconscious dimension, because it is always referred to the subject's projection elsewhere, that is, to the *construction of a gap* that is then to be bridged through action. The latter is solicited by or addressed to something, so that thought always has a predicative part and a content part.

I believe that the predicative aspect arises from affective verbalities, which are mainly dispositional and therefore of an emotional and motivational matrix, while the contents and circumstantials are the processing outcome of the information carried out by the individual's cognitive instruments. Emotional predications concern not only the aforementioned fearing-that, but also other basic emotions—for example rejoicing-if or getting irritated-at—and indicate propositionalities capable of flexing external conditions based on the subject's particular state. Motivational predications have to do with the tension of desire, which always has a propositional structure and never simply refers to the object. In practice, an animal wishes *for something to happen* or to be realised, so its wishing is projective and referred: in this sense it is propositional, because it refers to something that has not yet happened. Some propositionalities, such as believing-that or hoping-that, which seem to be indissolubly linked to rationality, are also to be traced back to the motivational sphere. The epistemic dimension is part of the exploratory and problem-solving aspect that is at the heart of the tendency to produce hypotheses about the world in order to understand correlations and causal relations.

As far as the contents of propositionality are concerned, they clearly arise through representational structures, thanks to processing instructed by the information available. Here are some examples of these references, taking as an example the propositional attitude of believing-that: (1) if the person picks up the bowl I will soon eat = anticipatory representation; (2) a certain goal requires a particular behaviour = operational representation; (3) to reach that target I must follow this route and observe those clues = mapping representation; (4) in order to recognise a target, I must focus on a particular character = extractive representation. These are only a few examples of the propositional contents that can be introduced by the predicative component. In a nutshell, the structure of behaviour, its descriptively enunciated phenomenology, looks more like a proposition than a mechanical stimulus-response reflex. I therefore consider it a moot point to assert that the mentalistic conception does not reflect the scientific precaution of sticking to what can be objectively ascertained.

A propositional explanation is undoubtedly more in line with what anyone can observe in animal behaviour: tracing it back to a reactive and atomic dimension is a stretch. But that is not all: this explanation is also more in line with the physiology and morphology of the brain, which is unequivocally organised: (1) on predicative levels, as shown by the use of neuromodulators or substances that inhibit reuptake, capable of activating specific propositional attitudes; and (2) on representational levels, as can be shown by stimulating specific parts of the neocortex. A propositional explanation is based on predictions, on strategies and operational planning, on attributions of meaning, on hypotheses and beliefs, on wishes and hopes, on imaginative projections and simulations. A propositional explanation is able to subsume different domains of behaviour—such as innate and learned aspects, simple and complex expressions, molar and molecular hierarchies—but also to resolve inconsistencies that the traditional model only explained through contrivances.

If we follow a mentalistic approach based on thought, we will see that in the end there is not much difference between a solution by trial and error and insight, no more than there is between doing a calculation mentally or using paper and pencil. Thought exceeds behaviour, because only a small part of the activations produce agency and not all of them find an expressive translation, as we can deduce by observing ambivalent behaviour. We must no longer think in oppositional terms, for example of instinct and reason or of innate and learned, because every content lends itself to being transformed in a dynamic representation to support the subject's predicative motive. Above all, however, we see that the mind is an emergent condition, but one that is closely connected to the body, because all it takes is a hormonal alteration, a storm of cytokines, a lesion in a part of the brain for the propositionality, that is the subject's thinking, to change radically.

Works Cited

Aristotle. 1961. *Metaphysics*. London: J.M. Dent.

Baars, Bernard J. 1997. *In the Theater of Consciousness. The Workspace of the Mind*. Cambridge, MA: Oxford University Press.

Barron, Andrew, and Colin Klein. 2016. What Insects Can Tell Us About the Origins of Consciousness. *Proceedings of the National Academy of Sciences of the United States of America* 113 (18): 4900–4908.

Benjafield, John G. 1992. *Cognition*. New Jersey: Prentice Hall.

Celentano, Marco, and Roberto Marchesini. 2021. *Critical Ethology and Post-Anthropocentric Ethics*. Cham: Springer International Publishing.

Dawkins, Richard. 1982. *The Extended Phenotype*. Oxford: Oxford University Press.

Dehaene, Stanislas. 2014. *Consciousness and the Brain*. New York: Viking-Penguin.

———. 2020. *How We Learn, Why Brain Learn Better than Any Machine*. New York: Viking.

Dennett, Daniel C. 1979. *Brainstorms*. Sussex: Harvester Press.

———. 1996. Kinds of Minds. *Toward an Understanding of Consciousness*. New York: BasicBooks.

Edsinger, Eric, and Gül Dölen. 2018. A Conserved Role for Serotonergic Neurotransmission in Mediating Social Behavior in Octopus. *Current Biology* 28: 1–7.

Fodor, Jerry A. 1987. *Psycosemantics. The Problem of Meaning in the Philosophy of Mind*. Cambridge, MA: MIT Press.

Fournier, Mat. 2016. *Quand la nature inspire la science. Biomimétisme*. Toulouse: Editions plume de carotte.

Godfrey-Smith, Peter. 2017. *Other Minds: The Octopus, the Sea, and the Deep Origins of Consciousness*. New York: Farrar, Straus and Giroux.

———. 2020. *Metazoa. Animal Minds and the Birth of Consciousness*. Glasgow: William Collins.

Grasso, Donato. 2018. *Il formicaio intelligente*. Bologna: Zanichelli.

Griffin, Donald. 1976. *The Question of Animal Awareness*. New York: Rockefeller University Press.

Hurley, Susan. 1998. *Consciousness in Action*. Cambridge, MA: Harvard University Press.

Karsten, Brensing. 2017. *Das Mysterium der Tiere*. Berlin: Aufbau Verlag.

Keijzer, Fred. 2015. Moving and Sensing Without Input and Output: Early Nervous Systems and the Origins of the Animal Sensorimotor Organization. *Biology and Philosophy* 30 (3): 311–331.

Kershenbaum, Arik. 2020. *The Zoologist's Guide to the Galaxy*. London: Penguin Books.

Köhler, Wolfgang. 1925. *The Mentality of Apes*. London: Routledge & Kegan.

Libet, Benjamin, Curtis A. Gleason, Elwood W. Wright, and Dennis K. Pearl. 1983. Time of Conscious Intention to Act in Relation to Onset of Cerebral Activity (Readiness-Potential). The Unconscious Initiation of a Freely Voluntary Act. *Brain* 106 (3): 623–664.

Merker, Björn. 2007. Consciousness without a Cerebral Cortex: A Challenge for Neuroscience and Medicine. *The Behavioral and Brain Sciences* 30 (1): 63–81.

Metzinger, Thomas. 2009. *The Ego Tunnel*. New York: Basic Books.

Montague, Read. 2006. *Why Choose This Book? How We Make Decision*. New York: Dutton Adult.

Morgan, Conwy Lloyd. 1894. *An Introduction to Comparative Psychology*. London: The Walter Scott Publishing.

Nagel, Thomas. 1974. What Is It Like to Be a Bat? *Philosophical Review* LXXIII (4): 435–450.

Pepperberg, Irene. 2008. *Alex and Me*. New York: HarperCollins Publishers.

Prigogine, Ilya. 1961. *Introduction to Thermodynamics of Irreversible Processes*. New York: Interscience Publisher.

Proust, Jöelle. 2005. Das intentionale Tier. In *Der Geist der Tiere. Philosophische Texte zu einer aktuellen Diskussion*, ed. Dominik Perler and Markus Wild, 223–243. Berlin: Suhrkamp.

Romanes, George. 1883. *Mental Evolution in Animals*. London: Kegan Paul, Trench.

———. 1897. *Darwin and after Darwin*. Chicago: The Open Court Publishing Company.

———. 1912. *Animal Intelligence*. New York: D. Appleton & Co.

Schrödinger, Erwin. 1944. *What Is Life?* Cambridge: Cambridge University Press.
Searle, John R. 1983. *Intentionality, an Essay in the Philosophy of Mind*. Cambridge: Cambridge University Press.
Taine, Hippolyte. 1888. *De l'intelligence*. Paris: Hachette.
Tolman, Edward. 1948. Cognitive Maps in Rats and Men. *Psychological Review* 55 (4): 189–208.
———. 1959. Principles of Purpositive Behavior. In *Psycology: A Study of a Science*, ed. Koch Sigmund. New York: McGraw-Hill.
Tye, Michael. 2016. *Tense Bees and Shell Shocked Crabs*. Oxford: Oxford University Press.
Uexküll, J. von. 1909. *Umwelt und Innenwelt der Tiere*. Berlin: Springer.
Wickens, Delos D. 1970. Encoding Categories of Words: An Empirical Approach to Meaning. *Psychological Review* 77 (1): 1–15.
Young, John Zachary. 1987. *Philosophy and the Brain*. Oxford: Oxford University Press.

CHAPTER 8

Leadership in Learning

Premise

One of the most striking characteristics of the animal world is its ability to learn from experience, to change one's expressive register through one's relationship with the world. Living is learning, as stated in *Leben ist Lernen: von Immanuel Kant zu Konrad Lorenz* (1981). This book, which stemmed from a conversation between Konrad Lorenz and Franz Kreuzer in 1980, places learning at the centre of attention as an implicit consequence of being alive. Today, many studies describe the incredible memorising capacities found in the animal world, not only among species with a great ability to remember past experiences—such as Clark's nutcracker (*Nucifraga columbiana*), which is able to remember around 10,000 supply sites six months after depositing its seeds there—but also in animals apparently stuck in an eternal present, such as insects or fish. Studying the mnemonic abilities of the frillfin goby (*Bathygobius soporator*), the biologist Lester Aronson (1911–1996) found that this small fish is able to memorise the topography of pools that emerge at low tide, so that it can escape predators by jumping safely into a neighbouring pool. Bees have also proven able to remember and therefore recognise faces, while ants use visual memory with extreme ease to find their way back. In particular, the Sahara desert ant (*Cataglyphis fortis*), unable to use the stigmergic effect of pheromones because the heat quickly evaporates these clues, uses visual-spatial memory to orient itself.

R. Marchesini, *The Creative Animal*,
https://doi.org/10.1007/978-3-031-07414-1_8

Learning abilities in the animal world are about problem-solving, the formulation of hypotheses to be tested, the tendency to reuse equipment to achieve a goal—in short, a series of activities that go far beyond random behaviours driven by the effect produced, as normally assumed by non-mentalist theories. Dogs can learn several hundred words, as shown by research carried out by Juliane Kaminski et al. (2004, 1682–1683), but not only: Rico, the dog from the experiment reported by the authors, was able not only to pick up the object whose name he knew, but also to return a new object if he heard a word he did not recognise. It has been amply demonstrated that many animals show surprise and attention-enhancing behaviour when, under experimental conditions, they are shown events that violate the laws of physics, such as the sudden disappearance of an item or the non-rectilinear movement of an object. This illustrates that experience is always based on hypotheses.

We must, then, take a broader view of animal learning, one that goes beyond the dualistic constraints of classical operant conditioning as described in the behaviourist tradition. Learning follows a multiplicity of paths, some of them very complex, which require us to go beyond the image of a passive exposure to stimuli or the metaphor of an empty vessel to be filled. A learning animal is never a *tabula rasa* exposed to stimuli, and learning does not take place *ex nihilo*. In learning, prior endowments, both innate and previously acquired, come into action: it involves a process of metamorphosis or reconfiguration of the resources available. Most of the time, a standard or virtual schema is converted into specific operational performances: this is how, based on the generic predatory model, through experience a cat develops the various hunting specialisations relating to different prey, such as lizards, mice or birds. The innate model is thus a sort of *generative master* that is modified to obtain new diversified competences. The variation can be (1) a specialisation of the resource, which makes it more adherent to the situation; (2) a generation, where starting from the resource, more dedicated specialisations are achieved.

This means that there are specific ways of organising experience and learning, which can be linked to the distinctive affective and cognitive endowments of a given particular species. Some species, in fact, tend to interact and play with objects, and are therefore very inclined to learn the use of tools, while other species have a more marked tendency towards acrobatic movements and kinaesthetic play, so their learning is more directed towards psychomotor performance. Some animals are very curious, and are prone to prolonged explorations, based on a strong capacity for concentration and attention to the structural requirements of a problem, while others prefer to patrol more extensive environments, through flyover reconnaissance, so they sharpen the mapping configuration of the territory through the definition of references. Loners generally approach learning through an exclusive and direct relationship with the problem, whereas collaborative animals tend to develop team styles. In stable, organised groups, social learning does not only last during the juvenile period of parental care but extends through the whole of life. On the other hand, if it is true that learning means modifying the heritage, it follows that the result of learning, and not only the didactic system, varies in different species.

Having said this, research has always also aimed to extrapolate didactic models that could serve as general laws of learning, obviously to be adapted to the specificities of each individual animal. It should be remembered, however, that there is not just a species-specific kind of learning, because the individual is also a primary variable in terms of predisposition and how learning takes place. There are countless factors that regulate this variability, such as motivational tendencies, emotional sensitivities, resourcefulness and stubbornness, attention and concentration skills, knowledge and memory, and cognitive flexibility. Indeed, animals have personalities which, as we shall see, also influence the events of macro-creativity which jump out at us as the peaks described in the literature. Such episodes were once swept aside by the accusation of anecdotality, but today we know that they have a fundamental role in the transformation of the styles of a population. In any case, the search for laws of learning has made it possible to carry out experimental research and to develop criteria for animal training. The best known of these is the law of effect, developed especially by Burrhus Skinner (1904–1990) and based on reinforcement, that is, the administration of a reward, usually food, every time the subject performs a useful behaviour.

This is a very convenient form of education, which suits the simple needs of experimenters and trainers, because it is able to select certain behaviours through a universal language (food pleasure) that can be stimulated by not feeding the trained animals. In reality, learning takes many more forms, and in this chapter I will discuss at least ten basic ones, which will illustrate the extraordinary phenomenon whereby an animal learns to include a new activity in its expressive repertoire. Another important issue concerns the interpretation of the learned, that is the endowment that is formed as a result of a learning process. In general, when talking about non-human species, the word *knowledge* is carefully avoided, as if it were an anthropomorphism, in favour of the term *conditioning*. My work aims not only to indicate the importance of creativity in animal ontology, but also to deconstruct the conceptual constraints that have been put in place to bring animal behaviour back to the automatism model. It makes no sense to talk about animal consciousness or the cognitive capacities of other species while maintaining the previous model, because this gives rise to epistemological inconsistencies.

As I have already said, it is not sufficient to attribute consciousness to an entity that is normally described through automatisms, in order to explain certain behaviours that clearly contradict this very paradigm. In fact, by doing so, one risks adding props to the model itself, as Thomas Kuhn pointed out in *The Copernican revolution* (1957) with regard to epicycles. Consciousness, used as a *deus ex machina* to explain what the paradigm cannot, does not pave the way for a full questioning of the paradigm itself, with the result of sticking to the Cartesian model revived by behaviourists in the twentieth century. The invocation of consciousness, therefore, serves as a convenient smokescreen to avoid facing the innumerable explanatory contradictions of the existing model. If animal behaviour escapes the parameters of this model, this means that the explanatory structure is inadequate. Although it is generally accepted that animals are not automatons and that they are capable of cognitive performance of different kinds—indeed, there are countless studies that highlight this—a serious revision of the explanatory model has not yet been carried out. Thus we continue to speak of classical and operant conditioning, implicitly assuming that learning produces automatisms. In this sense, the appeal to consciousness, on the one hand, has led to a reconsideration of animal ontology, but on the other, it has caused very serious epistemological damage, concealing the urgency of a global revision.

Learning: Between Ethology and Neuroscience

In order to understand the phenomenon of learning as opposed to other processes of behavioural modification, such as motivational phases or maturation, it is essential to refer to the characteristics of the neuroendocrine system, where the learning process takes place. Using Joseph LeDoux's metaphor in *Synaptic Self: How Our Brains Become Who We Are* (2002), it must be recognised that every endowment is ultimately embedded in the synaptic connectome. The influence of environmental conditions on synaptic education was highlighted in Daniel Lehrman's *A critique of Konrad Lorenz's Theory of Instinctive Behavior* (1953, 337–363), emphasising that an organism is never completely isolated from environmental influences. Lehrman's critique, however, fails to grasp Lorenz's evolutionary epistemology, which, like Donald Campbell (1916–1996) and Karl Popper (1902–1994), regarded a-prioris not as finished products, except in certain cases, but as knowledge operators. On several occasions in the aforementioned interview with Franz Kreuzer, Lorenz points out that the ability to learn is an innate faculty, an open programme with multiple possibilities. Both authors came to the conclusion that there is a relationship of *direct,* not inverse, proportionality between innate and learned: the more complex the innate, the greater the space for openness.

This principle is also fundamental to my own considerations of philosophical ethology. Having an innate heritage does not mean being closed off in phylogenetic—or even genetic—determination, but owning a kit that allows for many variation opportunities or, if you like, useful material for the construction of experience. For example, precisely because of their copulative character (the fact that we are interested in the world), motivations increase experiential possibilities. Moreover, innate representational contents have generative potentialities—as Jean Piaget (1896–1980) pointed out in the concept of genetic epistemology, described in his *Psychologie et Epistemologie* (1970)—so the greater the starting content, the greater the assimilative possibilities. Lorenz takes up this argument using the metaphor of embryology, where the more possibilities there are, the more articulated the programming must be. The importance of the comparison between a real fixed in the subject's a-priori and a real *after* the individual's cognitive experience is fully expressed in the Austrian ethologist's essay *Die Rückseite des Spiegels* (1973). It is prior *phylogenetic* experience (i.e. having previously been confronted with what lies behind the phenomenon) that enables the individual's *ontogenetic* experience, understood as an experimentation process.

The individual's dialogue with the world, using a-prioris, allows for those reshuffling processes—for example through play, imagination, discovery—that make it possible to create something new in the basket of individual behaviour. The innate and creativity are therefore strongly linked. Lorenz hypothesis, among other things, that the development of a *Gestalt* approach to phenomena, such as the assumption of constancy in the perception of the world, underlies the capacity for attraction. The innate, that is the phylogenetic nature of a species, can therefore be considered *a master of learning*. Many other authors move in this direction, emphasising the importance of innate tendencies as mental modules that predispose to learning and the construction of particular types of knowledge. In other words, we can talk about learning systems. This means, as LeDoux (2002, 115–120) reports, that the brain learns different things using different structures, so that different animals possess specific acquisition capacities, in contrast to the behaviourist assumption of a universal learning function. Already in the 1960s John Garcia was able to point out that nausea-inducing stimuli could be associated with tastes, but not with sounds, in contrast to the behaviourist idea that any stimulus could be associated with any other—see the essay *Relation of Cue to Consequence in Avoidance Learning* (John and Koelling 1966, 123–124).

As Aleksandr Lurjia (1902–1977) pointed out, remembering, far from being a simple passive process, is complex and active in nature. In fact, learning abilities are linked to very specific neurobiological features, as demonstrated by the process of *strengthening synaptic connections*. Examples include (1) an increase in the number and size of dendritic spines and the contact surface between neurons as a consequence of the repeated passage of the nerve impulse and the release of the neurotransmitter; (2) a greater ease of action potential by a *c* neuron on a *b* neuron if the latter is preactivated by repeated stimulation by *a* neuron. In 1966, Terje Lomo discovered a type of synapsis in the hippocampus whose repeated stimulation caused increased postsynaptic potentials that persisted for prolonged periods, known as *long-term potentiation*. Nobel laureate Eric Kandel's studies on the sea snail (*Aplysia californica*) were able to identify synaptic and molecular changes in processes such as habituation and sensitisation. As Stanislas Dehaene reports in the essay *How We Learn*, when a mouse learns to avoid a certain place because it receives a small electric shock there, there is a change in the synapses of the hippocampus and its connection to the amygdala (2020, 120–124).

On the other hand, the same author reminds us that the evolution of the mammalian brain already involves the organisation of the neural network, partly defined genetically and partly by epigenetic processes, during the embryofoetal phase. The author speaks (2019, 100–110) of veritable *highways of the brain*, mediated by the action of neurotransmitters, whereby each excited axon literally explores the surrounding areas and develops in a vector fashion on the basis of the chemical concentration present. In other words, these messengers act as topographical-directional labels. This means, in practice, that the initial scaffolding of the learning system is innate and is built up in utero.

As can be seen, studies in neurobiology do no more than confirm Lorentz's hypothesis: an animal is not a tabula rasa that is passively exposed to the world, but is predisposed to interacting with reality following preset guidelines. Admitting this does not mean limiting the possibility of revisiting the setting coordinates in a more or less revolutionary way. Some innate capacities are fundamental in the learning process: an example is working memory, which in some animals (chimpanzees, for instance) is particularly well developed if not superior to that of humans. Indeed, in 2007, the Japanese researchers Inoue and Matsuzawa showed that chimps were able to outperform any human competitor in the ability to remember the random order of digits from 1 to 9 shown for a moment on a screen (2007).

To understand the complexity of the learning system, one must keep in mind the concept of ontogenetic virtuality. In the essay *Neuronal man* (1986), the neuroscientist Jean-Pierre Changeux maintains that learning means eliminating, and this concept would also underlie Gerald Edelman's conception in *Neural Darwinism* (1987). In fact, one can envision the virtual starting condition in mammalian ontogeny as a redundant structure that is organised through processes of pruning and evolutionary differentiation between different alternative synaptic networks based on the activation gradient. Indeed, one does not only learn when confronted with something unusual, but also in regular agency. Every action is due to the activation of a neuroendocrine configuration in the animal organism and this implies a positive feedback in the network itself, for two reasons. The first is that, (1) by the law of exercise, first formulated by Edward Thorndike (1874–1949) and confirmed by neurobiological findings, each activation strengthens the existing synapses and the thickness of the myelin, making the subsequent evocation of that behaviour more likely. The second has to do (2) with the assembly theory, formulated by Donald Hebb (1904–1985), whereby each coactivation produces a conjunction of neural elements that will subsequently tend to act together, so that the activation of one point will cause activations in the whole pattern.

In short, the animal's own life produces the changes in the system's wiring that are responsible for the subject's expressive identity. Of course, there are times in an animal's life when the configurative plasticity of the connectome is particularly high: in mammals, for example, the development age is a phase where learning processes have stronger effects because the neural network has a higher virtuality gradient. Consequently, during this age the plasticity of the brain and the learning capacity are proportionally greater: this is why we speak of *formative* processes (the organisation of an individual's character), and not only of *informative* events (the acquisition of knowledge). From what has been said, it is clear that learning is never a passive process but, on the contrary, it is directly proportional to the individual's proactivity, curiosity, tendency to interact and make attempts and enquiries, exploratory motivation, propensity to observe and imitate, playfulness and expressive flexibility, ability to rearrange strategies and so on. Relying on a juvenile phase, during which the parent takes care of the most important survival activities, from foraging to protection, and in which the parental relationship itself becomes a training ground for knowledge, represents the very heart of this learning-centred proactivity.

There is a close relationship between acquisition and ontogeny, which in mammals and birds is achieved through social learning and experiential facilitation. The former refers to all the content transfers from parent to young, through modelling, emotional osmosis, target orientation or teaching, just to mention a few examples; these can be grouped together in the concept of *allomimesis*. The latter is the propulsive action of the parent's presence in terms of facilitating the experience through the *safe base* effect, as studied by John Bowlby and Robet Hinde, the creation of a *zone of proximal development* as noted by Lev Vygotsky, and the development of play conditions. The juvenile period is certainly one of the key learning phases. Even in insects, which have a larval phase and in some orders undergo a real reconfiguration of the body structure, it has been possible to highlight a persistence of the experiences acquired in the first period of life. In birds, in addition to imprinting in nesting species, there are a series of parental transmission processes that lay the foundations of identity expression—for example, chirp, orientation and preferences—as early as the hatching stage. This is the case with the zebra finch (*Taeniopygia guttata*).

Learning is an organisational process, so we must forget the Hellenic metaphor of the empty vessel or the empiricist one of the blank sheet of paper. At the same time, the inverse hypothesis of innatism must also be rejected, because ontogeny is not an already written parchment scroll or the mere reminiscence of previous experiences. Learning is a dialogue between three conditions of animal subjectivity: (1) being the bearer of prior content, innate or simply learned at a previous time, which defines the way in which one deals with critical issues; (2) being in a precise condition, in what is defined *hic et nunc* and exemplifies not only the environmental situation, but also the physical-psychological state of the subject; (3) being projected into a future condition and elsewhere, with the propositional attitude that emerges from the integrative condition of the mind. The subject is the master in this dialogue, the one who oversees the diachronic process—that is, the one that moves between the temporal dimensions of the past, *Dasein* and projection. In this sense, learning is always the result of a creative and singular process, which takes place at the very moment it occurs. Therefore, this event cannot be deduced from previous content: neither in terms of meaningfulness—as implied by the etymology of the word *development*, which recalls the act of unrolling parchment—nor in terms of deficiency, as posited by the purely acquisitive vision of the blank sheet of paper.

In other words, in learning one does not simply incorporate content, but translates an experience into an organisation of the system itself, into a wiring recipe or a packet of information; one does not simply remember something, but modifies what one has by giving it a new conformation. Learning is, therefore, a translation of external contents and a reconfiguration of the connectome—two events that can be achieved thanks to the heteronomic condition of the animal mind, its ability to acquire information from its own expression. One of the characteristics of complex systems, as I pointed out in the previous chapter, is that they present degrees of virtuality, phenomenological spaces that can be interpreted in different ways through external contributions. We can identify virtuality according to a number of parameters: (1) redundancy, the presence of several elements that can perform the same function; (2) modularity, being made up of elements that can change without altering the other functions of the system; (3) degeneracy, a high level of connectivity and therefore many paths that can be taken or exploited to implement new functions; (4) iterability, the repetition of modules that are identical in structure and can be adapted in different ways. A virtual system can evolve, as it has few constraints and many usable resources; at the same time, it is highly sensitive to the outside world and capable of taking in information.

The virtuality gradient indicates the phenomenological space of expression: that is, how much that system is able to introject what the subject is experiencing, converting virtuality into actualisation. The latter process is reminiscent of the sculptor's act of cutting marble from a slab to make a statue: the latter is ultimately the product—what has been acted upon—losing the other potentialities present in the slab. A newborn mammal has a certain *virtuality gradient* which, of course, varies in different species, but is in any case considerable, both because the animal has not yet had any experiences and because the neocortex is a formidable database. The number or, better, the conjugative availability of neurons, activatable synapses and receptors, just to mention a few examples, gives an idea—however approximate—of the introjection gradient of that particular system. There is no doubt that the learning capacities of a species can be linked to the complexity and performativity of the neuroendocrine system, especially the integrative ones we have seen when discussing the animal mind. We know that there are significant differences in the integrative conformation of vertebrates compared to arthropods and molluscs, but learning capacities are clearly related to system complexity.

Learning also means introjecting something external—in the form of organisational information or wiring—into the system itself, and there is no doubt that without a certain degree of virtuality this process would not be possible. Some authors, such as Gerald Edelman (1929–2014) in the above-mentioned essay, describe actualisation as basically the selection of neuronal groups through processes of neuroapoptosis and synapoptosis. At the same time, we have developmental phenomena such as neurogenesis, synaptic plasticity and synaptogenesis, as well as axon growth through the nerve growth factor (NGF) discovered in the 1950s by Rita Levi-Montalcini (1909–2012). The virtuality parameter is obviously also affected by the constraints of the system, which have a low level of interpretability because (1) they have less redundancy and degeneracy, a condition that decreases the range of convertibility; (2) they are more resistant to change, that is they have more stable and innately predefined configurations; (3) they have a reduced-scale conjugative capacity or a low modularity gradient, so that a change would mean compromising some overall functions.

The learning ability in the animal world is a predisposition of the behavioural identity to take on—at least in certain respects—a form and therefore a biographical recognisability, thanks to the information that the system has received from the outside (heteronomy). But it is not simply a matter of plasticity or virtuality, however essential they may be: rather, what matters is the animal's *status* as *being-conformed-to* the learning process, designed to favour internal organisation through the acquisition of information. In other words, introjection defines the type of wiring that the system assumes and, consequently, the individual's behavioural identity. The propensity to learn is a constant in an animal's life and gives rise to various modes of learning, which correspond to real laws. I have already mentioned the law of exercise and the law of composition, but I will examine others as well, in order to show that learning is never passive. Learning is affected by this introjective tendency of the subject, which is proportional to two parameters: (1) the already mentioned virtuality gradient, which establishes the room for expression of that particular animal at a certain time and varies through factors that are affected by the age of the subject and the species it belongs to; (2) the sensitivity to experience, which indicates how susceptible that system is to external influence.

Sensitivity to experience indicates to what extent a system is affected by events in the external world: it is an important parameter, because it shows the degree of learning that a subject can achieve in a certain experiential situation. If it is true that the virtuality gradient is a sort of prerequisite for learning, it is undeniable that sensitivity to experience is also an important factor, because what an animal is experiencing does not always have the same effect on introjection. Sensitivity can be assessed in terms of (1) elements to access world, for example sensory organs or endocrine receptivity and their particular responsiveness to available stimuli; (2) affective components in terms of projection into the world, such as emotions and motivations, and interest in what is occurring; (3) physiological parameters of attention, concentration, working memory, state of the sensory system in relation to conditions of illness, suffering, digestion, and so on; (4) the socio-parental dimension of a species, which influences the transmission of intergenerational content, or the development of cultural traditions and group diversification. Ultimately, sensitivity leads us to consider the interface dimension of an animal because that is where acquisition processes take place: the aforementioned sensory organs, for example, facilitate certain kinds of learning over others, just as parental care causes the transmission of models, target orientations, experiential facilitation and playful activities.

There is one aspect, however, which is not always taken into account: learning involves the subject's ability to be present in the here and now, by which I mean not only the obvious sense of being present in the situation, but four other abilities. One (1) is that of keeping the memory of oneself, of what one has already and just done. A second is (2) that of being active and proactive in the situation, while another is (3) that of being participatory and thus responsible for the unfolding. Finally, we have the ability (4) to be projective towards something that is not yet present, something not given but pertaining to a possible condition towards which one strives. The subject's presence is realised, thanks to the combination of these four points. This demonstrates that learning means a number of things: being able to use the past as a plastic and dynamic material or simply as a term of comparison; being proactive, for example by posing questions, exploring possibilities, diversifying points of view; being involved in the experience, feeling it both emotionally and through the drive of desire; and knowing how to project oneself into an elsewhere that is not a distraction from the situation at hand, but a possible evolution of it. Learning is therefore closely linked to creativity, because one does not learn a simple *condition of* the world, but a new *relationship with* the world, shifting one's point of view to new horizons.

Learning Means Being Present

Learning always involves the animal: no matter whether it is driven by simple curiosity or by the desire to explore, whether it follows an innate tendency to stir up the leaves in a serendipitous surrender, or the propensity to probe and test. Learning often requires asking questions, trying to understand the structural requirements of a problem, grasping the extent of a challenge and trying to solve it through targeted attempts or mental simulations. In other words, learning involves an active presence, acting on the environment, which is always much more complex than mere sensitivity to a stimulus or the tendency to respond mechanically through a reflex, in a knee-jerk fashion. An animal learns to recognise important clues to navigate the dynamic complexity of the world: it does so in order to follow a trail or anticipate an event, and in this case learning is always a journey within a temporal dimension. An animal learns about the different problems with which he will have to deal and learns to test the soundness of its hypotheses, like a researcher evaluating Bayesian probability gradients. Learning is about giving a meaning to things, referring them to oneself and one's needs, pulling them out of the landscape of simple presences and transforming them into externalisations of one's own body. One learns strategies and tactics, operational models and maps of the territory, but also targets on which to direct one's motivations and contingencies to mark through one's emotions.

Learning entails a mental condition: however conscious it may be, this skill certainly cannot be reduced to a mere reaction, to a blind orientation based on reinforcements, to an obscure proposition of random behaviour without a figurative objective, where actions are not chosen for their consistency or probability with respect to the challenge but are simply casual. Only a mind in action is capable of transforming a phenomenon into information, of pursuing a project through the whirls of the imagination: in a non-mentalistic hypothesis there is no room for intentions or problems. Clearly, if the animal were a puppet moved by strings or a machine driven by its cogs, there would be no place for the construction of goals, projects, possibilities and challenges. In order to learn, it is necessary to be motivated, attentive, focused and projected towards a condition that is not yet present. Similarly, if we regarded it as a cascade of reactions or a volitional fire, like a fruit falling from a tree or a whirlwind, we would never dream of speaking of a presence full of suggestions. An object does not have a critical and possibilistic stance in the face of a phenomenon: it can react, but it certainly does not have the power to learn or discover a causal relationship or a correlation between events, nor can it bring out problems to be solved or experience frustration. Unfortunately, even today the prevailing models for explaining animal learning follow the associative paradigm of the stimulus-response link, assuming simple reactivity and the blind tropistic orientation of reinforcement.

Behaviourism has stubbornly insisted on the animal's passivity in learning, but this has required quite a few explanatory crutches, which now appear as veritable exercises in logical acrobatics and cumbersome redundancies. In order to deny the mind and avoid any mention of mental activity, it has been assumed that animals have no goals, only unexpected consequences—the so-called reinforcements—with the exception of lure-targets. It was assumed that animals do not understand a problem, but blindly react to it. This forced behaviourists to believe in the model based on random attempts, contradicting the elementary rules of entropy. Denying a subjective, mind-based presence has led them to an explanatory contrivance that complicates rather than simplifying the explanation. Behaviourists had to appeal to tautological entities, such as variable reinforcement, according to which reinforcement, if administered on a regular basis, does not reinforce: yet they seemed not to notice the implicit contradiction! On closer inspection, the problem would be easily circumvented if, instead of the word *reinforcement*, we were to use the word *objective*, because it is understandable that an expectation decreases in value the more likely it is to achieve. But the word "objective" implies planning, understanding, the choice of heuristics: this would make the paradigmatic system fall apart, and its falsification could certainly not be ignored.

When I say that the assumption of consciousness has somewhat hindered the revision of the associative paradigm based on automatisms, it is because even today I hear indiscriminate talk of conditioning. And yet most learning processes do not take place in this way, and above all do not produce passive reflexes: they do not create switches in the animal, but give it knowledge and tools of which it has full ownership. We keep talking about conditionings, simple associations, problems faced at random: it matters little if this clashes with the evidence and with a number of simple observations. For example, (1) once it has learned something, an animal uses it with great flexibility in other situations, demonstrating its ownership of it, something that would not be possible in the case of conditioning. Or (2) most learning produces a complex representation of experience and a utility characterised by several functions: something that would be unthinkable for a simple association. Another fact is that (3) one cannot raise problems without having goals, just as one cannot tackle a problem without understanding its structural requirements, let alone solve it without resorting to purposeful attempts. But if we accept all this, we will inevitably end up with a systemic explanation, whereby the very premises of the behaviourist paradigm would crumble. Terms such as reinforcement, chain, stimulus, reaction and conditioning would lose the explanatory foundation they had when learning was to be explained without involving the animal's propositional mind.

Learning implies the animal's presence, as the individual must assess the situation at hand, understand the structural requirements of the problem, imagine potential results, remember its knowledge, choose behaviours that may be useful and evaluate the outcome of its actions. Learning cannot be done passively, but rather by expressing what has been learned in an integrated way, affectively and cognitively. It is necessary to be a protagonist, that is, to be present at the event to be introjected: this is to be understood in the sense of attending and participating, but also in the sense of giving meaning to what is taking place through the production of beliefs, hypotheses, hopes, desires, fears referring to the *mental images of the problem*. Being present also means introjecting the contents of the situation experienced and expanding them within a space of *self-referred* possibilities, whereby one always learns a type of relationship established with a certain phenomenon. The mental images that a situation evokes may concern a predictive, operative, orientative, evaluative or any other need, but they are always *referred*: they indicate the relationship between the subject-learner and the contents it will deduce from the experience itself. Experience always has to do with the propositionality of the subject, that is, with its being present: *it is clear, in fact, that there is a difference between a learning process based on fearing-that and one based on believing-that.*

The expression "being present", in its twofold semantic guise, indicates being present through superintendence, which means, in practice, not being totally immersed in the phenomenon itself. The resulting experience is thus shaped according to the specific proposition that the subject submits to it. As Shaun Gallagher and Dan Zahavi point out in *The Phenomenological Mind*, one can look at a car as a vehicle to reach a certain place or, in other circumstances, as an object to be cleaned or, again, as something to be sold or that does not work (2008, 10–13). It is propositionality, rather than a generic being present, that specifies how an entity is experienced and thus the type of learning that follows. "Believing-that" means approaching the phenomenon through hypotheses, hoping-that means illuminating it through the desiring condition, fearing-that recalls the need to anticipate and foresee the event and therefore learn to grasp the predictive clues. Being present in relation to a phenomenon means interpreting it according to the action that one intends to devote to it at a certain moment, acquiring new elements to bring back to the referred Self. An event takes place in front of us, but we transfigure it through internal images of resonance or recognition as well as with prospects referring to purposes or to something simply imagined in the future. Learning means intervening in the phenomenon.

The event experienced becomes so saturated with subjective attributions as to be only partially ascribable to its mere phenomenal dimension. Being present in relation to an event is what makes us experience the propositional condition that makes us fear, desire, want, believe something that is not explicit in the event itself. We are there (*Dasein*) precisely because we do not follow the passive flow of the phenomenon, but bring it back to *us*, to all those internal images that the predicative mind produces. *Dasein* has to do with the intentionality inherent in the subject, which arises not by virtue of being conscious of what one experiences in the here and now, but simply from experiencing it: that is, experiencing something Self-referred. The mind unfolds in its continuous production of propositional attitudes and implicit images: we find it in the processes of meaning attribution, in approaching phenomena through a hypothetical-deductive perspective, in going beyond appearances and seeking the deducible, in the choice between two possibilities and in decision-making. But presence is above all the innovative contribution that the subject makes to the order of things. The term "propositional synthesis" refers precisely to a capacity for superintendence, the ability to stand above the fact and beyond the simple causal convergence of influencing factors.

In learning, the subject is faced with events and participates in their unfolding—a diachronic condition that we could identify as accompanying the phenomenon—and at the same time anticipates them through prospection. In other words, one learns by moving between different temporal dimensions or, if you like, by narrowing chronological gaps. There is a close relationship between diachrony and learning, since clearly it is not possible to be a protagonist if one is totally immersed in what is happening, if one does not separate the phenomenon from the subject. An animal never just undergoes the situations and occurrences around it, but sheds light on them through an activity that we might call *thinking* about them, that is, placing them in a condition of interaction between past and future. Therefore, obviously, in most learning processes we cannot speak of a simple associative mechanism—the Pavlovian kind for which bell equals meal—because thinking about a situation means transforming it into a dynamic entity linked to the subject's expectation, the motive that makes it worthy of being learned. Learning requires recalling certain mental images, knowing how to strengthen attentional capacity (often by modifying the perception of time or altering the state of arousal), being able to concentrate one's interest and persist in the face of an unsolved problem, putting all of one's knowledge resources into motion.

We have seen that the emotional and motivational components produce a wide range of copulative modalities, but now we must add that each entity-event is linked to a multiplicity of mental images. For a chimpanzee, a stone can be thrown at someone or used to crack nuts: the difference lies in the propositionality that the subject attributes to itself and that leads it to address the stone in a given way—and the same goes for any other thing that can be used for the subject's purpose. In learning, one does not merely record a fact but transforms it into something significant, referred to one's own projections. The mental condition produces, in fact, not only expressive plasticity in the subject, who is thus able to change reality by breaking away from predefined tracks, but also a self-referential instance of events, a subjectivity of experience. Being present means transforming the stage of reality and one's own endowments based on the singularity of interpretation of the here and now. But in order to understand this process, which is intrinsically creative, it is necessary to leave behind the logic of triggers, associative mechanisms, drive expressions and reflexive chains. The mental condition indicates a presence that, while referring to the here and now, is not caught up in it, but produces a state of continuous learning. The subject is present in its unity which, despite originating from activations scattered throughout the somatic landscape, also yields a plan that shows the close relationship between learning and intention.

I would say that the concept of ***having intentions*** can truly enlighten the meaning of learning, better than any other reference. A learning individual sets goals that are not immediately attainable, so a special condition occurs where presence is not immersion (Heidegger's captivation): rather, this tension underlies the diachronic incentive. Learning is, therefore, the result of a prospection that increases the dynamic participation of the subject—its presence—multiplying its propositionality, so that the verbal predicates of believing, hoping, fearing and so on beat louder in it. In the course of learning, the individual imparts a certain order to things, consistent with its intentions, and in so doing produces an innovative condition. One learns because one is always faced with a gap to bridge, as Kölher intuited, but this gap can be measured in terms of propositionality. An animal that is startled by a noise may intersect it with the attitude of fearing-that: if so, it will try to understand what the danger is. Or it may react in other ways, and then the gap will have different connotations. At the same time, if one expects something from a certain behaviour, one clearly has beliefs, just as, if one has to solve a problem, one has to test one's hypotheses.

Being present means having an interior dimension that tackles reality in a process of reciprocal influences: the exterior modifies the state of the subject and the subject, in turn, transforms the very image of the world, going beyond appearance and the already given, bringing out possible conditions. Interiority and external reality, crashing against each other, give rise to representations that do not properly belong to either of these two worlds but are made available by their collision. This interaction gives rise, in fact, to the intentions that are at the basis of learning. An animal has recourse to absent images, referring to the past or to a future projection, and compares them not by holding firmly to the present order of things, but by balancing on the range of possibilities that this implies. Learning is therefore possible not only because the system of knowledge has its own gradient of virtuality, but also because every order of things (reality itself) has its own virtuality of development: using quantum mechanics, we could say that it has various states and ways of unfolding. An animal is capable of learning insofar as it is able to go both beyond its past (in other words, to reinvent itself), and beyond the order already present in the world: in this sense, all learning draws, albeit in different ways, on percentages of imaginative creativity.

We might believe, in fact, that images resulting from the past simply give rise to repetitive conditions, so that, with a sort of Laplacean demon, we can deduce the propositional attitude from the innate or learned past. However, this is not the case. Everything that is part of the individual's heritage is plastic material that is reconfigured according to the finalistic projection that the subject is experiencing. Learning is one of the most explicit ways of showing how each behavioural resource is an endowment available to be reused in other contexts, co-opted to new functions, used as a probable solution recipe and modified to acquire greater adherence to the particular conditions in which the individual is living. As I said, the mind is not a place where finite variables converge, but a productive dimension that, in integrating functions occurring in different parts of the body, generates new contents. For this reason, it is more correct to imagine the mind as a factory of thoughts rather than as a theatre or a working space; it's a sort of laboratory that does not merely put together the elaborative or dispositional specialisations coming from the limbic system and the cortical areas, but organises them. I have spoken of the mind as a new level of subjectivity, capable of relating to the world through propositionalities such as believing or hoping. Likewise learning, when it is not a simple reactive modulation, is possible insofar as there is a unity that learns in its relation to external reality.

As mentioned, thought, in its propositional dimension, is to all intents and purposes a singular product: a freshly made meal, not a pre-packaged one. The integrative process is therefore not an assemblage but a real emergent reaction of functions. This gives the individual a certain room for manoeuvre beyond mere local adaptation of established behaviour. The way an animal learns in nature is quite different from the way it learns in a Skinner-box. Indeed, it follows patterns that (1) in some ways have approximately universal characteristics, leading to genuine *laws of learning*; (2) in other ways are intimately related to species-specific characteristics, so one must always understand the given type of *interest in learning*. Presence implies an emotional and motivational condition, but this varies considerably between species and individuals. Some species are very competitive, learning in confrontational situations, others are more explorative, others yet have a problem-solving approach and are more reflective. Similarly, among individuals we find the stubborn and the calm, the curious and the disinterested, the fearless and the timid, the cooperative and the solitary, the trusting and the suspicious, the dominant and the submissive, the extrovert and the shy. All this shows how presence produces very specific learning processes.

The Laws of Learning

The term "learning"—see the *Dizionario di Etologia* edited by Danilo Mainardi (1992, 48–64)—refers to a specific process of information acquisition that meets three basic requirements. It must (1) be capable of inscribing information in the nervous system, (2) take place following an experiential event, (3) determine an adaptive modification of behaviour in a relatively permanent form. Thus the boundary separating learning from other physiological events seems quite clear. In reality, phenomena are not always so easily distinguishable from the continuum of other processes, as we shall see when discussing the law of exercise or mimesis. Learning, in fact, should diverge from transient conditions such as sensory adaptation or pathology, and yet it is not easy to separate processes such as forgetfulness or neurobiological evolutionary differentials from other forms of maturation or degeneration of the system. On the other hand, the nervous system is profoundly affected by relationships with endocrine and immune neuromodulators, so experiential changes may result in stable behavioural shifts without necessarily passing through the synaptic configuration. Moreover, it is not certain that an experience that has been retained, that is, that has led to learning, translates *ipso facto* into an immediately observable change in behaviour, as shown by the latent learning highlighted by Edward Tolman. In short, things are never so linear and any classification invariably has a margin of arbitrariness. There is, however, one aspect that needs to be stressed right away: there are many forms of learning, and believing that there is a single universal law or some sort of arithmetic of learning, as behaviourists thought at the beginning of the twentieth century, is a gross error, which requires explanatory complications and sometimes artefacts, ignoring the evidence and the relationship with the innate.

Behaviourist research has been very important and still sheds light on some of the mechanisms that indisputably regulate learning processes. My criticism essentially concerns three aspects. The first (1) is the explanatory model adopted, which, starting from the rejection of mentalism, has led to explanatory paradoxes and stretches to adhere to the reactive canon. The second (2) is the idea of a universal grammar of learning based on the two conditionings, reactive and operant, in line with the basic associationist model (S-R). Finally, there is (3) the lack of consideration of innate resources in shaping learning processes, which actually make up the starting point or the canvas from which to make the leap towards new knowledge. By abandoning the behaviourist explanatory model and reconsidering the role of the innate, the findings of these scholars can be used as valuable information to better understand the mechanisms of learning. What I would like to emphasise in this paragraph is the *plurality* of these mechanisms—I will mention a dozen by way of example—which gradually progress from very simple forms of learning, such as habituation or conditioned reflex, to incomparably more complex modes, such as learning by mimesis and by solution, where the associative canon can no longer be applied.

On the other hand, talking about basic mechanisms does not mean ignoring the species-specific diversity with which each animal faces learning. The innate, in fact, is a flywheel of experience, which brings the subject into a precise relationship with the environment; it is a tank full of petrol that makes the journey possible, but the itinerary is in the making and open to singular possibilities, to the creativity embedded in the specific relationship with the world. The innate is the astonishment that reality arouses by responding to the affective system of a certain species; it is desire expressed in acts directed outwards, like *chasing* for a kitten or *collecting* for a child. The innate is a body that the individual must learn to know through the stimuli of curiosity and surprise, which support searching and interaction, the enthusiasm and exuberance that immerses one in play. Each innate behaviour is like a magnet that leads the subject to have some experiences and not others; it is preference for particular targets, favouring the subsequent learning of their characteristics. Heritage is a flywheel that acts by positive feedback: the greater the innate content of a certain behaviour, the greater the individual's learning in this area. Innate and learned are therefore directly proportional to each other.

Every competence, constructed by the individual in the course of its particular experience, rests on the innate and does not arise from nowhere. On the other hand, the innate would have no meaning as a simple resonance or repetition of the past: its value lies in being available for the realisation of the singularity we call the individual, itself a tribute to the value of diversity, the true leitmotiv of life and its world-making. One of the main arguments of ethology, therefore, concerns individual diversity within a species, that is, the unique and unrepeatable pathway called ontogeny. It is a *viaticum* that, through processes of growth, learning and experience, creates a singular identity, which could not be repeated even if we cloned that particular animal—indeed, not even if the same story could be repeated in the same individual, by turning back the clock and subjecting it to the same developmental conditions. The reason for this is clear. Growth is like a pathway, where each step builds on the previous ones, so that over time the smallest divergences grow larger and larger, just like trajectories. Therefore, an infinitesimal difference in the first steps of development is enough to give rise to enormous identity shifts: hence the extraordinary adventure of the individual, who matures the grains of its personality moment after moment.

The keyword for understanding the success of living beings, even before adaptation and specialisation, is surely diversity. Ontogeny is the process that, starting from the phylogenetic heritage, creates an individual identity—which is, after all, always a singular manifestation of the species' behaviour. Individuality is not divorced from the identity of belonging: there is no contradiction between being *a* dog and being *that* dog. At the same time, it is not surprising that in many species belonging is also achieved through learning processes, such as the aforementioned parental apprenticeship in animals with parental care, and more generally through social learning. Phylogeny and ontogeny are not two separate processes, two opposing or independently regulated domains: the innate and the learned are not to be regarded as separate elements, like water and oil poured into the same glass. Indeed, the learned arises *from* the possibilities provided by the innate, from the phenomenological spaces it provides. This means that both the way of approaching learning and the results that can be obtained from a learning situation vary between species. This is true in a broad sense but, *a fortiori*, it changes from individual to individual or from moment to moment. Having made this appropriate consideration, it remains to be asked whether there are general conditions which, while maintaining the species-specific diversity of approach and outcome, can offer a meta-representation of the learning processes.

From my point of view this can be done, or at least it is a feasible course of analysis, provided that we do not restrict the field of possibilities too much, as has been done by behaviourist psychologists. This means *highlighting* the multiplicity of learning mechanisms and the gradual complexity of each, without reducing all forms of learning to a simplified criterion.

In my opinion, there are ten general laws of learning, subdivided as follows: (1) non-associative learning, characterised by the simple modification of the thresholds of reaction to a stimulus; (2) associative learning, like the Pavlovian conditioned reflex; (3) law of effect, that is learning from the consequences of the actions carried out; (4) law of mimesis, that is learning linked to the presence of a demonstrator; (5) law of exercise, that is learning by evolutionary differential between the parts; (6) law of composition, whereby, according to Donald Hebb's indications, coactivating means joining; (7) law of assimilation, based on the reuse of prior knowledge in new fields; (8) law of solution, based on the understanding of the problem and acquisition of solution strategies; (9) law of engagement or ludic learning, referring to the field of play and motivational activation; (10) law of experience, that is the construction of a body of knowledge relating to a certain type of behaviour.

As far as non-associative learning is concerned, a stimulus can activate the subject's attention and its persistence can cause a modification in the reaction threshold. In general, there are two opposite situations: (1) if the stimulus is not too strong and occurs continuously, there is a decrease in the reactivity referred to it; (2) if the stimulus is sudden, surprising, intense or strikes the subject in a moment of particular vulnerability, it produces an increase in sensitivity or reactivity which is called *sensitisation*. We have seen that the component involved by stimuli is attention, but it is necessary to distinguish interest from alertness, two antithetical forms of paying attention to a stimulus. The repetition of a stimulus that arouses a modicum of alertness produces a phenomenon that is called *familiarisation*, while the persistence of a stimulus that initially aroused interest produces *habituation*: both processes lower attentional reactivity to the stimulus, but in the first case there is a reduction of alertness, in the second case of interest. Another quite simple learning process in its purely reactive form is the Pavlovian *conditioned reflex*, whereby providing a neutral stimulus (e.g. the sound of a bell) in anticipation of an unconditioned stimulus (food) produces an association between the two, so that the neutral stimulus is able to produce the salivation reflex normally due to food. On the other hand, the studies conducted by Robert Rescorla (1940–2020) in the 1960s showed that the predictive relationship is more important than temporal contiguity and that a central role is played by expectation in the learning process (Rescorla and Wagner 1972).

From the law of effect, first illustrated by Edward Thorndike and used by behaviourists in the operant conditioning scheme, we thus shift to a more complex system. The basic structure of such learning is quite simple and could be reduced to the observation that we learn from the consequences of our actions. But unlike behaviourists, I find it useful to make a clear distinction between three consequential factors: reinforcement, lure and goal. Reinforcement is an unexpected and unintended consequence, which can be traced back to a kind of serendipity, whereby the subject, while doing a certain activity or absent-mindedly exploring a context, suddenly comes across something pleasant or unpleasant. In this case, it is chance that governs the whole mechanism: this factor is by no means irrelevant in an animal's life—at most, it can be influenced by probability gradients—but certainly, in this situation, we cannot speak of oriented behaviour, so the consequence acts unexpectedly. In the case of a visible target, otherwise defined as lure, behaviour is never random, but always oriented and regulated by prior experience, so that the rewarding consequence consists in getting closer to the target. In the case of a non-visible (imagined or *expected*) target, the behaviour is even more complex because, in addition to being based on useful solution recipes, it relies on anticipation and forecast resources. This explains the mechanism of variable reinforcement used by behaviourists: it has to do with expectation and that is why it is harder to quench. So, considering learning by consequence only in terms of operant conditioning is an unacceptable reduction.

The law of mimesis is fairly superimposable on so-called social learning, although the latter includes social play and active educational interventions by the parent, such as closures or punishments, as well as engagement incentives. First of all, *allomimesis*, that is the ability to learn by observing a conspecific, is predominant. In many animals, there is evidence of the mirror neurons discovered by Giacomo Rizzolatti: specific encephalic structures capable of activating neural configurations in an observer that overlap with those of the individual performing the behaviour. This means that seeing an action take place activates the same areas as performing it in the first person, so the neurophysiological pattern can be transmitted directly to the observer, without going through a phase of personal predisposition and training. The process of mirroring is all the more likely the greater the isomorphy between demonstrator and observer, and this is why *allelomimesis*, that is imitation of a heterospecific, is less probable. On the other hand, this latter type of learning is not strictly constrained: in fact, interspecific socialisation events may promote allelomimesis, whereby a dog may imitate its human partner in certain actions. In animals, allelomimesis is facilitated by the following factors: (1) the parental relationship, as Paul Leyhausen (1916–1998) points out with respect to the behaviour of cats, which learn mainly from their mothers; (2) the hierarchical position of the demonstrator; (3) the young age of the observer.

The law of exercise, also formulated by Thorndike, tells us that the more a behaviour is performed, the more likely it is to be repeated. A behaviour, in fact, involves a certain pattern of synchronic and diachronic activation, but the repetition of an activation also induces a synaptic reinforcement, not only at the level of the dendritic junction but also because of the thickening of the myelin. The Latin saying *repetita iuvant* thus finds a neurobiological explanation. But what causes exercise in nature? Habits are a fairly universal feature of animal behaviour, so if an individual starts to prey on a particular species, it is likely that, for a thousand reasons, including a certain seeking image, it will continue to do so. The law of exercise can be explained in terms of the probability of expression of a behaviour K given by a set of variables, whereby: $K = E x A / C$, where E is the recall on that behaviour, A is the ease of expression and C is the cost that the subject has to bear. In reality, each of these parameters has an intrinsic and an extrinsic component, so that (1) E can be divided into the subject's sensitivity to recalling that behaviour and evocation, that is to what extent the environment calls for that behaviour; (2) A can be divided into the subject's skill with respect to that behaviour and the environmental conditions that allow or favour it; (3) C can be divided into an objective burden linked to that behaviour and the subjective cost of putting it into practice.

The law of composition says, in short, that everything that is learned or realised simultaneously or sequentially is linked together, so that it will be enough to activate one point of the network to turn on all the others. As said, the first person to hypothesise *neuronal assemblies* was Donald Hebb (cfr. in particular *A Textbook of Psychology* (1958)), for whom coactivating means connecting. We all know that having a dog wear certain things calls up a set of behavioural styles, a real script of behaviours and styles referring to a certain context—civil servants and research workers know this well. After all, all training activities are based on the law of composition because they configure behavioural packages referring to a particular situation. Much of what is known as procedural memory, for example what is triggered when riding a bicycle or playing table tennis, is based on a pattern of elements. In sporting and disciplinary activities, motivations are channelled through a specific expressive framework, that is a certain order of composition. The construction of a topography of a territory, the so-called cognitive map, is an example of a compositional configuration of space. Emotional marking of situations also has to do with composition, because unlike sensitisation it is not a relationship between a stimulus and a response, but an overall picture of the situation, so it does not work by common denominator but through the patterns of family resemblance, following the model illustrated by Ludwig Wittgenstein in *Philosophical Investigation* (1953).

The law of assimilation is based on Piaget's genetic epistemology, whereby, when faced with a new object to which it must relate, an animal tends to generalise, using behaviours that have been already tested in dealing with similar entities to the target in question (1970). This learning model is based on an assessment of similarity that, in certain situations, has to do with conceptualisation, as demonstrated by Richard Herrnstein's (1930–1964) research on pigeons' ability to formulate abstract concepts based on images. This is indeed assimilation, so, for example, the first thing a child does when approaching the bottle is to treat it as if it were the mother's breast. Then, of course, she realises that her behaviour is not perfectly adaptive and, starting from the first assimilative approach, she adjusts through trial and error until she reaches a good level of use efficiency. The life of animals is full of similar situations and the process of assimilation-accommodation, more than any other form of learning, shows how new specialisations can be built through metamorphosis and generation starting from innate endowments. In this case, too, it is difficult to explain this behaviour without embracing the mentalistic assumption, since assimilation involves a comparison by similarity with other known entities, and accommodation entails a selection of useful strategies and a continuous evaluation of the results achieved. In this learning mechanism we already recognise a certain level of creativity that arises by enlarging the space of use of available equipment.

There is a similar situation with regard to solution learning, which involves understanding the structural requirements of a problem, developing solution strategies and checking the results achieved. Once again there can be a number of different situations, which can be divided into (1) objective or explicit problems, like the situation where bananas were placed out of chimpanzees' reach, illustrated by Wolfgang Köhler; (2) subjective or implicit problems, when an animal is called to find a solution without having an objective as a reference point. In the first case there is a clearly visible but unattainable goal, so behaviour is based on trying to bridge the gap or remove the constraint. Kohler's banana problem is challenging because bananas hanging two metres high require a means of reaching them, like stacking crates, whereas if they are two metres outside the cage they require a means of bringing them closer, like a stick. In any case, the problem is strongly target related. In other situations, for example when using experimental practices where an animal is rewarded for solving a problem, the goal is not visible and what motivates the behaviour is expectation. In these cases the animal is more focused on the problem because it is not distracted by the goal. The solution can be carried out through oriented attempts (*heuristics*) and a gradual approach to the challenge, or through internal simulation and immediate solution (*insight*).

The last two laws concern the relationship between learning and activity. In the first case we speak of learning through playful engagement, where play is understood as a behavioural dimension in which motivational expression prevails over the objective to be achieved. The pleasure of the game lies in the expression of one or more motivations that, freed in some ways from the burden of achievement, allow the subject to try freely all possible directions, covering a very wide phenomenological spectrum. There is a very close relationship between play and creativity, so it will be necessary to return to this subject in future chapters. Play certainly has many factors that reinforce learning: (1) first of all it relies on motivational copulas, so the subject is very involved and interested in what it is doing; (2) then the presence of opening emotions, such as joy and a sense of security, undoubtedly favours acquisition and retention in long-term memory; (3) finally, the expressive freedom of play allows us to cross different behavioural areas and explore the space of possibilities through the innovative use of components. However, talking about play in a generic sense is only partially correct, because in reality every game relies on particular motivations and these produce different copulative effects, so we should rather talk about experiential dimensions. The competitive motivation, for example, produces playful (and therefore learning) experiences that are very different from the exploratory one, just as a predatory game teaches and exercises different skills than an athletic one. If it is true that play as a meta-condition encourages expressive freedom and creativity, it is evident that the various motivations open up different ludic experiences.

Something similar can be said about the last law of learning, the one linked to an individual's most common experiences, so that simply acting, following the precious teaching of Maria Montessori (1870–1952), means learning. But it is clear that different animals, producing different behaviours and living in different environments, have non-overlapping experiences. The sensory organs interfacing with the world are different, as are the type of sociality, the niche inhabited, the way of moving, the type of foraging, the kind of reproduction and so on. Let us consider the differences between an animal that runs and one that flies or swims: what changes is the very relationship of one's body with the world, which can be in contact with an open space and a solid surface under one's feet, or surrounded by the liquid medium of the ocean or the current of a stream or, again, swept by the upward currents of air. Life means constant experience, a body that unfolds in action and finds its singular expression in it—an animal relies on implicit creativity in giving meaning to its own expressive here and now. Considering learning as a particular phase in an animal's life is a legacy of the behaviourist school that still features in psychology textbooks today, insisting on the animal's reactive passivity in the face of the *primum movens* of the stimulus. If, on the other hand, we consider the individual as the protagonist of the action, then we discover that learning is the direct consequence of this proactive dialogue with the world.

So what do these ten laws of learning tell us? The answer is simple: they show us that learning is indeed the construction of something new or a non-temporary modification of the behavioural set-up. It is indeed the result of an individual's interaction with the environment, but the new endowments that are created do not respond to a universal and non-specific grammar: they are not constructions that emerge from nothing, but represent the very expression of subjectivity, the internal contents of the subject. In other words, in its interaction with the environment, an animal uses the resources it possesses to construct new opportunities: and this is the process we usually call learning. Notwithstanding the possible configurations described above, we cannot ignore that each species learns according to its own style and not in a universal manner. We must consider the innate endowment as a semi-finished product, but I would go further and say that the innate, rather than an operational tool still to be perfected, represents a lathe that allows each individual to build the tools it actually needs, given the extreme variability of each environment. This is why, in many species, parental care has become not only a tool for optimising the success of the reproductive effort through protection, feeding and care, but also an important source of information for the configuration of the species and for apprenticeship in order to limit, through standard experiences, any excessive individual range. For this reason, a complex innate not only allows for articulated learning, but also requires and provides for complex learning sessions to realise one's own species canon.

Learning and Ontogeny

Some animals, such as mammals and birds, have embarked on a path that is articulated and tortuous, to say the least: by delaying the moment of reproduction, they give rise to a particular period known as the developmental age, a phase in which the organism matures, preparing itself for the sexual and social maturity that will mark the beginning of its reproductive career. This is a complex process that affects all systems, especially the neuro-immuno-endocrine system, which acts as an interface. The developing subject is in a condition that can be defined by two terms: (1) virtuality, since it has redundant interface networks that will then be pruned according to its experiential specification, only actualising some of the possible options; (2) vulnerability, since it is small and with few experiential endowments, which makes it easy prey or prone to accidents, but also, precisely because it lacks basic information, more exposed to alterations in its behavioural profile. Parental care exists to compensate for these problems and to take full advantage of the opportunities made available by this phase.

The developmental age, a period of sexual and social preparation and maturation, represents a sort of apprenticeship aimed at structuring most of the social styles that the subject will exhibit as an adult. In this sense, among mammals the interactive-relational habit is always perfected through (1) maternal teaching, which defines the basic models; (2) playful litter relations, which establish the modes of interaction; (3) the first affiliative and competitive confrontations in the peer group, which in some species determine the bases of social standing. Social training is referred to as *social learning* or primary socialisation, since the organisational information of innate propensities is, so to speak, externalised in parental care. This is an onerous path of development that takes place by virtue of the emotional vitality and motivational exuberance of the young animal, which engages in various ways with its own ethnographic dimension. It does so first through its mother and then through its litter and peer group, thus entering society gradually and without trauma, after having received the appropriate preparation. The parent is a fundamental figure in both mammals and birds, although its contribution is slightly different in nidicolous species compared to nidifugous ones.

Parental care is not only about food, protection and care for the young, but, when it comes to mammals and birds, it is also an irreplaceable apprenticeship that allows the newborn to fully assume its species profile. The first form of learning aimed at identity-building is called *imprinting*: observed by Douglas Spalding (1841–1877) and then studied by Oskar Heinroth (1871–1945) at the beginning of the twentieth century, it was then defined as *pragung* by Konrad Lorenz in 1935. Lorenz described it as an innate predisposition, defined by a precise time window, in which the subject assumes the image of the parent almost like an imprint. Imprinting is therefore a sort of identity print left on the animal which, in this way, will not only seek guidance from the adult—following it, seeking refuge in it in the event of difficulty, assuming its emotional states by osmosis, focusing on the targets it pursues—but also assume the behavioural style and the parameters of belonging that the parent expresses in social and sexual interactions. For example, collared doves (*Streptopelia risoria*), when reared by hand, have shown a sexual interest in human hands as adults. In galliformes, such as chickens (*Gallus gallus*) and turkeys (*Meleagris gallopavo*), as well as in anseriformes, such as ducks (*Anas platyrhynchos*) and geese (*Anser anser*)—species that have precocious offspring, able to follow their mothers immediately after being hatched—the phenomenon is very evident, as there is an immediate attachment to the parent, which is easily noticeable in the behaviour of the hatchlings.

The newly hatched chick immediately tends to seek refuge in one of the entities present in the context: this is usually the parent. The choice is made by means of certain preference criteria, for example the fact that the entity moves increases its attractiveness, as does vocalisation. After having had a certain experience with the chosen entity, which paradoxically could even be an object, the chick tends to remain faithful to it and to be reassured by its presence. If the entity on which it is imprinted is removed, the chick immediately becomes restless and emits insistent chirping calls. Imprinting is a predisposition that, however, relies on a process of learning or experience. There is a sensitive period in which the young can be influenced by imprinting, a sort of timer set at different intervals in each species but which, even within the same species, can vary from a few hours to a few days. On the other hand, as subsequent studies have shown, this process is both innate and learned: in fact, in the presence of several potential attachment referents, the chick is led to recognise the adult of its own species, demonstrating that it has a pre-experiential image of it. Cases of re-imprinting have also been found: if the newborn chick sees, for example, only a human being, it can imprint on the latter; but if, within a short time, it is shown a subject of its own species, it suddenly changes its orientation. In practice, it is as if the chick were saying, "Oops, I made a mistake!".

It is important to distinguish the act of following the parental referent and taking refuge in it, termed *filial imprinting*, from the subsequent pattern of sexual preference, the so-called *sexual imprinting*. These are two different types of orientation that in early studies were considered to be the result of the same process. The former takes place over a much more limited period of time, while the latter not only takes longer, but often needs post-weaning experiential consolidation and, if carried out on a subject of another species, can very easily be reversed on the species of belonging. This phenomenon of species redirection has been demonstrated in Australian zebra finches (*Taeniopygia guttata*) bred by white-rumped munia (*Lonchura striata*). Male finches reared by munia were able to court females of the latter species once they became adults but, if they came across females of their own species, they changed their minds about the correct partner. Studies on sexual imprinting by ethologist Patrick Bateson (1938–2017) on Japanese quail (*Coturnix japonica*) have shown, among other things, that sexual preference is played out in a complex dialectic between belonging (being in the same macrocategory: the species) and diversity (differing a little from the parent), so that orientation also refers to individual characters. According to Bateson, this preference would facilitate exogamy, preventing inbreeding, which, as we know, produces homozygosity and is highly inadvisable.

The tendency to avoid endogamy can also be seen in the Tundra swan (*Cygnus columbianus*), whose individuals have a distinctive face mask per family, a sort of pedigree that establishes their lineage. However, the members of pairs have dissimilar facial masks, which means that they carefully avoid inbreeding. A very different and in some ways opposite situation occurs in the phenomenon of homogamy, known as *assortative mating*. In practice this means non-random assortment between partners, whereby the sexual choice is oriented towards a partner with similar characteristics to the parent. Two or more phenotypes are often found side by side in the same species: for example, in the snow goose (*Anser caerulescens*), which breeds in colonies, there are two plumage colour morphs, white and dark grey. It was noted that 90% of geese born to white or grey parents tended to choose a partner displaying the same plumage colour. Supposing that plumage colour is in line with an allelic difference, assortative mating increases the divergence between phenotypes; in the long term, it may be one of the mechanisms underlying sympatric speciation, which takes place in the same territory. Preference for the similar, which demonstrates a precise orientation in sexual choice, may concern size, vocal repertoire, type of behaviour or smell. In the late 1970s, Patrick Bateson proposed the formula of optimal outbreeding, whereby the balance between these opposing tendencies produces the maintenance of certain characters while avoiding homozygosity.

But what are the most important stimulatory elements in imprinting? In the 1980s, neurobiologists Johnson Mark and Gabriel Horn (1986) discovered that a chick's most important reference point is its head. Orientation therefore seems to be favoured not only by movement and vocalisation, but also by a precise *Gestalt* structure made up of eyes and beak. A particular form of imprinting is learning the parent's voice during hatching, as has been observed in some species of ducks, for example in the wood duck (*Aix sponsa*): the young do not need to see the mother immediately after hatching to follow her calls. In the Australian zebra finch (*Taeniopygia guttata*), the particular harmonic structure of the song, typical of each male, is learned by the young while they are still in the egg. Discovering this was not easy, but it was possible thanks to the great parental disposition of this small estrildid: ethologists exchanged the eggs between two pairs of finches, in such a way that they were brooded by adoptive parents, but restored the newborn chicks to the legitimate parents as soon as they hatched. They discovered that the song of the hatchlings mimicked the notes of the foster parents, the ones who brooded them, and not of the natural parent, even if the latter had welcomed and nourished the chicks from birth to flight, through all the phases of growth. The cuckoo (*Cuculus canorus*) presents another variant of this phenomenon. Since the cuckoo has adopted the strategy of parental parasitism, it would be disastrous if it were to direct its sexual preference towards the parental host: this species is not susceptible to sexual imprinting, which is why sexual choice is determined by an innate *Gestalt*.

In mammals, parental training is even more complex. First of all, the young very often do not realise the potential danger around them: for example, they may follow their parent when it is preparing for a dangerous hunt and then, as happens for example in lions (*Panthera leo*), the mother must intervene; or they may swallow dangerous substances, as in the case of young gorillas (*Gorilla gorilla*), which tend to put any vegetable in their mouths, even toxic ones, so that the mother is forced to literally snatch the food out of their hands. At other times, the parents' task is the opposite, that is, to calm the young or push them towards an experience that terrifies them, as is the case with sea lions (*Neophoca cinerea*): pups put up strong resistance to entering the water and it is up to the parent to help them overcome the first moment of fear. Likewise, many primates flee in panic in front of any snake, and it is the mother who teaches them to distinguish the really dangerous ones. Other times, the relationship with the mother serves to teach the young some very complex behaviours, for example how to use tools or how to face a hunting session. Mother cats bring live mice to their kittens for the sole purpose of teaching them the difficult art of hunting.

In this last case we have a real lesson, as the kitten does not merely observe its mother while she hunts, but witnesses an action carried out expressly for training purposes. Other members of the felid family behave in a similar manner: for instance, the cheetah (*Acinonyx jubatus*) mother brings her cubs a slightly knocked-out gazelle to train them, because in the future they will have to compete with these top sprinters. The otter (*Lutra lutra*) mom must not only give swimming lessons but also teach its young the art of underwater fishing. An even more complex task is that of the sea otter (*Enhydra lutris*), which has to pass on to her pups the art of using stones to open bivalves. Teaching the use of tools is common practice among chimpanzees (*Pan troglodytes*), whose young learn not only how to fish for termites with specially prepared sticks, but also how to use stones to crack nuts. Ontogeny, therefore, is a fundamental phase in animals that go through a developmental age, where learning is never a passive exposure to stimuli but the expression of an accompanying relationship with the world and towards the emergence of one's own personality.

Works Cited

Changeux, Jean-Pierre. 1986. *Neuronal Man*. New York: Oxford University Press.

Dehaene, Stanislas. 2020. *How We Learn, Why Brain Learn Better than Any Machine*. New York: Viking.

Edelman, Gerald M. 1987. *Neural Darwinism*. New York: Oxford University Press.

Gallagher, Shaun, and Zahavi Dan. 2008. *The Phenomenological Mind*. New York: Routledge.

Hebb, Donald O. 1958. *Textbook of Psychology*. Philadelphia: Saunders.

Inoue, Sana, and Tetsuro Matsuzawa. 2007. Working Memory of Numerals in Chimpanzees. *Current Biology* 17 (23): 1004–1005.

John, Garcia, and Robert A. Koelling. 1966. Relation of Cue to Consequence in Avoidance Learning. *Psychonomic Science* 4: 123–124.
Johnson, Mark, and Gabriel Horn. 1986. An Analysis of a Predisposition in the Domestic Chick. *Behavioural Brain Research* 20 (1): 108–109.
Kaminski, Juliane, Josep Call, and Julia Fischer. 2004. Word Learning in a Domestic Dog: Evidence for "Fast Mapping". *Science* 304 (5677): 1682–1683.
Kuhn, Thomas. 1957. *The Copernican Revolution.* Cambridge, MA: Harvard University Press.
LeDoux, Joseph. 2002. *Synaptic Self: How Our Brains Become Who We Are.* New York: Viking.
Lehrman, Daniel. 1953. A Critique of Konrad Lorenz's Theory of Instinctive Behavior. *The Quarterly Review of Biology* 28: 337–363.
Lorenz, Konrad. 1981. *Leben ist Lernen: von Immanuel Kant zu Konrad Lorenz.* München: Piper.
Lorenz, Konrad, and Franz Kreuzer. 1973. *Die Rückseite des Spiegels.* München: Piper.
Mainardi, Danilo. 1992. *Dizionario di Etologia.* Torino: Einaudi.
Piaget, Jean. 1970. *Psychologie et epistemologie.* Paris: Gonthier.
Rescorla, Robert A., and Allan R. Wagner. 1972. A Theory of Pavlovian Conditioning: Variations in the Effectiveness of Reinforcement and Nonreinforcement. In *Classical Conditioning II*, ed. Abraham H. Black and William F. Prokasy, 64–99. Appleton-Century- Crofts.
Wittgenstein, Ludwig. 1953. *Philosophical Investigation.* New York: Macmillan.

CHAPTER 9

Animal Orientation

Premise

Creativity belongs to the subject. In a sense, it is inherent in all its expressions, and indicates that spontaneous vocation to interpret the script of one's existence in a singular way in the here and now. The individual is the result of various stages of ontogeny as well as the bearer of an agglomeration of stories: such is, after all, the phylogenetic equipment that somehow sums up very different dimensions of life—think of Neil Shubin's essay *Your Inner Fish* (2008) with respect to the expressions of the human body. Against this background, the subject soon becomes the full protagonist of its own journey through the world. Creativity is like the air we breathe: not something exceptional, but a fundamental element in our dialogue with reality, which calls for constant adjustments as well as the ability to innovate, revise, experiment, but above all to seize the unexpected. Be it a challenge, an opportunity or a danger, what happens does not give the subject time to fit it into a preformed canon, but requires instant inventiveness, though obviously starting from the information material available. I repeat: creativity is not a mere reiteration of heritage, but neither is it a construction *ex nihilo*. As Henri Bergson put it in *L'évolution creatrice* (1907), a living being is not a machine that always repeats itself, but continuous and incessant change. Hence a magmatic image of the *bios*: life is always innovating; it systematically brings novelty to its very manifestation and expression, and it grows, renewing the past by projecting into the

R. Marchesini, *The Creative Animal*,
https://doi.org/10.1007/978-3-031-07414-1_9

future. In line with the concept of duration, so dear to Bergson, time becomes invention, the creation of forms, the continuous elaboration of the new, as we read in *La pensée et le mouvant* (1934, 19–20).

Animal creativity manifests itself in the tiniest things, and no great elaborative skills are required to put into practice this innovative, interpretative, poetic vocation of animal expression. What is needed is not consciousness, but rather the capacity to emerge from the phenomenal depths through distinction, self-reference, ownership and unity, as mentioned above when discussing the four levels of subjectivity. Phylogeny equips the animal with orientation tools that are ready for action, in the form of stable or evolutionary schemes, but above all cognitive resources to elaborate new strategies and glimpse opportunities hidden from the surface of tradition. Creativity, therefore, manifests itself in a plurality of ways as a process of idea formation that transfigures the subject's intersection with external reality. Creativity undoubtedly belongs to the subject, to its capacity for *poiesis*, but also to reality, which presents itself as a space of possibilities to be configured into a form. An animal living in a certain environment is not faced with a table explicitly laid out with opportunities to be seized and risks to be avoided. It is more like carving a slab of marble, which certainly presents constraints but otherwise leaves the artist the honour/duty of creating the statue they wish. The poetic animal navigates its way along this journey through creativity.

Not all individuals have the same creative skills. However, by saying this I do not wish to refer to some hierarchical scale of ingenuity, but rather to a diversity in terms of flair, so that even to speak of creativity in the singular is arguably wrong. Creativity is the ability to bring out something new in the warp of reality, most often through daring connections, at other times by discovering unexpected room for manoeuvre. One of the first to explore the territory of creativity was the Scottish Presbyterian William Duff (1732–1815), who published *An Essay on Original Genius* (1767), but I also wish to recall Cesare Lombroso's *Genio e Follia* (1864) and Francis Galton's *Hereditary Genius* (1869). In *Science and method* (1914), Henri Poincaré defined creativity as the ability to find new and useful connections between distant elements. On the other hand, the different personalities within a population, with more timid and traditionalist subjects and more open-minded and innovative ones, undoubtedly have an evolutionary value. Besides, when we talk about the way in which an animal navigates the world, it is evident that we cannot speak of total predictability, as in a chain of dominoes, but of probability, given the presence of strange attractors that influence the outcomes.

Animal creativity is often understood as the ability to find new solutions to the trivial problems of survival. This is the case with the insight illustrated by Wolfagang Köhler in 1917 with reference to the admirable solving skills of the chimpanzee Sultan, kept at the Tenerife zoological station, or the discoveries of the Japanese macaque Imo (*Macaca fuscata*) from the island of Kojima, studied in the 1950s and described by Kinji Imanishi. We find it in the amazing technical skills of Betty, the New Caledonian crow (*Corvus moneduloides*), which devised a hook to retrieve food, studied at Oxford University, or in the ingenious strategies of dolphins that hold a sponge in their mouths to search the seabed without injuring themselves.

These examples of inventions, use of tools, modification of the environment and implementation of cultural traditions will be discussed in the next chapter. In the numerous essays on animal cognition, there are frequent references to surprising behaviours, developed by talented subjects and observed in their individual specificity. The importance of the individual factor is analysed in *Animal Personalities*, edited by Claudio Carere and Dario Maestripieri (2013). The authors underline that "the concept of personality has been used to refer to the existence of behavioral and physiological differences among individuals of the same species, which are stable over time and across different contexts or situations" (1). The expressive, ethnozoological and even evolutionary significance of the animal personality factor is central to explain how cultural traditions and shifts in the phylogenetic trajectory of the species develop through the actions of particularly creative individuals.

However, animal creativity, although less surprising or less impactful on the development and evolution of a population, represents the most ordinary character of animal behaviour. Highlighting some of its high-profile aspects is certainly very useful, but I worry that—as with the reference to consciousness—it may be used as a smokescreen, an excuse to not question the basic paradigm used to explain animal behaviour. Creativity is common currency, extremely widespread and devoid of the aura of speciality that some would like to attribute to it. We could say that creativity is nothing but the expression of the cognitive functions that enable the subject to work on situations instead of simply applying ready-made solutions. Therefore, the first thing to do is highlight the inconsistencies of the explanatory model in force with regard to this C-factor (creativity), which has long been misunderstood until the emergence of striking manifestations put it on the table. On the other hand, just as with animal consciousness, it is necessary to get out of the dead end of exceptionalism and bring creativity back to the realm of ordinary animal functions.

The concept of marginality, which relegates creativity to a kind of exception or deviation from the rules, unable to challenge the current mechanistic model, no longer holds. Creativity is not an occasional expression, an extraordinary wave in the calm sea of repetitiveness and automatisms. I believe, therefore, that a reformulation of the model in the Bergsonian sense is overdue, admitting the statutory role of creativity in terms of a natural manifestation of the living. If we continue to explain animal expression through reactive outputs activated by stimuli, we have no prerequisites for exceptional performances. Creativity calls for a paradigm shift involving the entire spectrum of animal behaviour. It is not possible to cling to the automaton model by merely describing these unusual illuminations as if they were the exception in a boring repetitive flow. On closer inspection, without a serious admission of *ownership of choice* on the part of the individual in its orientation, we could not even understand the ordinary aspects of its life. Over the last twenty years, research has focused on personality and individual interpretation as strategic factors in assessing the expressive space of the species. The anecdotal, therefore, can no longer remain on a descriptive level, but requires a reformulation of the explanatory model.

In fact, we can no longer ignore the evidence accumulated over the last few decades, which shows that the species-specific repertoire is not a standardised and homologated habitus, but rather a space of expressive virtuality, a range bordered by constraints but which can be interpreted in an almost infinite number of ways. The individual acts within that window and inevitably relies on creativity to orient itself in the labyrinth of possible choices. On the other hand, the image of the Heideggerian animal—captivated in the fruition of the world, stunned by trivial needs, relegated to a perspective corner, limited to survival and reproduction—is slowly disappearing even in the humanistic environment. When evaluating the concept of *openness*, we should go beyond both Rainer Maria Rilke's (1875–1926) Duine poetics and Martin Heidegger's interpretation. Animality is a dimension that encompasses us and involves an open perspective on the unpredictable, built on the creation of the future. The emergence of the C-factor in twentieth-century primatology turned the tables, exposing individual protagonism—something that was soon found in all subsequent research with respect to the animal's orientation. Driven by a canon of preferences and by the ability to project itself into aims and objectives, the individual has regained its ownership.

Creativity is not a means of survival, but a way of life. When zoologist Desmond Morris handed Congo, a two-year-old chimpanzee, a piece of paper and a pencil in 1956, he could hardly have imagined Congo's pictorial flair. Congo's work was praised by internationally renowned painters such as Miró and Dalí, and Picasso even bought one of Congo's paintings. In 1957 there was even an exhibition at the Institute of Contemporary Arts in London curated by Morris himself, and in 2005 some of Congo's paintings were sold for almost £15,000. Morris published two essays on this subject: *The Biology of Art* (1962) and *The Artistic Ape* (2013). It is not surprising that the artistic sense has a common animal origin, and this view is not only found in this Morris' work. In the second half of the twentieth century, the German biologist Bernhard Rensch (1900–1990) carried out research showing that certain animals had preferences for certain forms, comparing certain primates, such as capuchin monkeys and guenons, and corvids, in particular crows and ravens. As Giorgio Celli points out in his essay *Arte e biologia: una scommessa evoluzionista* (2004, 217), the fact that one can speak of a preference for forms, that is an animal aesthetics, supports the hypothesis of "an emancipation from utility, of a perception that matters as such".

Moreover, Charles Darwin himself, in his 1871 essay *The Descent of Man and Selection in Relation to Sex*, had invoked a species-specific conception of beauty to explain sexual dimorphism, and was harshly criticised for this. On the other hand, the aesthetic orientation of bowerbirds in decorating their patios is unquestionable. As Donald Griffin reports in *Animal Minds* (1992, 110–121), the eighteen species of these extraordinary birds construct their arbours using aesthetic criteria of symmetry, colour combination, colour preference and the use of natural paints applied with tools. Bowerbirds use feathers, berries, leaves, snail shells and flowers to decorate their habitats; Jared Diamond's studies (1982, 297, 99–102; 1986, 17, 17–37) of populations of *Amblyornis inornatus* showed that decorations of the same colour tend to be grouped together, suggesting the birds' preference for certain ornamental patterns over others.

These behaviours show great capacities for improvisation and creativity, so it is not possible to speak of creativity simply by referring to practical innovations and concrete problems to be solved, for creativity has to do with a system of preferences. Perhaps a revision of the explanatory paradigm should start from this aspect, bearing in mind that the classic question of why animals express themselves can no longer remain confined within Tinbergen's classic explanation, which refers to how they conform to their environment and the underlying physiological mechanisms. This requires serious reflection on what an animal is feeling, how it is oriented with respect to its system of preferences and what objectives it wants to achieve. Cognitive characteristics, like all other expressive components, have a conformative history and a physiological substratum—no one doubts this—but at the same time they place the subject in a particular condition: a point of view on the world which gives rise to creativity. This paradigm shift can no longer be postponed.

The Four Questions to Explain Expression

The subjective characteristics of animals have led to the emergence of protagonism in the exercise of an individual's presence in the world—an existential character that does not fit in well with the traditional mechanistic image. As mentioned above, the explanatory paradigm that took shape in the first decades of the twentieth century, based on the concepts of instinct and conditioning, was reluctant to accept any form of mentalistic or psychological description of animals and relied on concepts that carefully avoided referring to expressions of preference or purpose in animal behaviour. A good example of this explanatory principle is the Four Questions model formulated in 1963 by Niko Tinbergen. This involves (1) two remote causes relating to the evolutionary history of the species and the function of the given behaviour in adaptation and (2) two proximate causes relating to the functional mechanisms underlying that behaviour and the ontogenetic developmental process in the individual. What I am asking is whether this explanatory model is still valid in the light of new discoveries in terms of animal cognition. Indeed, I have noticed a certain resilience of the traditional paradigm, which is forcibly trying to fit the new findings, which clearly indicate the mentalistic character of behaviour, into a conceptual framework that remains anchored within a non-mentalistic vision. This is the task I have set myself as an analysis of *Etologia Filosofica* (Marchesini, 2016), namely to reform the ontological paradigm of animality.

The first two remote causes recalled by Tinbergen in the Four Questions show how animal behaviour is affected by Darwinian evolutionism: firstly, because each species is the result of a specific phylogenetic trajectory, so that the steps taken by a taxonomic ramification influence subsequent ones; secondly, because each character is subjected to the scrutiny of selective pressures rewarding the phenotypic expressions that make the individual more replicative. The characteristics of living things, elective sensitivities like preferences and canonical behaviours like cognitive abilities all result from a specific history and are affected by the meticulous work of selective pressures, creating an admirable conformation to the species' environment and lifestyle. Obviously, variables do not only concern environmental parameters, for which we usually speak of adaptation, but also concern a large number of factors. Smaller species often need high-quality sustenance to cope with their high metabolic rate, requiring a greater ability to locate food and sometimes to make food reserves. In such species, good spatial memory is of greater importance. Species living in complex social groups, on the other hand, need better episodic memory, capable of remembering all the relationships within the group.

Often, particular performances derive from a different development of body parts (allometry). This is also true for the brain, whereby certain cognitive endowments are due to an increase in the hippocampus or telencephalon compared to what we would expect from the overall size of the body. Explaining why certain features are present within a species risks falling into circularity or tautology, not to mention the Panglossian bias of Rudyard Kipling's *Just So Stories* (1912) denounced by Stephen J. Gould. Besides, although Dennett's idea of evolution as an algorithm, as expressed in the essay *Darwin's Dangerous Idea* (1995), is unacceptable, it is clear that the characteristics of animal behaviour cannot be understood outside the Darwinian evolutionary framework. This also applies to cognitive traits, which are endowments available to the individual and tailored to the specific challenges that the species or, better, the evolutionary trajectory had to face. In *Natural Selection and the Emergence of the Mind* (1978), Karl Popper points out that, since Charles Darwin views the evolution of mental faculties in the animal world as a result of natural selection, it follows that such powers are of central significance in behaviour.

The two hypotheses commonly used to explain the evolution of cognitive endowments in the animal world are (1) the foraging intelligence hypothesis, according to which brain development is directly proportional to the species' difficulties in obtaining food, either generally in the living environment or at specific seasonal times; (2) the social intelligence hypothesis, which suggests that animals living in large organised groups with complex role articulations and hierarchies must keep track of individual relationships. I believe that these two hypotheses represent only one aspect of the evolutionary mechanisms of animal cognition. Other aspects may concern (3) the need to monitor living and moving spaces, as is the case with migrants; (4) the need to modify the environment to make it suitable to one's needs, as builders do; (5) the need to make tools to increase the extent of one's intervention in the world, not necessarily in terms of food; (6) the need to pass on knowledge in a transgenerational way and (7) the strategy of colonising new territories and therefore modifying one's behaviour by adopting new styles. More generally, we could say that every expressive characteristic is favoured by cognitive skills.

What is certain is that we cannot speak of intelligence in a universal sense, because each species has a cognitive set-up that is shaped by particular needs. There are countless examples of this, because the challenges faced by the various species are very different. Sara J. Shettleworth, in her essay *Cognition, Evolution and Behavior*, highlights this by showing that: "birds that migrate might be expected to use memory and spatial learning more than relatively sedentary populations, not necessarily because they actually need learning to migrate, but because they need to acquire spatial and other information about each of the places where they spend a few months at the ends of their travels, and perhaps at stopovers along the way" (2010, 41). Different cognitive abilities may therefore relate to (1) the type of learning specialisation, whereby there can be mapping skills rather than comprehension abilities; (2) the type of memory, which includes working, procedural, spatial, episodic and conceptual abilities; (3) the type of discernment, such as the ability to detect even stylised targets, to recognise shapes, to follow pattern generalisations; (4) the way of interacting with the environment and the gradient of niche construction; (5) social and empathic capacities, that is the ability to navigate with ease the network of complex relations in large groups; (6) problem-solving or operational capacities, the construction of strategies and tools; (7) the type of creativity or innovation, that is in which areas of the repertoire there is the greatest flexibility.

When we speak of cognitive capacities and, more generally, of the physiology of behaviour, we must never forget that in nature the principle of diversification applies, so that the evolutionary trajectories in the various taxonomic levels—right up to the last level, that of individual personality—are always under the banner of conformation. This produces a particular correlation that synthesises in the body the adaptive path of that individual. The phenomenon of specialisation is incredibly explicit in certain physiologies, such as the mechanisms of preparation for migration in the bird world. As Scott Weidensaul reports in the essay *A World on the*

Wing. The Global Odyssey of Migratory Birds (2021), in the different phases of sedentary and migratory flight, birds drastically modify many areas of the body. For example, they can (1) increase the neuronal density in the hippocampus, to improve spatial perception, as in the case of the dark-eyed junco (*Junco hyemalis*); (2) increase the area of the caudolateral nidopallium, which oversees executive thought, in the case of diurnal migrants, such as turtle doves or pigeons; (3) increase the number of red blood cells so as to acquire more oxygen to compensate for the rarefied atmosphere, as in the case of the godwit (*Limosa gen.*); (4) contract digestive organs like the gizzard and the intestines, which will not be needed during travel, and double the mass of pectoral muscles; (5) burn fat reserves ten times more efficiently than mammals due to high levels of mitochondrial enzymes.

It is clear that each species has its own repertoire linked to different variables, such as the type of diet, social density, living environment, type of predator or competitor, sedentary nature or tendency to move or migrate in different seasons, type of reproduction, tendency to colonise or maintain a balance in a certain ecosystem and so on. An animal is born already predisposed to lean towards specific food stimuli, to feel particular motivational calls, to perform certain behaviours on the basis of seasonality, to express certain choreograms during courtship, to complete the entire reproductive cycle from copulation to the full autonomy of its young and to respond emotionally to certain events. Cognitive functions are also linked to these innate predispositions, so that some species have a great capacity for spatial memory, like animals that stockpile food or move in intricate tunnels, while others have a remarkable capacity for problem-solving, like all the great foraging soloists; some have a particularly acute social intelligence because they are part of groups with complex relational systems and dynamics, like baboons and chimpanzees, while other species easily use objects as tools, like corvids and psittacidae. Cognitive endowments therefore respond to the same adaptive mechanisms as the other traits in a species.

Phylogeny, as the development of a certain trajectory—for example that of birds or mammals—and likewise the refinement of a species through the action of particular selective pressures, provides the individual with a certain basic equipment that provides the first set of tools to embark on its journey in the world. In 1950, Tinbergen's elegant experiment with variously coloured beaks showed how seagull chicks responded innately to certain colour configurations with a greater number of begging pecks

(Tinbergen and Perdeck, 3, 1950, I, 1–39). As I pointed out in the previous chapters, one should not think of innate endowments as constraints or automatisms, which would reduce the expressive variability of the subject. Indeed, this phylogenetic equipment represents the raw material both of learning and of the expressive flexibility (non-repetitiveness) that marks creative behaviour. The innate, therefore, creates phenomenological room for individual action, which can be defined as the range of creative freedom available to the subject.

Let's now turn to the proximate causes of behaviour in the Four Questions. These show that every expression is the result of a particular physiological state and an ontogenetic process, factors that involve the whole body—understood as a system that develops an internal unity based on its relationship with the world. For this reason it is no longer possible to think of intelligence as an incorporeal entity because, as Dick Swaab points out in *Wij zijn ons brein* (2010), there is no gap between how we think and the underlying neurophysiological mechanisms. When we wonder why an animal at a certain time and in a certain situation tends to perform a particular behaviour or attitude rather than another, we are also wondering what mechanisms underlie this predisposition. And the mechanisms inevitably refer back to bodily functions, whether they are affected by internal phasic processes or called upon by external factors. Can we compare these mechanisms to those at play in our man-made machines? No, and the reasons are linked to the plurality of levels, whereby physiology influences the psychological condition and the other way around. Moreover, these mechanisms are anything but linear and additional, but rely on integrations on several levels, so that it would be difficult to extrapolate a Laplacian diagnostic or predictive account even if we had all the variables at hand. We can only work on probability indices, approximate pictures, propensity conditions and general indications.

For example, studying the various neuromodulators and claiming to give them a precise meaning in terms of their effect on behaviour lead us to make blatant errors, such as the idea that oxytocin is the hormone of love and dopamine is the hormone of gratification. It is true that, in mammals, oxytocin stimulates affiliation, encourages grooming behaviour and reduces social distance, but it is also recognised that this neuromodulator increases an individual's sense of security and, in combination with dopamine, can produce social aggression. Similarly, dopamine can determine exploratory and industrious behaviours, but it can also underlie compulsive and psychotic behaviours that lock the individual into repetitive attitudes and stereotypes. A similar situation applies to serotonin, which is considered a mood regulator, but which can increase glutamate release resulting in the paradoxical effects of arousal, hallucination and disinhibition. The release of endorphins tends to produce drowsiness, but when associated with adrenaline they cause states of arousal, just as, while it is true that endocannabinoids promote states of play and playful learning, in association with endorphins they lower hippocampal capacity for episodic memorisation—see Neil R. Carlson's *Physiology of Behavior* (1977).

In the same way, if it is true that the brain is characterised by different areas with precise specialisations—in mammals, for example, lateralisation provides further functional specifications in the right hemisphere with respect to the left—as well as precise pathways with receptors for particular neurotransmitters, nevertheless the system presents a strong global integration. A nucleus such as the amygdala undoubtedly has its own functional specialisation, as does the insula or the *locus coeruleus*, but it is equally true that in association with other nuclei it gives rise to changes in functional significance. On the other hand, the activation of the neural connectome is always networked, to the point that, to avoid epileptic outbursts, each activation of a functional pattern also involves the release of the GABA inhibitor on the surrounding areas. And this neuroactivation through synchronic or sequential patterns has a precise value in the economy of behaviour. In social activities, for example, animals need to have a good knowledge of their mates, the relationships between them, the overall style of the group, as well as semiotic, interactional and cultural knowledge. When a chimpanzee or baboon wants to interact with another individual in order to obtain something or strengthen the bond, several areas of its brain must be activated in a conjugative way, for example: the prefrontal components of decision-making react with the hippocampal components of episodic memory, with the dopaminergic pathways of expectation and with the temporo-parietal area (rTPJ) of the social acceptability of an expression.

Behaviour is always the result of a developmental process, and ontogeny, seen as a proximate cause in the Four Questions, shows both embryogenetic and experiential aspects. After all, embryonic formation also occurs in very different versions in different animals, such as metamorphosis in insects, hatching in birds or internal gestation in mammals. Species with parental care must have developed great capacity for social learning during development and a strong tendency to play as a metamotivational condition. Learning, moreover, is linked to working memory, which varies between species. This is demonstrated, for example, by the chimpanzee Aymu, which proved able to memorise an image with nine numbers on the screen and tap them in ascending order after they were hidden by white squares (Inoue and Matsuzawa, 17, 2007, 23, 1004–5). In tests of this kind, the chimpanzee has always resoundingly beaten its human opponents. What significance, in terms of fitness value, does working memory have for an animal that navigates in a forest habitat moving from one tree to another? This is the question we should ask ourselves, avoiding the obsessive comparison with the human being, whose innate endowments and way of learning are different.

Similarly, there are significant differences in the development of the hippocampus, responsible for spatial and biographical memory, in different species and even at different times in the same species. Attention capacity is a cognitive factor that characterises different species on the basis of discernment, categorisation, development of seeking images, priming effect of interaction between stimuli and habituation. Some species are stubborn and competitive when confronted with a problem, others work better in collaboration and others yet appear to give up in certain situations, such as social ones. In certain conditions, the degree of attention and concentration is made difficult by environmental criticalities: for example, many animals need to spread their attention over perspectives of interest, such as searching for a territory or food, and vigilance, to watch out for possible dangers. As Shettleworth notes, "An animal that is foraging cannot wait until it is satiated to check for predators but should continuously divide attention between foraging and vigilance" (2010, 91). The difficulty shown in this attentional breakdown explicitly demonstrates the cognitive-systemic character of the process itself.

One of the relevant factors in the learning process is the activation of emotional and motivational affective systems that strengthen the experiential copula, creating a higher level of involvement, and retention in long-term memory. There is a long tradition of active involvement of animals in learning-based experiential processes, ranging from Maria Montessori's pedagogical applications at the beginning of the twentieth century to early cognitive science research from the 1960s onwards—such as the carousel experiment conducted on cats by Richard Held and Alan Hein, which demonstrated that active exploration of the world is essential for the development of vision. As Stanislaw Dehaene points out in the essay *How We Learn*, one learns when one is motivated by the outcome to be achieved (2020, 215–220). Knowing the motivational diversity of different species, it is understandable that the ontogenetic pathways of learning must take this into account. The emotional system also plays a very important role in the development of memories, just as some related proteins—for example CREB-1—are known to form new synaptic connections through the activation of genes that induce the synthesis of the necessary proteins (McGaugh, 1992; Milner et al., 1998). There is a close relationship between ontogeny, understood as a developmental process based on the relationship with the world, and neurophysiological conformation. As Daniel Siegel points out in *The Developing Mind* (1999, 47–50), repetitive and emotionally intense experiences have a major impact on the formation of neuronal circuits.

At this point, however, I wonder whether Tinbergen's Four Questions are able to fully account for behaviour and the reference to cognitive activities. In my opinion, the four explanatory matrices are fundamental and indispensable, but not exhaustive. In particular, they are not able to address the specific orientation of subjectivity, which, although affected by this explanation, appears as a psychological dimension in its own right, based on preference and projection. The subject, in fact, is characterised by a particular background which, while resulting from several aspects and often bearing ambivalences, is translated into a singular presence in the here and now. The psychological state defines what the subject is feeling at that moment, its viewpoint and prospection. If I ask myself why my cat is playing with a ball at the moment, I can certainly answer by sticking to the phylogenetic configurations that have sanctioned its predatory tendencies and note that its experiential pathway has given it the ball as a surrogate for prey, just as I can hypothesise a certain geography of neuroendocrine activation and perhaps visualise it with neuroimaging techniques. Yet I feel that doing so would circumvent my cat's motives, perhaps out of a need to treat it aseptically as a phenomenon that requires no psychological explanation.

The psychological state of the subject is the joint result of physiological functions but also emerges as an expression of a self that navigates the world according to preferences. As noted in reference to mammals by John Z. Young in *Philosophy and the Brain*, the cerebral cortex is certainly subdivided into distinct parts with different functions, but the strong interaction between these individual parts and the rest of the brain produces an overriding condition (1987, 101–109). When we observe play in baby mammals, we notice the ability to integrate different attitudes and hybridise affective states, motivational and emotional alternations, and attributions of meaning that change in the course of play. What stands out is their singular protagonism in action, following their flow of preferences like a surfer follows the waves. This shows the condition of individual presence: feeling something, leaning towards something because it is pleasant or aesthetically satisfying, because it expresses the Self—an aspect that is not part of Tinbergen's Four Questions. An animal's behaviour is a continuous dialogue with the outside world, an active participation in action where pleasure acts as a compass and planning as a navigation system. If I were asked why an animal at a certain moment performs a certain behaviour, I would have no doubt in answering "because it takes pleasure in it and because it has planned it". Whether a certain activity lowers a condition of unpleasantness or produces additional pleasure, the main helmsman guiding the expressive ship of the individual is the hedonic system.

The explanatory considerations of the Dutch ethologist remain fundamental to understand the conformation and functioning of animal predicates, but in my opinion there is something missing: an explanatory Fifth Factor accounting for the individual's orientation, choices and expectations, based on a criterion of preference. We humans are also the result of conformative processes, and we too have precise physiological conditions underlying our behaviour. And yet, if I were to say that I am afraid or in love because my physiological systems respond in a certain way, by virtue of an evolutionary and developmental conformation, I would feel that my explanation, though correct, would be insufficient. At the same time, if we admit that the subject, as propositional, projects itself into a future condition, we will inevitably require a Sixth Explanatory Factor, based on purposes and plans, on imagined, believed, feared or yearned-for conditions that together produce an intrinsic purpose. I believe, therefore, that it is not possible to understand animal behaviour by sticking to Tinbergen's Four Questions and that this has been done in order to preserve the paradigm of the animal as a machine governed by automatisms. Tinbergen's explanatory considerations, I think, are important but not sufficient because, when the psychological condition of the subject comes into play—its feelings and projections—we are faced with a Self that is the protagonist of its choices.

The Fifth Factor: Having Preferences

Accustomed to thinking in mechanistic terms, we believe that the psychological condition has no reflection on physiology, but is merely an expression of it, just as we think ontogeny is nothing but the assumption of new automatisms, in defiance of any form of choice-based orientation. In reality, the subject is the master helmsman of its journey through the world, so the explanation must be brought back to the basis of experience. It must be traced back to a dimension that has been avoided at all costs, as if it were a taboo: *what the animal feels*. I do not doubt that the psychological condition depends on the physiological state, but these are two different levels of reality, and they influence each other as such. In his *In Defense of Mentalism and Emergent Interaction* (1991, 221–226), Roger Sperry points out that mental properties emerge from the activity of the nervous system, but also hold causal control over the brain. The emergentist conception, otherwise defined as property dualism, therefore does not envisage a *res cogitans*, but neither does it negate the presence of a mental realm capable of acting on the levels below. It should also be remembered that the psychological realm descends from the whole body and not just the nervous system, so the influences of the other apparatuses should not be considered mere internal variables, but co-factors responsible for the integration from which the mind emerges. They too are influenced by the psychological orientation of the subject.

An animal's life is based on continuous choices that must necessarily be based on a system of preferences. We have seen what produced this criterion when discussing phylogeny and ontogeny; what remains to be explained is what it causes in the subject. The aesthetic canon puts the individual in direct contact with a changing world that can confirm or refuse its preferences, or perhaps encourage it to look deeper. The research conducted by Bernhard Rensch and Jared Diamond has shown that there are numerous examples of aesthetic orientation in the animal world. The fact that these preferences can be traced to evolutionary meanings does not change the focus of my argument, because there is no doubt that each character has been subjected to the test of adaptive compatibility. I speak only of compatibility and not of the expression of a specific need, in accordance with the considerations made by Stephen J. Gould and Richard Lewontin in *The Spandrels of San Marco and the Panglossian Paradigm: A Critique of the Adaptationist Programme* (1979). I am specifically interested in how the individual feels about the expression of preferences. After all, we too are attracted to sweet, salty and fatty foods, because from an evolutionary point of view these substances had a high nutritional value, but this does not mean that we do not follow pure pleasure cues when eating such foods.

It is this feeling, the orientation based on what one feels, that influences an individual's choices—not adaptive calculation. The subject is an entity with preferences and judgement, drawn by the compass of pleasure, a judge that comes into play in all of the animal's expressions. One does not eat for nutritional needs, but because one is hungry and craving is an unpleasant sensation that demands to be compensated through the act of eating. This applies to all animal behaviour, from courtship to mating, from parental care to affiliative or competitive behaviour. Evidence of neurobiological activation of pleasure centres and the release of particular neuromodulators such as dopamine and endorphins confirm the hedonic significance of behaviour. Allogrooming in primates, for example, performs precise social regulation tasks, but we also know that during this behaviour both subjects are flooded with endorphin flows. As mentioned above, for Darwin the sense of beauty is related to utility; we also get this from Amotz Zahavi's analysis *The Handicap Principle* (1997), which sees aesthetic examination as a test of performativity. At the same time, the appreciation of beauty should be regarded as a *genuine aesthetic sense*, as it was defined by the father of evolutionism. Wolfgang Welsch, in his essay *The Animal Origin of Aesthetics* (2013), argues that aesthetics originated in the animal world, so that the cultural evolution that occurred in humans must be traced back to a root already present in other species. The author emphasises how the analysis of animal aesthetics can help us to better understand the aesthetic orientation in human culture, avoiding the prejudice that humanity can and should only be understood from within.

In contrast to Darwin, who spoke of a genuine sense of aesthetics, there is often a tendency to relate every preference to a fitness value, when instead it can only define the style of a species later on, by means of selective screening. Obviously, not all orientation criteria are compatible, but sometimes a character can be neutral without causing an adaptive disadvantage. In this sense, according to Darwinian principles, sexual selection could exploit these spaces of freedom and give the evolutionary trajectory of a species new directions. Again, according to Darwin, aesthetic judgement is based on the pleasure caused by a certain feedback. In this regard, Welsch makes an interesting point (53) when he emphasises that even if beauty and fitness completely overlapped, the fact remains that females, when faced with the male display of song, plumage and choreography, must first of all perceive beautiful appearance as such. Any attempt to evade the aesthetic phenomenon, in the author's opinion, is doomed to failure because, whatever adaptive advantages or physiological components are in place, these are achieved through the perception and evaluation of beauty. In a certain sense, we can say that even the phenomenon of so-called supernormal releasers, which causes many birds to prefer a larger or spotted egg, can be traced back to an aesthetic criterion.

I do not agree, however, that animal aesthetics is exclusively a matter of sexual selection, although the latter is certainly relevant, and above all I do not believe it is confined to visual and acoustic sensation. Perception, in fact, does not merely provide a glimpse of the world; it does not merely give the subject an objective frame of reference—*what is out there*, the sense data of experience—but is always related to a first evaluation of perceptions, an orientation that indicates a criterion of preference. Aesthetic feedback characterises (1) the choice of food, often in terms of simple gustatory pleasure, such as the particular preference for hemp seeds in many fringillidae; (2) the olfactory orientation towards particular molecules, such as indole and skatole in dogs; (3) the tendency to stroke the coat in grooming behaviour. We can say that aesthetics is a first form of judgement that occurs in all the interactive activities between the subject and external reality. The aesthetic canon clearly has a species-specific dimension, as shown by the animals' different preferences, but at the same time there is a certain degree of variability, so that the physiological state predisposes a psychological condition. The latter could be defined as *feeling something at a certain time* or in a phase of the individual's life, especially in relation to the hormonal condition.

In humans too, for example, greater sensitivity to music has been observed during the adolescent period and during ovulation. According to some research on birds, there is a reciprocal influence (basically positive feedback and therefore amplification) between singing and the production of sex hormones, a circularity that in turn sharpens both the areas involved in singing and those of receptivity. Seasonality affects aesthetic orientation as well as the estrous cycle. As Henkjan Honing reports in the essay *Aap slaat maat* (2018, 85–90), it is possible to measure differences in women's hearing based on phasic factors, such as seasonality and menstruation, so that when oestrogen levels are high they perceive the male voice as fuller. The individual manifests itself through a continuous unravelling of orientations that have to do with its preferences, and each choice produces effects that inevitably reflect on its psychological state, also influencing its physiology. Essentially, these are recursive processes that can act in a homeostatic way to keep the system in a state of confirmatory balance or, instead, in an evolutionary dynamic. In other words, preferences not only generate choices but also induce experiences, and these pave the way for indeterminacy. Experience is always a gamble, a non-deterministic process, a contamination with the unexpected that produces a unique identity, so having preferences means immersing oneself in the world and getting dirty with it.

The mind is a relational system or, in the words of Daniel Siegel in his *The Developing Mind* (1999, 1–10), a condition that is formed and expressed in relation to the world. There is no doubt that the behavioural basis of this aesthetic copula is to be ascribed to the affective dimensions that, therefore, are the first regulators of orientation through the system of preferences. It follows that a species' aesthetic evaluation must be correlated with the hedonic feedback experienced by the subject. I do not think I am attributing excessive qualities to heterospecifics when I speak of an animal aesthetics, an orientation compass that influences choices and involves not only perception but the whole body, thanks to the intermediation of the limbic nuclei and the action of certain neuromodulators such as dopamine, serotonin, endorphins and endocannabinoids. Aesthetics is certainly about sensory feedback, for example images or choreographies considered pleasant or interesting, sounds or rhythms that may appeal or excite, tactile and gustatory sensations regarded as pleasurable. But, more generally, aesthetics applies to any feedback that the species-specific system is able to detect. The aesthetic response is predominantly emotional, with reactions that can evoke joy, excitement, curiosity, surprise, annoyance or fear, in a sort of archetypal collection of the categories of beautiful, ugly and sublime. Thus, the whole body participates in this aesthetics of life, which is at the very heart of existence.

Taking up the Four Questions, we can say this: it is true that phylogeny predisposes preferences—so that each species presents its own canon—and it is also undeniable that ontogeny refines this criterion through relational experiences with the world. However, we must recognise that the protagonist of this feeling is always the individual. Aesthetics gives the subject a point of view that is never neutral, cold or detached, but always participatory and involved. The aesthetically open individual is therefore the true nerve centre of the peripatetic search for satisfaction in the world—a fundamental requirement for self-ownership. The subject displays a fullness that cannot be reduced to mere gratification or reaction to a stimulus. It is undeniable that pleasure has a physiological basis, but aesthetic orientation is firmly rooted in a psychological condition of the hedonistic subject. Aesthetics concerns the evolution of living beings, because it induces or facilitates appropriate orientations through pleasure, but above all because it opens the subject to new experiences. The aesthetic subject therefore has a fundamental orientation compass: the canon of preference, in fact, is unlikely to go wrong, precisely because it has been specifically developed by phylogeny. However, the aesthetic canon does not produce predefined experiential effects, but a large number of possible epiphenomena: our strong parental tendency, for example, also predisposes us to feel tenderness towards the young of other species.

The relationship between aesthetics and evolution has been analysed by various authors, some of whom were interested in the forms of the living following in the tradition of Johann Wolfgang Goethe, such as Ernst Haeckel (1834–1919) or D'Arcy W. Thompson (1860–1948). But it was especially Adolf Portmann (1897–1982) in the collection *Aufbruch der Lebensforschung* (1965) that gave rise to the concept of the living as self-representation. Already in the essay *Selbstdarstellung* (1958), the author complained that, when observing how biology deals with the phenomenal appearance of animals, the explanations given were disappointing. Portmann's vision, instead, focused on aesthetic self-display referring to an evaluating other. Henry Walter Bates (1825–1892) and Alfred Russel Wallace (1823–1913) had already examined *allaesthetic* qualities—that is related to perception at a distance—as important factors in evolution, in line with Charles Darwin's insights. At the beginning of the twentieth century, the entomologist Johannes T. Oudemans (1862–1934) had noted a correlation between a butterfly's part exposed to light and the patterns that these lepidopterans—in particular *Pheosia tremula*—display with aesthetic intent. This argument was taken up and expanded on by the French entomologist Roger Caillois (1913–1978) when he discussed the mimetic characteristics of the butterfly *Kallima inachus*, which at rest, when folding its wings, takes on the shape and colour of a leaf, disappearing from view.

What would later be referred to as the "Oudemans rule" shows how heterogeneous parts of the body and wings come together to form a harmonious whole and combine with such extraordinary concordance as to imply a unitary role, rather than a generic random variation of individual patterns. Unlike Oudemans, Portmann emphasises that the body's configuration is not simply exposed to light, but to the eye. Phenomenal appearance therefore becomes central to the analysis of interactions between living beings, with conformations that are as likely to attract the eye as to divert it. It is no coincidence that internal organs generally do not have symmetrical shapes or patterns related to self-presentation, confirming the aesthetic significance of the animal form. An interesting case is the shrimp *Periclimenes amethysteus*: its musculature is as transparent as glass, so that the body cavity is visible. Because of this transparency, the base of the body cavity has specific patterns, as is usually the case with the dorsal surface in other crustaceans. As Portmann notes, there is a correspondence between the patterns on the surface of the body and the internal structures of transparent organisms, which are manifested in coherence with the laws of Gestalt. We must therefore recognise that one of the fundamental needs of the animal world is the self-presentation, which implies that there is someone on the other side who is oriented according to an aesthetic criterion. In birds, this need becomes central in the phenomenon of courtship through the redundancy of manifestations.

In fact, observing some courtship expressions, as in the Gouldian finch (*Erythrura gouldiae*) and the diamond firetail (*Stagonopleura guttata*), the singing display corresponds to specific jumps and choreographic patterns, based on rhythms, exposure of particular parts of the body and lifting of feathers in certain areas. This presentation takes on a certain relevance among birds of paradise, with amazing effects. In *Cognition, Evolution and Behavior*, Sara J. Shettleworth explains this redundancy through two possibilities: "multiple sexual signals (…) may signal different aspects of a male's quality or different features important in mate choice (…) Different signals are used at different points in the mate choice process, as when a female is attracted by song from a distance and then responds to visual signals as she approaches the singing male" (2010, 171). The use of different modalities produces an accentuation of the visual image, such as chromatic contrast, the Tyndall effect, texture or different colouring in neighbouring parts of the body. The individual's self-display, which Adolf Portmann and Wilhelm Roux (1850–1924) define as an activity of relating to the world, giving rise to a kind of interiority, manifests a coherent unity of exhibited forms. This *phanerology*—as Portmann defines it—evidently dialogues with the animal's aesthetic sense, that is, with the property of being a bearer of preferences. If there is a displayer, it follows that there is a spectator watching.

In this respect, in relation to the singing of male canaries in terms of the preferences expressed by the females, Eric Vallet and Michel Kreutzer describe certain trills as "sexy syllables", that is singing elements capable of producing pleasure in females (1995, 49, 6, 1603–10). Interestingly, female Australian zebra finches (*Taeniopygia guttata*) also become excited when they hear males singing and exhibit a higher-than-normal level of dopamine in the mesolimbic tract. We now know that dopamine has to do with the seeking, interest and reward system, and is also released during sex or eating. We must not consider behaviour only in terms of its physiological and adaptive significance, but also in terms of the pleasure it produces. For example, with regard to motivations, it has been said that cats' predatory tendency is the result of remote causes, related to (1) their specific phylogenetic trajectory and (2) the action of precise selective pressures. Moreover, it takes on a connotation: (3) on the basis of the individual's experiences, (4) and with respect to its particular physiological state. There is no doubt, therefore, about the validity of Tinbergen's Four Questions, but it is equally clear that a cat's act of chasing a ball also responds to (5) the pleasure principle, because the presence of a character in a species implies an expressive *hedonic label*.

These preferences are what we should reflect on when we observe animal behaviour. While resulting from an evolutionary process that has shaped it in a certain way, even if an animal is caught up in a vortex of physiological flows as a clash of internal and external currents, it is still the owner of its movement in the world, because it is intrinsically devoted to pleasure. Through aesthetic pre-judgement, pleasure allows it to navigate the labyrinth of reality. We find this explanation in Irenäus Eibl-Eibesfeldt's monumental essay *Der Mensch, das riskierte Wesen* (1988), which is about specific human orientations.

If we ask why an individual performs a certain action at a particular time, the answer is very simple: because it has seen an opportunity for enjoyment. Giorgio Celli (1935–2011) reminds us that bees like jagged shapes rather than regular ones and prefer a star to a square, while ants prefer vertical lines to horizontal ones (2004, 216). Rensch's experiences were later confirmed by observing the behaviour of other species. Celli himself found that house mice tend to stay longer in diverticula marked by certain figures compared to others, just as it was already known that human babies as young as a few months old explore certain shapes with greater interest. Research into primate aesthetics began in the 1930s, when the zoologist Nadia Koths investigated the graphic skills of the chimpanzee Joni. The same period saw the studies by Mr and Mrs Kellogg, followed by the aforementioned research by Desmond Morris, whose work on Congo attracted the attention of the art historian Ernst Gombrich.

Natural aesthetics had the *imprimatur* of the naturalist Edward O. Wilson (1929–2021), who considered art to be an expression of our biological origins. In his essay *Biophilia* (1984), he posited an innate human aesthetic tendency towards natural landscapes, especially when referring to the characteristics of the African savannah. Moreover, the idea that aesthetics has to do with psychophysical sensation had already developed in the nineteenth century through the work of Gustav Theodor Fechner (1801–1887), inaugurating experimental aesthetics, and was supported a century later by the psychologist Daniel Berlyne (1824–1976), who sought to show how aesthetic feedback has deep roots in the physiological structure of the individual, which can be measured through orientation responses. As we have seen, in *The Question of Animal Awareness* (1976), referring to the prolonged flight of seagulls, Donald Griffin hypothesised that hovering for hours, enjoying the upward currents, is an aesthetic pleasure. Certainly, when in springtime we observe the joyful carousels of swifts (*Apus apus*) swooping around church towers at full speed, we cannot deny that they experience a certain amount of pleasure; the same is undeniable in the playfulness of baby mammals. As Roger Caillois wrote in *Méduse et Cie* (1960), there are aesthetic correspondences between what we see in nature and our own tastes, and this reveals a general character of animality aimed at orienting itself based on what it feels.

Being able to repeat a habit acquired at a young age also provides a certain amount of pleasure. On the other hand, it is precisely this bodily availability to pleasure that underlies the freedom of the individual, who is capable of going beyond all predefined conditions. Pleasure and its affective declinations give rise, in fact, to copulative needs, making the individual more open to the occasionality of the world and the singularity of situations. The psychological condition turns the tables because it introduces circumstantial, unpredictable, chaotic factors, leading to new behaviour. Pleasure is a factor of amplification but also of unpredictability because, starting from a particular yearning, a situation can give rise to anomalous translations or epiphenomena. The aesthetic experience, therefore, is not limited to re-proposing the heritage but foresees its precise evolution in the subject. And here the close relationship between creativity and aesthetics is evident because, if it is true that aesthetics favours orientation and experimentation, it is also probable—I am personally strongly convinced of this—that the creative product also causes a shift in the animal's aesthetics. As soon as a new aesthetic orientation develops, the rules of the game of sexual selection change, just as happens with niche construction, leading to modification effects on phenotypic fitness. Creativity, through the shifting of aesthetic canons, is thus able to play a central role in the alteration of selective parameters. I will return to this when analysing the subject's protagonism in phylogenetic history.

The Sixth Factor: Orientation Through Knowledge

Aesthetic orientation has to do with affectivity, a condition that is not based on consciousness or even cognition, but on feeling something, having preferences, following pleasure as the compass of one's choices. Aesthetics is a pre-judgement of what the world has to offer, because it is not based on a specific representational framework, let alone a conscious evaluation. I said that I want to temporarily suspend judgement on consciousness, neither because I doubt the wide presence of such states in the animal world nor because I ignore its functional importance, but because I believe that the animal's subjective ontology requires a paradigm revision that has been overshadowed by the appeal to consciousness in recent years. Affectivity, with its copulative and self-referential bearing—this is only an apparent oxymoron, because it means going in search of something in the world to bring back to oneself (*for-Self*)—lays the foundations for an intrinsic finalism. If we put together (1) the individual's ability to stand out from the background, distinguishing between sensibility and agency and thus bringing out a *self*, with (2) the affective and aesthetic tendency to navigate the world through preferences, giving rise to a *for-Self*, we will come to consider the individual as endowed with self-ownership. I think this is the prime cornerstone of a new explanatory paradigm for animal orientation.

On the other hand, an animal is not just a copulative and self-referential entity, but a presence that builds its identity through choices, projects and plans—that is, a projective effort and a protagonism that gives it full ownership of action. It is true that each species lives in a certain environmental context and is equipped with well-defined endowments of knowledge, but the individual's identity journey is neither fatefully restricted to an a-priori destiny nor completely influenced by what the subject encounters along the way. Through its experiences, an animal ***dialogues with*** the world and follows a precise learning path, which cannot be reduced to a mere *exposure to* the world. In the same way, the information it possesses, be it inherited or acquired, does not *impose a behaviour* on it, but *enables its cognition*, that is a title of use.

I must now point out the difference between learning and cognition, starting from the beginning. A subject cannot only project itself affectively onto external reality: it must also have the tools to do so. In other words, it needs elaborative resources to transform its landscape into a meaningful horizon, so that it can make predictions, extractions, selections, hypotheses and operations. Both the phylogenetic process and the various learning experiences allow the individual to accumulate these resources, which I have (1) called *knowledge*, (2) modelled as *schemas*, that is maps capable of functional plurality, (3) considered as *representations*, that is entities capable of standing for entities-events-performances. Overcoming the idea of expressive automatism found in the instinct-conditioning dyad is the *Second paradigmatic foundation* of the new ontological model I want to put forward.

Considering innate and learned endowments as usable resources, that is tools of which the subject has ownership of use, means focusing on the operating subject, who is not determined by its endowments, but rather enriched by them. This is an important step, because it establishes a close relationship between resources owned and available freedom: in this case, the greater the knowledge, the wider the *phenomenological space of possibilities*. By this I mean the virtual field of unnecessary and non-repetitive choices in which the subject can operate. If we were to imagine—to make a thought experiment—an animal without any innate endowment, there would be no room for freedom available in its life. And here lies the immediate difference between considering the innate as a mechanism whose activation gives rise to a predefined response, and regarding it, as we see especially in Lorenz's epistemological writings, as an endowment that enables the formulation of hypotheses.

Similar considerations must be made with regard to the concept of conditioning, because the term itself refers to the subject's closure as opposed to its acquired freedom. Do forms of conditioning exist in animal learning? I think so, just as there are innate resources with little, if any, scope for arbitrary use in every species. But these are particular mechanisms that cannot be generalised transforming all animal learning into an accumulation of switches or conditionings. Learning produces knowledge, that is, tools that enrich the individual's life insofar as they increase the space for freedom, arbitrariness of choice, application to the singularity of the situation and finally creativity. At this point, we can move on to the *Third foundation* of the paradigm I intend to propose: the emergence of cognition, which is not to be understood as the overlapping of resources—because then we would fall back into a vision based on the determinism of endowments—but as the free use of the phenomenological space of possibilities. Animals own their endowments as instruments, not automatisms, but they can use them repetitively or by shifting very little in the space of possibilities—a bit like I did myself when I used the computer as a typewriter in the mid-1980s, without resorting to the possibilities that were actually available in the new medium.

The copulative experience based on pleasure makes individuals *unstable* in their heritage condition rather than captivated and stunned: the affective components favour contamination precisely because of the heteronomy they imply and the restlessness they provoke. This is what I have defined as the peripatetic condition of animality. This copulation is not limited to producing a kind of fruition of the world, which is of course undeniable and implicit in the thermodynamic condition of the living, but underlies the subject's evolution as a biographically defined entity. Of course, there are situations where learning produces a response so simplified as to resemble a reflex, but in most cases learning generates utilities and spaces of identity expression. Knowledge is not a switch in the animal's body, conditioning it or imposing a precise reaction on it, but a tool that enriches and defines the individual's singularity. In this sense, changing the term with which we describe the learned—abandoning the indiscriminate use of the word *conditioning*—means radically changing the way we consider subjective experience. Knowledge, in fact, always makes the subject freer, because it gives it greater depth, applicability and adherence, but mostly greater *room for interpretation* in its relationship with the world. The ability to process information and compare it with an inner perspective, generating knowledge, that is increasingly personal representations of reality, creates the dimension of subjectivity that I define as *cognition*.

Cognition is therefore *the mode of use of the knowledge possessed*. It has been said that knowledge results from the metamorphosis and generativity of the innate as a consequence of individual experience. Here we already find individuals who have had a great deal of experience or who have been forced in some way, for example by occasional habitat transformations or life vicissitudes, to expand and reorganise their innate equipment. In such individuals, the phenomenological space of possibilities relative to the knowledge possessed is greater. In this sense, I maintain that there is direct proportionality between learning processes and expressive freedom in the mode of cognition: the more complex and articulated the knowledge, the greater the range of possibilities for its use, that is its *intrinsic virtuality*. On the other hand, cognition also depends on other factors, generally referred to as *cognitive metacomponents,* such as working memory, attention and concentration, discriminative, categorical, correlative and causal functions, reflexivity and metacognition, as well as the subject's flair or imagination. In addition to this, there are also aspects that concern the individual's overall personality, which may also relate to emotional and motivational aspects. There can therefore be higher levels of cognition than the canonical or adaptive use of knowledge, such as (1) reuse, that is the simple application of knowledge to new contexts; (2) recycling, that is the mobilisation of a resource for co-optative use; (3) extension or synthesis, that is the creation of new functions by assembling several resources.

When we speak of (4) invention-discovery in the use of an endowment or (5) functional innovation, we are referring to a very high level of cognition that we could identify as the ability to travel the entire space of virtuality, or at least to come up with new itineraries. Cognition, therefore, can be expressed at a basic level, because even the canonical use of knowledge always entails a margin of discretion, but it can also reach surprising peaks of creativity, especially when they are manifested by animals far removed from us. What I mean is that very few people are surprised by the cognitive heights reached by anthropomorphic primates or cetaceans, or even by corvids and spittacids, but when we are faced with expressions of a high intellectual level in arthropods or molluscs, the astonishment is obvious. I am convinced that these researches have been and are still fundamental in shattering the anthropocentric arrogance that would like to elevate the human being to the status of unique and special entity, the only one who can access the upper echelons of arbitrary behaviour. I believe that books like Frans de Waal's *Are We Smart Enough to Know How Smart Animals Are?* (2016) or Colin Allen and Marc Bekoff's *Species of Mind* (1999) are extremely useful in deconstructing such prejudices. However, I think it is necessary today to start from the basic behaviour, the kind that does *not* surprise us, in order to change an explanatory paradigm that is outdated, full of corrective epicycles and far from being in line with Occam's razor. I am sure that even Lloyd Morgan would agree.

Cognition has various fields of application and can be used in learning processes themselves through its ability to produce reasoning about the state of things and possibilities. We have seen, in fact, that the integrative action of the mind places the subject in a propositional position with respect to the situation being experienced: the subject is a diachronic entity capable of moving on several temporal fronts, remembering something from the past and projecting itself towards the future. Propositionality, for example believing or hoping, is not only a syntactic function, capable of putting together an affective attitude and images of the world, but also solicits reasoning, moving within the space of possibilities. Instrumental reasoning, for example, stimulates the subject to apply its knowledge of the world in a pragmatic sense in order to obtain goals, while inferential reasoning allows the individual to derive consequences, conclusions or generalisations from the data available. There is also referential reasoning, such as when the subject takes into account the behaviour of others in order to orient itself, or selective reasoning, which highlights only useful clues—not to mention problem-solving, when the individual understands the structural requirements of a problem. I will speak of *matching to sample* to indicate the ability to recognise something as similar or of *perspective taking* as the ability to take on someone else's viewpoint: all this is possible if, and only if, the individual has its endowments and information at its disposal.

Let us return, then, to the theme of this chapter, namely the individual's way of navigating the world. The psychological condition, by bringing out the self, produces the state of feeling something, and we have analysed this by talking about animal aesthetics or system of preferences. Now we have to consider the fact that being endowed with knowledge and cognition brings out a *sixth factor* not mentioned in the Four Questions. I am referring to an animal's tendency to *project itself into something*, to imagine situations that have not yet come true or to set goals to be achieved. These images of the future, which arise in the animal's mind through various forms of propositionality and reasoning, act as irresistible magnets that influence its orientation. It is not possible to explain behaviour without these images, which represent aims, eventualities, possibilities, solutions that the individual uses as a target for its action, a guideline to follow, a reference itinerary. I therefore agree with the rejection of extrinsic finalism, which regards the entire evolutionary phenomenon or the trajectory of a species as an undefined goal or as an a-priori project that is realised as it unfolds. However, this does not mean that the individual is devoid of purpose in the course of its life. Planning is a cognitive activity that I would even call trivial, because it is the inevitable consequence of the animal's tendency to project itself into a future condition. Animals do not live entirely in an absolute present, with no memory or imagination.

Obviously, given the species-specificity of the innate endowments of knowledge and dispositional copulatives—that is of emotional and motivational characteristics—the experiential outcomes also vary, so that the phenomenological space of resources changes in different species. If we consider that cognitive metacomponents also present marked differences between species and between individuals, it follows that we must speak of cognition not only as an arbitration field but also as a functional multiplicity, so that, for example, one species might be more inclined to instrumental reasoning and another to inferential reasoning. This allows us to speak of *multiple intelligences*, that is of a marked peculiarity in the various species to work on their own projectivity. When we talk about goals, in fact, we are not simply referring to objectives to be achieved, but to a certain way of looking to the future, referring to conditions that are not present. That is why, in my opinion, intelligence is not a way of solving a problem, but a way of *raising* a problem, that is of creating or hypothesising the conditions of a possible future. A performance, whatever it may be—from the capuchin monkeys' ability to use a tool to an archer fish's skill in learning the trajectory of a throw by observing another and correcting the refraction—can be utilised through endowments structured by phylogeny, through abilities obtained by training, or by specifically using one's intellectual capacity. This means that the complexity of the performance is not what indicates a particularly active mind.

Of course, the mind is always in action, but in certain situations the reflectivity quotient is higher, such as when an animal makes a discovery, finds a new solution, goes against a predefined behaviour, moves away from the objective to reach it or to obtain the means to achieve it. The tendency towards instrumental reasoning leads an individual to focus all the knowledge it possesses on a pragmatic dimension, to achieve its objective. In this case, even the observation of the object loses some of its evocative meaning (what it heralds or arouses) and its feasibility meaning (what it allows to be expressed), developing a dialogue with the body for the purpose of extension and externalisation. In such cases, for example, a stick becomes an extension of the body, so the subject mainly analyses two aspects: (1) the affordances, that is the possibilities offered by the object to relate to the body; (2) the performativity, that is the result that can be obtained with that tool. The subject, therefore, orients itself according to a future image: in this case that of its own body hybridised with the object, grappling with the task, in view of a goal. In the same way, the classical structures of inferential reasoning—syllogism, induction, deduction, abduction—are always based on a content that is not present, but are essential to achieve the objective. Projection can often create a temporal distance between the immersion in the problem, or even just its presentation, and the solving action: we have seen this in the insight studied by Köhler, or in the accumulation of useful tools among orangutans, to escape from the cage in the cases reported by Benjamin Beck (1980), or to extract honey from hives as shown by van Schaik et al. (1999).

As de Waal (2016, 110–115) reports, it is easy to observe that primates think before they act: there is a reflection gap that also aims to assess the consequences of an action, also in terms of social relations. The work of Elisabetta Visalberghi (1994, 108, 15–22) at the Bioparco di Roma is significant in showing how problems present a multiplicity of situations, not all of which can be easily tackled by a species, precisely in relation to the type of projection the latter makes on the unfolding of the problem itself. When studying sphecidae, solitary hymenoptera, Tinbergen discovered that before flying away from the paedotrophic nest, the wasp did a reconnaissance flight to find useful references so that it could easily come back. Here we see once more how animal behaviour uses memorisation not to live in the past but to project itself into the future. When we look at how a species orients itself for migration or to return home, we see that it never relies on a single source of navigation but integrates several systems when necessary. We are used to thinking that navigating the world is essentially a sensory matter, and the great discoveries of the twentieth century—for example, echolocation in microbats, magnetorhexis in birds or electrorhexis in fish—have made a fundamental contribution to understanding the biodiversity of sensory inputs.

On the other hand, perception is not only about the ability to access a certain physical and chemical repertoire, but also about placing it within a certain time frame. We, too, experience this variable perception of time, for example when an adrenaline rush seems to slow down the unfolding of events. Today we know that the sense of time changes across species, for example, a fly or a pigeon may see us slowed down while a tortoise sees us speeded up; but there are also phasic oscillations, especially in crepuscular predators, as is the case with cats, whose timing is accelerated in the evening by the combined action of dopamine and catecholamines. A bee communicates not only the direction of the food source but also its distance, just as a fly in a tunnel has a sense of the distance travelled based on the speed of its flight. The Sahara ant (*Cataglyphis fortis*) uses the number of steps taken on the way out to orientate itself in terms of distance, for example when homing in and returning to the anthill (Wittlinger et al. 2006, 312, 1965–67). As Emmanuelle Pouydebat points out in *L'intelligence animale* (2017, 105–116), in order to orient oneself, a sensory compass is not enough: one needs a tachymeter and a stopwatch, that is, one needs to master time as well as space. Orientation is based on numerous aspects relating to the perception of the space-time relationship, to working and episodic memory, but also to vector recognition and calculation. There is no perception without time awareness.

This means that one of the prerequisites of orientation is the ability to stay in a condition of diachronic experience: this is the most important aspect that I would like to point out in order to dismantle the thesis or, better, the prejudice that sees the animal captivated in a sort of eternal present. Animals have been thought to be incapable of programming, that is of setting long-term goals and targets, and of planning, that is of mentally constructing a sequence of actions useful for achieving what has been set in the programme. In the essay *Le culture degli altri animali* (2008, 41), Michelangelo Bisconti points out that herbivores and frugivores must not only have a topographical idea of where they forage, but also remember how the foraging areas are distributed (which is never the same), and which areas have recently been used. The same can be easily seen in bumblebees, which avoid visiting already-foraged flowers to collect nectar. We must remember, however, that a fundamental imperative in the animal world is the ability to minimise search times in different activities, which is why episodic memory is a basic endowment. In vertebrates, episodic and spatial memory are linked in the hippocampus, and it is no coincidence that the most performative animals in this sense, such as Clark's nutcracker, have a more developed hippocampus than other related species, and that at certain times this structure is enlarged and has a greater neuronal density. After all, the hippocampus has a function not only in retrieving previously acquired information but also in projecting it into the future.

Here I do not wish to go through the list of animal performances that disprove their alleged captivation in the present moment. Just to make a few examples, consider the ability of elephant matriarchs to remember the location of water holes to direct their thirsty herds, the planning of routes in orangutans, who communicate at dusk the direction they want to take the next day, and the tendency of chimpanzees to return to food resources by moving through the forest with great ease and in different directions. Frans de Waal, in the aforementioned book (2016, 260–295), reports a lot of research disproving the widespread belief that animals are stuck in the present. The first daring studies in this sense were carried out in the 1920s, first by Edward Tolman and later by his student Otto Tinklepaugh; the two scholars investigated purposive, that is, goal-driven and expectation-driven characteristics in reference to the future. The fact that animals have expectations is amply demonstrated by tests of attention maintenance or expressions of surprise. That chimpanzees show a tendency to collect tools in different places from where they will use them, that they are capable of self-control over a resource with a view to greater gain, that they strategically strengthen friendships a few days before a confrontation, that they assess the consequences of their actions in advance should no longer surprise us. Unfortunately, the anthropocentric need to continually posit an insurmountable gap between humans and other species risks leaving these clear refutations within the community of ethologists, while humanists continue to draw unfounded differences.

After all, it is not just a matter of admitting our taxonomic cousin to the high table of programming and planning, because evidence is constantly accumulating that one of the fundamental characteristics of animality is precisely this unavoidable *diachronic dimension*. Being an animal means, in fact, bringing out the present by collapsing the past onto the future, that is, creating a temporal space (a window) in the continuum. I might add that there is no objective present, since there are different time scales in different species and even in the various moments of an individual's life. In short, the present is not an objective and homogeneous reality, but a variable construct that animals produce in their diachronic expression. This means that the subject is not immersed in the present, but emerges from the temporal flow through the construction of the present that allows it to produce a referred image (i.e. a certain image) of the past, consistent with its intentions, and an expression of a possible future. Well, creativity lies precisely in this emergence of the present, the opening up to a *not-yet* that lies within the space of virtuality in which the individual operates.

Therefore, animals do not live in an isochronous condition: their time is not a constant but a space of use, so their existence is certainly projective but also productive, that is, creative. Their behaviour is not a sequence of automatisms that follow one another in a necessary, domino-like way, but involves facing up to the space of possibility, *taking one's time*. I would like to underline this aspect, because I believe that the invention of the present is just a matter of taking time in order to be able to construct one of the possible futures. It is neither the senses nor memories that impose an orientation, for it is the individual that uses its resources to create its own future. Obviously, in order to do so, it needs a temporal space that allows it a margin of creative manoeuvre. But at this point one must inevitably admit the intrinsically anecdotal condition of animals, who are always designing and planning, that is reinventing their future. The present is a personal and intimate space that allows one to reflect on one's possibilities; it is an antechamber to action available to the individual, which is why the present never has the same duration in the life of an animal. The present is what allows the subject to set goals, to evaluate different options, to develop plans and strategies, but above all to get lost in the imagination. The animal dimension is a creation not only of possible worlds but also of different times. Creativity makes it possible not only to initiate new behaviours, that is to escape from the mere repetition of acquired skills, but also to escape from the present in order to imagine possible conditions.

Works Cited

Allen, Colin, and Marc Bekoff. 1999. *Species of Mind*. Cambridge, MA: The MIT Press.

Beck, Benjamin. 1980. *Animal Tool Behavior*. New York: Garland STPM Pub.

Bergson, Henri. 1907. *L'évolution Creatrice*. Paris: Presses Universitaires France.

———. 1934. *La pensée et le mouvant*. Paris: F. Alcan.

Bisconti, Michelangelo. 2008. *Le culture degli altri animali*. Bologna: Zanichelli.

Caillois, Roger. 1960. *Méduse et cie*. Paris: Gallimard.

Carere, Claudio, and Dario Maestripieri. 2013. *Animal Personalities. Behavior, Physiology, and Evolution*. Chicago: University of Chicago Press.

Carlson, Neil R. 1977. *Physiology of Behavior*. Boston: Allyn and Bacon.

Celli, Giorgio. 2004. Arte e biologia: una scommessa evoluzionistica. In *Il Bello e le bestie. Metamorfosi, artifici e ibridi dal mito all'immaginario scientifico*, ed. Lea Vergine and Giorgio Verzotti, 215–223. Milano: Skira Editore.

Darwin, Charles. 1871. *The Descent of Man and Selection in Relation to Sex*. New York: Merrill and Baker.

De Waal, Frans. 2016. *Are We Smart Enough to Know How Smart Animals Are?* New York: W.W. Norton & Company.

Dehaene, Stanislas. 2020. *How We Learn, Why Brain Learn Better than Any Machine*. New York: Viking.

Dennett, Daniel. 1995. *Darwin's Dangerous Idea: Evolution and the Meanings of Life*. New York: Simon & Schuster.
Diamond, Jared. 1982. Evolution of Bowerbirds': Animal Origins of the Aesthetic Sense. *Nature* 297: 99–102.
———. 1986. Biology of Birds of Paradise and Bowerbirds. *Annual Review of Ecology and Systematics* 17: 17–37.
Donald, Griffin. 1992. Animal Minds. *Beyond Cognition to Consciousness*, Chicago: The University of Chicago Press.
Duff, William. 1767. *An Essay on Original Genius*. London: E. and Dilly.
Eibl-Ebesfeldt, Irenäus. 1988. *Der Mensch, das riskierte Wesen*. München: Piper.
Galton, Francis. 1869. *Hereditary Genius*. London: Macmillan.
Griffin, Donald R. 1976. *The Questions of Animal Awareness*. New York: Rockefeller University Press.
Honing, Henkjan. 2018. *Aap slaat maat*. Amsterdam: New Amsterdam.
Inoue, Sana, and Tetsuro Matsuzawa. 2007. Working Memory of Numerals in Chimpanzees. *Current Biology* 17 (23): 1004–1005.
Kipling, Rudyard. 1912. *Just So Stories*. New York: Garden City Pub. Co.
Lewontin, Richard, and Stephen J. Gould. 1979. The Spandrels of San Marco and the Panglossian Paradigm: A Critique of the Adaptationist Programme. *Proceeding of The Royal Society B. Biological Science* 205 (1161): 581–598.
Lombroso, Cesare. 1864. *Genio e follia*. Milano: Chiusi.
Marchesini, Roberto. 2016. *Etologia filosofica. Alla ricerca della soggettività animale*. Sesto San Giovanni: Mimesis.
McGaugh, James L. 1992. Affect, Neuromodulatory System, and Memory Storage. In *The Handbook of Emotion and Memory: Research and Theory*, ed. Sven-Åke Christianson, Part III. New York: Lawrence Erlbaum Associates.
Milner, Brenda, Larry R. Squire, and Eric R. Kandel. 1998. Cognitive Neuroscience and the Study of Memory. *Neuron* 20: 445–468.
Morris, Desmond. 1962. *The Biology of Art*. London: Methuen.
———. 2013. *The Artistic Ape*. London: Red Lemon Press.
Poincaré, Henri. 1914. *Science and Method*. London: T. Nelson and Son.
Popper, Karl. 1978. Natural Selection and the Emergence of Mind. *Dialectica* 32 (3-4): 339–355.
Portmann, Adolf. 1958. Selbstdarstellung als Motiv der lebendigen Formbildung. In *Geist und Werk. Aus der Werkstatt unserer Autoren. Zum 75*, ed. Geburtstag von Daniel Brody, AA VV, 139–173. Zürich: Rhein.
———. 1965. *Aufbruch der Lebensforschung*. Zurich: Rhein-Verlag.
Pouydebat, Emmanuelle. 2017. *L'intelligence animale*. Paris: Odile Jacob.
Shettleworth, Sara J. 2010. *Cognition, Evolution, and Behavior*. New York: Oxford University Press.
Shubin, Neil. 2008. *Your Inner Fish*. New York: Pantheon Books.

Siegel, Daniel J. 1999. *The Developing Mind. Toward a Neurobiology of Interpersonal Experience*. New York: Guilford Press.

Sperry, Roger W. 1991. In Defense of Mentalism and Emergent Interaction. *The Journal of Mind and Behavior* 12 (2): 221–245.

Swaab, Dick F. 2010. *Wij zijn ons brein. Van baarmoeder tot Alzheimer*. Amsterdam: Contact.

Tinbergen, Nikolaas. 1963. On Aims and Methods in Ethology. *Zeitschrift für Tierpsychologie* 20: 410–433.

Tinbergen, Nikolaas, and Albert C. Perdeck. 1950. On the Stimulus Situation Releasing the Begging Response in the Newly Hatched Herring Gull Chick (Larus argentatus argentatus Pont.). *Behaviour* 3: 1–39.

Vallet, Eric, and Michel Kreutzer. 1995. Female Canaries Are Sexually Responsive to Special Song Phrases. *Animal Behaviour* 49 (6): 1603–1610.

Van Schaik, Carel P., Robert O. Deaner, and Michelle Y. Merrill. 1999. The Conditions for Tool Use in Primates: Implications for the Evolution of Material Culture. *Journal of Human Evolution* 36 (6): 719–741.

Visalberghi, Elisabetta, and Luca Limongelli. 1994. Lack of Comprehension of Cause-Effect Relations in Tool-Using Capuchin Monkeys (*Cebus apella*). *Journal of Comparative Psychology* 108: 15–22.

Weidensaul, Scott. 2021. *World on the Wing. The Global Odyssey of Migratory Birds*. London: Pan Macmillan.

Welsch, Wolfgang. 2013. The Animal Origin of Aesthetics. *Rivista di Estetica* 53 (3): 181–205.

Wilson, Edward O. 1984. *Biophilia*. Cambridge, MA: Harvard University Press.

Wittlinger, Matthias, Rüdiger Wehner, and Harald Wolf. 2006. The Ant Odometer: Stepping on Stilts and Stumps. *Science* 312: 1965–1967.

Young, John Zachary. 1987. *Philosophy and the Brain*. Oxford: Oxford University Press.

Zahavi, Amotz, and Avishag Zahavi. 1997. *The Handicap Principle*. New York: Oxford University Press.

CHAPTER 10

Niche Construction

Premise

Creativity is one of the most important drivers of evolution. Life is in fact an incessant flow of inventions and discoveries, a pilgrimage into the land of possibilities, a continuous innovation on the stratigraphy of the phylogenetic heritage. The very variability of individuals within populations creates the conditions for exploring new opportunities, for testing the value of one's own attempts, for venturing into the meshes of the unknown, of what is not already given. We have seen how the animal dimension plays on individual personality, regardless of the dimension of belonging: the subject's singular projectivity is the basic condition for completing something that is still immersed in virtuality. Therefore, we shouldn't measure creativity based on the result achieved, but rather in terms of slight deviations: a small innovative shift that produces, in the unfolding of time, a new morphopoietic trajectory. This is how Charles Darwin envisioned it.

The productivity of individual existence does not yield monotonous repetition generation after generation, but inaugurates events that unfold through the metamorphosis of styles and the layering of discoveries. Like clothes, these inventions envelop a population until they become recognisable in a fashion beyond somatic boundaries, giving rise to a niche. A new species is a different kind of appearance on the stage of the world, manifested in a particular pattern of development, a symphony of fabrics that adapt to each other by exploiting modularity and architectural planes. The

R. Marchesini, *The Creative Animal*,
https://doi.org/10.1007/978-3-031-07414-1_10

result is a morphological and behavioural phenotype, an identity, however fleeting in the passage of biological time. In addition to anatomical features, there is an expressive make-up, a *modus vivendi* that can be identified within the taxonomical continuum, at least as a range of behavioural variables. Living beings are threshold systems, which do not develop through morphopoietic linearity but through global rearrangements; bodily modelling is always systemic, the result of complex interactions in the process of the organism's development.

Creativity introduces constant novelties into the score of life, because every living creature follows its own trajectory, partly as a protagonist and partly driven by contingencies—fortuitous events that the animal is able to accept precisely because it is free from determinism, order and simplification. But, as said, the individual is also the protagonist of this symphony of serendipity: the single animal is an actor in phylogeny. The mistake of considering every subject's existence a mere transfer of replicators, translating the history of life into an algorithm much like the ghost of the Laplacian demon, falls apart when we realise that a random mutation of the genome and the subsequent work of selection is not enough to explain the complexity of living forms.

The claim that mechanistic determinism applies to living beings, supposedly locked in the ahistorical order maintained within the tracks of monodian necessity, only reveals our need for mathematical control over the chaotic turbulence of life. Jean-Jacques Kupiec, in his book *Et si le vivant était anarchique* (2019), reminds us that while it is true that order is the obsession of researchers, it is also undeniable that this eagerness represents an epistemological obstacle to understanding the living, because variability and disorder are the cornerstones of biological organisation, whose phenomenology is inherently stochastic. Each living organism brings to the theatre of life its own grain of creativity, a minimal deviation from uniformity, so that (perhaps) we should focus more on the genetic and epigenetic mechanisms called upon to contain the individual on the orbital of the species—namely the mechanisms devoted to regimenting the morphopoietic inspiration—rather than wondering about the origin of variability. Life is openness and virtuality, a constant quest to extract something new, to cross a frontier, to see what lies beyond.

Most of this individual creativity is certainly lost in time, washed away by the great dissipation of geological eras. Sometimes, however, its resonance endures, either by chance or through unexpected success, as when a mass extinction leaves open immense colonising prairies, with partially pre-established niches, as happened in the great transition of the Cenozoic. Behaviour is the great protagonist of this occupation and inauguration of new expressive and adaptive styles, through the construction of worlds that take shape from the undergrowth of minimal attempts. It is strange, and frankly absurd, that there has been so little room for ethology in twentieth-century evolutionism, betraying the legacy of the great founding fathers of the previous century. We must not view phylogeny as a passive process, as if animal populations were being dragged through derivative motions into different habitats: the origin and transformation of species has been guided by precise orientations.

Specialised trajectories stem from the cumulative action of individuals, correlated with parameters of preference, exploratory events, opportunities for discovery and even episodes of inventiveness. The development of innate behaviour is, therefore, the expression of accumulative processes of innovation, each bringing a small but significant deviation from the starting base, which are realised through the experiences, learning processes and inventions developed by each individual. Just as environmental induction on developmental factors, that is phenotypic construction, precedes the selection of the most suitable alleles for that particular adaptive morphotype, so the behaviour assumed by a certain population through learning and creativity creates the basis for coherent genetic selection, assimilating experience into the innate.

The individual's behaviour, in its bringing novelty to the interpretation of environmental possibilities and new styles of interface with the world, represents a niche shift, which can be more or less consistent and suitable for being maintained over time: it is therefore a change in the organisation of the species-specific morphotype, in all its aspects. The theory of niche construction shows how, through the innovation of different individuals, a *modus vivendi* is achieved that is capable of acting on environmental parameters and selective pressures. Hence a process of phenotypic channelling that has repercussions on the evolution of the species and on the consequent definition of recognisable traits.

It is useful, then, to reflect both on innate behaviour and on the subject's ability to own and update it: the inherited expressive pattern can be innovated through creativity, giving rise to new behavioural dimensions. In *Animal Architects* (2007), James and Carol Gould highlight the importance of recognising the heterogeneity of animal artefacts. Within a standardised (i.e. species-specific) design and operating method, having to respond to the demands of ecological singularity, animals inevitably require cognitive superintendence over their life project. According to the authors, if we want to understand animal architecture, it is essential to admit the possibility that they are capable of incorporating a certain degree of creativity or intuition into their work.

How important was the description of animal behaviour in the formulation of evolutionary theories? At first, very, because innate knowledge in the animal world—not attributable to experience, not least because of its executional perfection with no need for apprenticeship—was the battlefield on which to disprove the creationist fixationism of natural theology. It was therefore necessary to start from behaviour in order to justify and support evolutionary theories. On the other hand, from the twentieth century onwards, the role of behaviour in the evolution of living beings strangely disappeared almost across the board, in a slow erosion of the individual as well as the phenotypic significance of the organism, reduced to a simple container of genes.

Not all scholars did this. For example, in *Le comportament, moteur de l'évolution* (1976), Jean Piaget emphasised the importance of individual behaviour as a factor in evolution. Other authors, such as Waddington, Baldwin and Lewontin, also supported the importance of behaviour in the evolution of living beings, but they were isolated voices crashing against the Modern Synthesis, based on population genetics and the adaptive action of natural selection. For Piaget, behaviour should not be considered simply a *result*, but rather a *promoting* factor of evolution, through the choice of environment and the circumstantial strategies that organisms implement. According to the French biologist, individuals integrate the environment via processes of relation and modulation, in a logic of constructivist interactionism that goes beyond the concept of passive determination from the outside. Piaget's vision would find a new potential for development in the theory of niche construction.

But what is a niche and what is the relationship between ecology and behaviour? In order to answer this question it is important, in my opinion, to refer to von Uexküll's concept of *Umwelt*, starting from the idea that every animal is immersed within a context-world where perceptive and operational characteristics are closely connected to each other in an interpretative bubble in the here and now. The *Umwelt* is, in fact, one of the possible worlds, a precipitate constructed by the species by converting, as it were, a condition of virtuality into a possible form, very much as a statue is extracted from a marble slab. The worlds of a tick, a bat, a bee or a human being are not total inventions, but neither are they mere expressions of partiality: they are more like emergences or constructions-extensions of possible dimensions. It is true that a tick perceives the chemical stimulus of butyric acid, but at the same time it transforms it into a meaning, that is into a way of constructing its own relationship with the world.

A spider's web, therefore, is not simply a product of the animal, but a specific *dimension of its life*, and the same can be said for any microenvironmental construction: it should not be considered an external entity, but an integral part of the spider's body. In this sense, the web is introjected into all ontopoietic processes, both phylogenetic and ontogenetic. As far as the so-called human technosphere is concerned, it is also completely implausible to regard it as an external entity, simply instrumental to our needs and ergonomic with respect to our bodies. Every technology is somatised, because (1) it intervenes in the dynamics of selective pressures and thus modifies the phylogenetic trajectory of the population; (2) it modifies developmental processes, operating *developmental differentials* of a quantitative type related to the timing of growth sequences.

A species does not merely inhabit a portion of the environment, taking advantage of the opportunities it finds there, nor does it objectively occupy a preformed niche, as if it were a simple rental process: a species *constructs* its niche through its own action. Each niche is something new, something that did not exist before. It can emerge in various ways, for example through circumstantial situations, such as applying an old behaviour in a new context by bringing out novel possibilities and habits, in a kind of ontogenetic exaptation. It can be the result of a discovery, an invention or the side effect of another behaviour. In all these situations the innovative action of the individual and the introjective capacity of the population play a major role in the affirmation of the new. It is a fact that each niche brings with it effects of great importance in terms of selective dynamics. Innovative behaviour, in fact, gives rise to new pressures which, over time, will reward progressive adaptation to the inaugurated performance. A new use of a behavioural pattern—as well as its establishment as a tradition within the new population—does not follow the pattern of inheritance of acquired traits, but that of selective shifting to new functions, exactly as in the classic case of *phylogenetic exaptation* introduced by Stephen J. Gould and Elisabeth Vrba (1982).

In this case, however, the new behaviour does not arise from a random mutation, but from an application based on individual creativity. Once again, it should not be assumed that creativity always operates on macro levels, because it is enough to interrupt a repetitive pattern to obtain new expressions; creativity is a rather common ingredient in the animal world since, as we have seen, it is not possible for animals to rely on mere repetition. Whether it is co-optation, reuse or an epiphenomenon—each of these definitions has small differences in meaning—of previous behavioural patterns, a shift is introduced into the expressive system of that population. This novelty, just like a genetic mutation, can be lost—which I believe happens in most situations—or it can form the core of a niche construction, with a different landscape of selective dynamics. When innovation is introjected into a tradition, in its first steps, it should not be seen as a revolutionary event, but as a reworking of previous patterns.

Environments change, and a species often finds itself chasing a favourable environment by moving along different territories, but in doing so it may be faced with selective counterparts, both in terms of risks (competitors, predators, parasites etc.) and opportunities, such as to redefine the selective matrix. At this point, even a small innovation, combined with other factors, for example a change in foraging, defence or reproduction style, produces a profound transformation of the niche and consequently of the selective terms. That is when quantum effects can occur on the species profile, from both a morphological and a behavioural point of view. A change in the selection matrix, that is the set of selective pressures acting on the population, inevitably entails an overall reshaping of the species. This could mean, for instance, a reshaping of the population, partly due to the interruption of genetic mixing with the species of origin (exactly as in allopatric processes), and partly because the conformation of the morphotype is always the result of the selection matrix. We must not, therefore, think of speciation only in terms of genetic disruption, but also as the result of a change in selective pressures. The novelty introduced into the population, if/when interacting with other environmental factors, determines the opening of a new niche, also due to positive feedback, so that each change leads to others.

Creativity, in its different degrees and scopes, therefore plays a central role in the evolution of species, seeing as the subject is not a passive entity undergoing selective pressures, but contributes in some measure to the definition of the selective matrix through its own behaviour. Ethology is a relatively recent discipline, with little more than a century of structured research using scientific methodologies, yet in these few decades we have witnessed new behaviours developed by animals based on established patterns, such as the tits' practice of breaking the foil wrapping of milk bottles to reach the cream, as reported by John Fisher and Robert Hinde in 1949. The behaviour of anthropoid apes in zoos showed how extremely flexible these animals are in developing new behaviours, taking advantage of the opportunities they are presented with. Today, we know that a species' behaviour is not just the repetition of established patterns, because alongside these, we find continuous explorations of the space of possibilities. Even when using inherited patterns, there is a frequent tendency to apply them out of context or for other functions: individual behaviour is a sort of workshop that is always active and open to possibilities, with incredible cognitive dissipations as well as discoveries destined to mark the advent of new niche dimensions.

Niche Construction Theory

The relationship between an animal and its environment is much more complex than we are accustomed to imagining according to tradition, which still interprets it in an essentialist way, as the occupation of a predefined niche. The fixist heritage is still present and leads us to view the niche as something that precedes the species: the latter would simply occupy the space allocated to it. According to William Paley's (1743–1805) *Natural Theology* (1802), there is a universal harmony, and each species was designed by a divine creator to occupy a precise place in nature. This approach still influences our understanding in terms of the rank occupied by each animal in the cosmos. With evolutionism we moved from the top-down view of the divine plan to the bottom-up view of adaptive correlation with the environment, but the interpretation of the animal-environment relationship remained unchanged. As we read in Ernst Haeckel, the environment defines conditions to which the organism must submit, so it is commonly believed that animals are subject to selective pressures that derive primarily from the conditions of their habitat. This view is certainly correct in some respects: we can often deduce the morphological characteristics of an animal from the climatic, pedological, ecosystemic characteristics of its life. However, this leaves the species totally passive in the face of adaptive challenges.

The concept of niche has undergone a considerable metamorphosis since the first formulations offered by zoologists at the beginning of the twentieth century—take, for example, the work of Joseph Grinnell (1877–1939) who, while studying bird populations in California, first considered the niche as a spatial reference point where a certain species can be found more frequently (J. Grinnell 1914, 1917), hypothesising that there could be unoccupied spaces, that is empty niches. Later, Charles Elton (1900–1991) came to define the niche in a more relative way, as a mainly trophic function performed by the species in the biocenosis to which it belongs. As Saverio Forestiero notes in *Ambiente, adattamento e costruzione di nicchia*, the prevailing difference between the two biologists lies in Grimmel's greater emphasis on the environment's action on the species or, conversely, in Elton's insistence on the species' action on the environment (2009, 258). With the studies of George Evelyn Hutchinson (1903–1991) on the fauna of freshwater lakes, the concept of niche was developed as a space occupied by a species, but also as the expression of an animal population, in a multidimensional vision defined on the one hand as a set of environmental parameters and on the other as the result of the action of the species.

In spite of the avant-garde works of Elton and Hutchinson, aimed at emphasising the animal's non-passivity in niche setting, the idea of objective space, or rather of an empty niche that a species merely occupies, has increasingly asserted itself, especially since the second half of the twentieth century. A turning point in this debate was the work of Richard Lewontin (1929–2021) who, starting from a general critique of the adaptationist view and with a co-evolutionary approach to the relationship between populations and the environment, rejected the principle of passivity, arguing that the environment is not autonomous in the process of a species' insertion. For Lewontin (1983), a species contributes to constructing the environment through its transformative action; he therefore opposes the idea that a species' niche pre-exists it, that is it can be considered an empty pocket waiting to be occupied. In *The Triple Helix: Gene Organism and Environment* (2000), Lewontin addresses this argument in four points that place organisms at the centre of niche definition. Animals (1) choose external elements useful for the construction of their environment; (2) actively affect their surroundings; (3) alter environmental parameters through their action and (4) produce modulations of external conditions through organismic introjection. Lewontin's work is very important precisely in emphasising the individual's role in evolution: an animal is not only an object of the phylogenetic process but also a subject, an actor capable of setting a trajectory thanks to its reactive capacities to environmental stresses.

If we associate von Uexküll's vision of the elective *Umwelt* of organisms with Lewontin's active conception of modifying and adapting the environment to the animal's needs, we obtain an image of the niche that is no longer a preformed space, to which the species must adapt, but a *dimension* of the species itself. This brings us to the theory of niche construction promoted by John Odling-Smee in *Niche Constructing Phenotypes* (1988, pp. 73–132): the author refers to the ability of organisms to construct, modify and choose the most important elements of the environments in which they live. Basically, this is the action of species-specific behaviour on the evolutionary mechanisms of natural selection, a topic touched on only marginally by authors such as Leigh Van Valen (1935–2010) with the Red Queen hypothesis and by Richard Dawkins in *The Extended Phenotype* (1982). For Odling-Smee, instead, animal behaviour plays a major role in the evolutionary process, as it is capable of modifying the ecological factors underlying selective mechanisms. To use the words of Forestiero (2009, 269): "By constructing their niche, organisms can alter the environment, not necessarily in a way that is favourable to them, but always contributing to producing the conditions that will affect their local adaptation". The theory of niche construction speaks of a developing correspondence between the organism and its living environment, so that the individuals of a given population eventually reach a kind of equilibrium of congruence between the expressed phenotype and the altered environment. Subsequent generations thus acquire a double inheritance, because in addition to the genetic inheritance they have the ecological inheritance of the environment transformed through the niche construction.

The relationship between animal behaviour and evolutionary mechanisms had already been indicated by Charles Darwin, especially in his second essay *The Descent of Man and Selection in Relation to Sex* (1871). Authors such as James M. Baldwin (1861–1934) with the concept of *organic selection*, Conrad Waddington (1905–1975) with the idea of *genetic assimilation* and Jean Piaget (1896–1980) would later point out in various ways how the expressive phenotype, through its responses to environmental stimuli, could direct the phylogenetic trajectory. However, these authors did so in a profoundly different way from Lamarck's hypothesis of the inheritance of acquired traits, recognising the role of natural selection in the definition of the morphotype. Ernst Mayr himself, in the essay *Animal Species and Evolution* (1963, 604), had stressed that: "A shift into a new niche or adaptive zone is, almost without exception, initiated by a change in behavior". In this case we are talking about a shift in selective pressures, which does not imply a suspension of natural selection—considered, in any case, the outcome of competition over reproductive resources and therefore never avoidable—but rather a modification of fitness parameters. This puts the relationship between creativity and evolution back at the centre, bearing in mind that every behavioural innovation affects the niche construction and consequently the selective pressure factors or morphogenesis.

The niche construction undoubtedly has a number of repercussions on the species, which are complex and certainly not linear. In fact, niche construction can support a stable polymorphism in the species, which would otherwise be drastically reduced by environmental factors, just as it can stabilise phenotypic homogeneity by eliminating polymorphism (Laland et al. 1996). However, the phenotype is the product of the genotype and the environmental paratype, so acting on the environment not only modifies selective pressures, favouring outcomes that would otherwise have very different trends, but also affects the developmental environment. Therefore, it modifies the factors that, by interacting epigenetically with the heritage, are responsible for the phenotype, which in turn is the real actor in the processes of natural selection. It is clear, therefore, that the animal population is not a passive entity in the face of the selection process operated by the external environment: this image is not only approximate but above all misleading. I do not doubt the central role of environmental conditions in influencing the morphogenetic agenda of organisms that, by necessity, must come to terms with the demands that each condition places on them, but I wish to reiterate that the niche is something far more complex than a space to be occupied and above all that each species does not merely inhabit but constructs its own living environment.

Reflection on the niche and on the active role of populations in its construction, even if correct as regards modification of selective pressures and influence on the phylogenetic morphogenesis of the species, clearly requires further explanation through ethology. The first question that arises is what behavioural factors are involved in niche construction and how they relate to creativity: it is not enough to affirm the centrality of the individual as an evolutionary subject, but it is essential to understand its title. The second question concerns the relationship between niche construction and the processes of speciation, in terms of (1) the mechanisms of anagenesis, understood as the progressive refinement of the morphotype, and of cladogenesis, as the formation of new species by disjunction; (2) the events of sympatric and allopatric speciation, that is, how the niche can intervene by facilitating or encrypting these possibilities.

In an extended conception of evolutionary theory, it should be reiterated that, as pointed out by Stephen J. Gould's *Punctuated Equilibrium* (2007), it is not enough to dwell on species transformation or speciation, but it is necessary to explain the long periods of stasis in the general morphotype of the species, as observed by palaeontologists. I believe that here, too, the theory of niche construction allows us to understand the role of the niche as a factor capable of buffering environmental fluctuations, stabilising selection. This was explained by the Ukrainian zoologist Ivan Schmalhausen (1884–1963): in a species, intermediate characters tend to prevail over extreme ones if conditions remain largely unchanged. Niche construction tends to create a microenvironment where parameters are kept under control, whence a calming effect on fluctuations.

The relationship between individual and population in niche construction can be facilitated by the processes of parental transmission and social learning, whereby an invention or discovery emerging in a single individual may end with the lifespan of the individual or a single group, but may sometimes spread within the population and constitute an initial aggregative niche nucleus. In fact, a particular invention can have amplifying effects—a tool may make a certain food accessible or facilitate the use of another tool—shortly inaugurating a complex tradition in a population, perhaps even diverging from the original species by giving rise to a new adaptive niche. Speciation is a process that requires biological timescales ranging from a few tens of thousands of years to even more substantial spans, but in any case it is based on an interruption of gene flow between two populations that suspends the mechanisms of cross-reproduction. Typically, this can occur through geographical isolation, for example through group confinement, or through migration as in the founder effect. The disjunction is not necessarily produced by a geographical boundary since, as proposed by Hugh Paterson in *A Comment on "Mate Recognition Systems"* (1980, 34, 330–331), each species has a kind of recognition signature that allows an individual to recognise its conspecifics, ensuring the correct formation of pairs. A niche divergence might produce two different recognition signatures, constituting, in fact, a reproductive barrier even in the absence of concrete geographical isolation. This would also explain the processes of parapatric speciation, when a new group enters an adjacent and therefore non-isolated niche, giving rise to the first core of a new species.

Thus, some cultural traditions that are observed in various animals, such as chimpanzees and cetaceans, and that in some cases, such as killer whales, produce reproductive segregation effects, may represent workshops for parapatric speciation that could, over time, result in significant morphotypic differentiations. Cladogenetic evolution is often the outcome of reproductive isolation, for example when a geographical factor tends to separate two populations (allopatric) or when a population migrates to a new environment resulting in a founder effect (peripatric). In the case of niche innovation by a single population, it is not a case of isolation caused by geographical barriers but of different ethological habits. We have seen a similar condition when discussing sympatric speciation by assortative mating: among geese, for example, species dimorphism tends to progressively increase segregation as a result of the propensity to choose a partner with plumage similar to the parent's.

On the other hand, niche construction should not be seen as a finite expression, the phenotypic manifestation of a precise genotype, but rather as a laboratory that develops on the basis of contributions that are partly innate and partly transmitted intergenerationally. It is clear, then, that in both cases expression oscillates around a certain constitutive barycentre, but cannot avoid showing degrees of flexibility and, I would say, evolutionary transformation. In a word, a niche is also an open construction site, so even in the presence of small changes that accumulate over time (*anagenesis*), refining the species' morphotype, there is a stability gradient. When we speak of innovative behaviour or established habits within a population, we cannot ignore the circumstantial nature of these factors, whereby a species maintains a certain level of niche variability within itself, so that it is easy for progressive changes to occur over time in relation to the overall evolution of the environment, alongside oscillations that fall into stasis.

The occurrence of behavioural factors is just as chaotic: a habit can be the expression of an epiphenomenon, of a co-optative process, of learning and even of an ambivalent mechanism—think of the stickleback's courtship dance. In *Behavior and Systematics* (1958, 341–362), Ernst Mayr points out that an expression can have several origins, such as redirective or dislocation behaviour, and can even be learned from heterospecifics. Sometimes a fortuitous event can bring about a change in a bird's courtship song, and this event can end with the individual or give rise to a new tradition. Animals are active entities, moving in search of new opportunities, chasing a certain habitat when it mutates geographically due to weather changes, inventing new strategies to expand their foraging space, modifying their environment to make it more congruent with their needs.

Creativity, in its different forms, enters deeply into these mechanisms of niche construction, ushering in veritable revolutions in the selective matrix of a population or buffering environmental fluctuations over long periods of time. Niche construction theory helps us understand the mechanisms of episodic evolution and long periods of stasis highlighted by Stephen J. Gould and Niles Eldredge in the 1970s.

Today, there is increasing talk of epigenetic mechanisms. This may not fully re-credit the hereditary nature of acquired traits, since these do not involve actual structural changes to the genome; however, through marking processes like methylation, these mechanisms can influence transcription—by silencing or emphasising a gene, intervening in regulatory genes, modifying expression times—and thus affect the phenotype. Palaeontology had been somewhat sidelined from genetics in the evolutionary explanation of the first half of the twentieth century, as Gould rightly notes with respect to the neglect of the stasis phenomenon. Similarly, developmental sciences had made little contribution to the debate until the emergence of Evo-Devo. Brian Hall's essay *Evolutionary Developmental Biology* (1992), indeed, paved the way for a more fluid conception of the construction of the body, emphasising the genetics of development and no longer just the genetics of character transmission, thanks to the action of regulatory genes on the definition of the somatic *baunplan*. Today, ethology must also be given a place in this debate.

Instinct and Intelligence

The study of innate abilities in the animal world has a respectable tradition in the history of biological thought. Erasmus Darwin, Charles' grandfather, already addressed this topic in his *Zoonomia* (1794). Moving away from empiricist thinking, which considered the mind a *tabula rasa*, he recognised four fundamental characteristics in animals—irritability, sensibility, will and associability—while maintaining a sceptical attitude towards instincts. Erasmus believed that animals have a great capacity for learning and were therefore more inclined to attribute to them a vivid intelligence, capable of acquiring behavioural skills through experience, rather than instinct. He cited, for example, the case of nightingales whose songs vary in different regions of Europe, and argued that even their migratory movements "are not produced by a necessary instinct, but are accidental improvements, like the arts among mankind, taught by their contemporaries or delivered by tradition from one generation of them to another" (1794, 167).

Erasmus Darwin's reasoning stood in contrast to the idea of perfect instinct proposed at the time by natural theologians, showing that animals are very often wrong in their responses. But what made Charles' grandfather most sceptical concerned one of the central aspects of niche construction: the creation of nest sites. In his reflections, Erasmus pointed out that when making their nests, birds used variable solutions, in terms of the end result, the material used and the designated site. The modulation of nesting behaviour in relation to environmental specificities, such as how to camouflage a nest according to the characteristics of the site, fascinated him, bringing to the fore the birds' evaluative capacity and flexibility rather than the presence of instinctive automatisms. For Erasmus Darwin, animals constantly adapt their behaviour, urged to a continuous adaptive exercise.

The expressive plurality in the construction of nests demonstrated by birds (which also use sites made available by other species, such as rabbit holes or the roofs of bell towers) demonstrated an intellectual capacity rather than a large number of instincts, each assigned to a particular situation. Erasmus Darwin's work then went on to analyse other animals, showing how the intelligence of a species is revealed precisely in its ability to operate in unusual circumstances, for example, at the edge of the niche or in situations in which an animal has to rebuild its comfort zone. He paid particular attention to the way in which worms cover their holes with leaves and straw, an aspect which was to influence Charles Darwin's last work on this subject, *The formation of vegetable mould through the action of worms, with observations on their habitats* (1881). Ultimately, Erasmus' reflections led to a strong downsizing of the notion of instinct in favour of the recognition of the individual's adaptive intelligence: a subject is capable of finding flexible solutions to its needs in relation to environmental variables. An animal is a sort of carpenter who must necessarily rely on acquired knowledge and experience, but must also be able to constantly evaluate his own work. In *Zoonomia* we thus find a great many references to the observational and experiential capacities displayed by animals, so that rather than perfect instinct it was necessary to speak of a constant adaptive exercise on the part of the individual.

A somewhat similar approach was taken by William Smellie (1740–1795), in a more gradualist perspective that, while recognising instincts, admitted their individual adaptability. In *The Philosophy of Natural History* (1791),

Smellie reiterated that instincts are general capacities that are applicable, that is adaptable, to specific circumstances and in some cases perfectible through experience. This author, therefore, differentiated between the *general sphere of species-belonging*, characterised by a specific style, given by instincts and placing a species in a precise rank—today we would say a niche—and the *individual sphere*, whereby the subject is the main actor in the process of adapting to a given situation. But the most interesting aspect of this thoroughly modern view of the concept of instinct was the idea that there is a direct proportionality between the number of instincts possessed and expressive plasticity. This was equivalent to affirming that a richer menu of options from which to choose corresponds, in the end, to greater expressive freedom. This aspect was to find Charles Darwin's agreement, in contrast to those who maintained the opposition between instinct and intelligence, in a relationship of inverse proportionality.

A great contribution to the study of instincts was made in the eighteenth century by French entomologists: consider especially René-Antoine de Réaumur's (1683–1757) *Mémoires pour servir à l'histoire des insectes* (1734) and Charles-Georges Le Roy's (1723–1789) essays in the *Encyclopédie, ou dictionnaire raisonné des sciences, des arts et des métiers, par une société de gens de lettres* (1782). The skills of insects—in particular bees' ability to construct perfect hexagonal cells or solitary wasps' ability to anaesthetise with surgical precision the nerve centres of their victims on which to lay their eggs—were undoubtedly overwhelming proof of instinctive capacities. Indeed, it seemed counterintuitive to think that these animals had the capacity to achieve such results through reasoning. Nevertheless, de Réaumur, while lamenting the custom of attributing human qualities to insects, at the same time refused to regard them as mere machines. Again, he considered the ability to vary behaviour in relation to circumstances and the tendency to make mistakes as evidence of animal intelligence. We find a similar attitude in Le Roy, who rejected the thesis of automatism, although he avoided making a clear separation between instinct and intelligence, considering them as two complementary faculties. According to Le Roy, instincts are general operative dispositions that have to be applied to particular circumstances through intellectual faculties such as memory, judgement and reasoning.

On the other hand, the most interesting aspect expressed by Le Roy in his essay *Lettres sur l'intelligence et la perfectibilité des animaux, avec quelques lettres sur l'homme* (1802) is the idea that instinct is an ancient acquired knowledge, transmitted intergenerationally until it assumes the character of spontaneity in the subject, rejecting Etienne Bonnot de Condillac's (1714–1780) idea that each generation is forced to start afresh. On the contrary, he believed that the individual can rely on the achievements of previous generations. This theme was taken up by Jean Baptise de Lamarck (1744–1829) in his *Discours d'ouverture* (1800), later included as an introduction to *Système des animaux sans vertèbres* (1801), where the author launches his transformist theory as the tendency of living beings towards complexity and the inheritance of acquired characters. The metamorphosis of species, according to the French zoologist, is due to environmental changes, but above all to the habits acquired by populations, capable of fortifying certain faculties through use and therefore preserved and propagated in subsequent generations. In *Philosophie zoologique* (1809), Lamarck recognised two factors in the process of transformation: the first is *endogenous*, intended to make the organisation of living beings more complex; the second is *exogenous*, linked to circumstances and relating to the influence of habits on organs, whereby a faculty, if not used, eventually deteriorates until it disappears altogether.

In *Histoire naturelle des animaux sans vertèbres* (1815), Lamarck divided animals into three categories: (1) *apathetic*, that is with no true sentience, but only irritative capacities, such as polyps and worms; (2) *sensitive*, that is other invertebrates, incapable of making true choices and driven by entirely instinctive actions, but devoid of cognition; and (3) *intelligent*, basically all vertebrates, capable of behavioural variability. According to Lamarck, there is a clear opposition between instinct and intelligence, to the extent that the presence of the former compromises the capacity for reasoning: intelligence is a different faculty from both instinct and habit, and the latter two risk limiting the former. An interesting contribution to this debate between instinct and intelligence was made by John MacCulloch (1773–1835) who, in considering the construction of bee cells and spiderwebs, as well as bird's nests, reiterated the importance of the *convergence* of instinct and adaptive intelligence. According to the Scottish geologist, the property of instinct was by no means sufficient in the implementation of projects; memory, evaluation and even inferential faculties were needed as well. Since animals adapt their projects to the circumstances, it is reasonable to assume, according to MacCulloch, that they are aware of the purpose of their actions, by which they evaluate the adaptation process moment by moment.

The skills of insects have often been at the heart of the debate on animal intelligence, for two reasons: (1) on the one hand, these animals have an obvious capacity to adapt their behaviour to particular circumstances; (2) on the other, it is clear that some of their expressions are entirely instinctive. But this made the explanation of their appearance particularly problematic, for it was impossible that these qualities could have been acquired by previous habits, whereas, according to natural theology, they were evidence of a divine design. In this respect, I would like to cite a study by William Kirby (1759–1850) and William Spence (1783–1860): *An Introduction to Entomology* (1828), which had a strong influence on Darwin. Spence, in particular, considered the instincts of insects to be much more elaborate than those of mammals and birds, but he rejected the rule of inverse proportionality between instinct and intelligence, showing how, in interaction with artificial objects, insects still demonstrate marked abilities. To support this view, he reported Francois Huber's (1750–1831) experiments on bees and glass—a subject that Darwin would take up again in his observation of earthworms to test their ability to use new materials like paper.

The evaluation of the behavioural characteristics of domestic animals was particularly influential in considering the evolution of instincts and dispositions. This subject had already been considered by both Erasmus Darwin and Jean Baptise de Lamarck and was to find further development in the analyses of Thomas A. Knight (1759–1838) and John Sebright (1767–1846), becoming central to Darwin's work. According to Knight's *On the Hereditary Instinctive Propensities of Animals* (1837), the training of hunting dogs over several generations consolidated in the form of instincts. For example, pointer dogs are driven to "point" birds and small game and retrievers, indeed, to retrieve the prey. Similarly, in his essay *Observations upon Instinct of Animals* (1836), Sebright claimed that breeding introduced new instincts into domestic animals by deleting others found in their wild state. Sebright pointed out an aspect that Erasmus Darwin had already discussed, namely that only birds living in areas frequented by humans are afraid of them. In these authors as well as in the early writings of Herbert Spencer (1820–1903), instinct appears to be the result of repeated experience in previous generations, which slowly became consolidated within the population. In *Principles of Psychology* (1855), Spencer considers the expressions common to all individuals of a species, explaining their innate character as a kind of ancestral experience that predates the individual's acquisitions. He hypothesised an intergenerational accumulation of experience so that, taking up George Lewes' (1817–1878) view that instinct is nothing more than lapsed intelligence made automatic by heredity, Spencer referred to this process as lapsing of memory into instinct. Not only Kantian a-prioris, but also basic emotions, such as fear, were for Spencer the manifestation of an ancestral memory.

In Charles Darwin we find echoes of all these authors, including those oriented towards natural theology, who allowed the English naturalist to test his evolutionary theory. Research on domestic animals and the feedback from breeders play a central role in his work, alongside, of course, the experience gained during the five-year voyage of the Beagle. The inheritance of traits acquired through breeding and training practices, emphasised by authors such as Sebright and Knight, was therefore held in the highest regard by him. However, alongside this, he believed that just as random mutations found a positive response from breeders, who rewarded them through backcrossing, so in nature the struggle for resources operated in a similar way. The work of the economist Thomas R. Malthus (1766–1834) *An Essay on the Principle of Population, as it. Affects the Future Improvement of Society* (1798), which showed a gap between the geometric growth of population and the arithmetic growth of resources, ignited in the English naturalist's mind the principle of the struggle for survival.

The characters of species were shaped by natural selection, as they were subjected to the continuous scrutiny of performative survival tests: the slow accumulation of small variations over the long course of the planet's gradual transformations, following the theory developed by Charles Lyell (1797–1875) in *Principles of Geology* (1830), made the species increasingly adapted to environmental challenges. The idea that animals were born equipped with instinctive equipment suggested to Darwin that the foundational ideas attributed to the pre-existence of the soul in Plato's *Phaedo* should instead be assigned to an ape-like ancestor. While Darwin could accept that the inheritance of acquired characters also had a function, alongside that of natural selection, in the definition of the behavioural *habitus*, he could not agree with Lamarck on the idea that extinction was an apparent phenomenon, precisely because it was clearly at odds with the process of natural selection. On the other hand, it was clear to him that not all instincts could be explained through habits acquired by ancestors because, as mentioned above, some of them could hardly have been learned and built by habit. Rereading Kirby and Spence's *Introduction to Entomology*, a text Darwin particularly appreciated, he had to grapple with the problem of sterile individuals in insect communities and their clearly non-heritable instincts.

Another important text in the formation of Darwinian theory was *Nouvelles Observations sur les Abeilles* (1814) by the aforementioned Francois Hubert, particularly with regard to intelligence in insects. Hubert pointed out that many gradual stages can be found in nature, so that the astonishing characteristics of bee cells are merely apparent: observing the structure of the cells of the genus *Melipona*, one can see that it would take only a few innovative measures to achieve the performance level of domestic bees. Darwin reiterated his opposition to Georges Cuvier's (1769–1832) idea that instincts exist in inverse proportion to intelligence. In his essay *The Descent of Man and Selection in Relation to Sex*, he argues: "Those insects which possess the most wonderful instincts are certainly the most intelligent (…) and amongst mammals the animal most remarkable for its instincts, namely the beaver, is highly intelligent" (1871, I, 37). Darwin considered the effects of culture and prosociality on human evolution with a conception close to today's niche theory. His approach was a consistently continuist one: this differentiated him from Alfred Russel Wallace (1823–1913) and his *The Origin of Human Races and the Antiquity of Man Deduced from the Theory of Natural Selection* (1864), which posited a clear discontinuity.

In *The Formation of Vegetable Mould through the Action of Worms* (1881), Darwin took up the gradualist approach borrowed from Lyell, whereby great things could be achieved through gradual and uniform action. The activity of earthworms, capable of producing vegetable mould through continuous digestion and expulsion of matter, was in line with his view of slow accumulation of small adaptive mutations. Despite the doubts expressed by Thomas Huxley (1825–1895) himself, for Darwin it was fundamental to explain the evolution of characters through the continuous and gradual action of mutations and selection. In this essay, Darwin pointed out that the earthworms' daily work brought man face to face with the need to change his way of reasoning, an inability that had previously delayed progress in geology and now was hindering the understanding of the origin of species. Furthermore, observing the earthworms' behaviour of dragging leaves into holes, he insisted that it should not be explained as a simple reflex, but by admitting the ability to couple attention with purpose-related behaviour. The attribution of mental qualities to earthworms, which Darwin tried to demonstrate through experiments with pieces of paper, influenced his disciple George J. Romanes (1848–1894).

Romanes saw instinct and intelligence as converging in the idea that *evolution itself is driven by mental activity*, with the individual as a bearer of innovative capacities responding to environmental stimuli. Following Spencer's perspective, Romanes saw the mind as a magmatic force necessarily pushed towards increasingly complex forms of adaptation, which makes the individual capable of reconfiguring hereditary endowments through innovation. For Romanes, these instinctive endowments have a dual origin, as the result of the inheritance of acquired habits—we might say in a Lamarckian way—and as the accumulation of chance variations rewarded by natural selection. In *Mental Evolution in Animals* (1883), Romanes considered such endowments in terms of plasticity: instinct, he claimed, is modifiable through intelligence. Romanes' view of the inadequacy of instinctive behaviour was also strengthened by the studies of Douglas A. Spalding (1841–1877). The latter had used the term "imperfect instincts" to describe behaviours such as the imprinting of chicks, which require particular experiential acquisitions in order to work correctly. There is no doubt that Spalding's work had blurred the distinction between innate and learned behaviour, and this was to have an influence on the work of George Lewes who, in *Problems of Life and Mind* (1874), emphasised the importance of experience and above all of the individual as a factor in evolution. It is surprising that all this attention to the centrality of behaviour in evolutionary processes was not adequately followed up in the twentieth century.

But there might be a reason for this. After the counter-reformation of Lloyd Morgan and his canon of parsimony, any focus on individual behaviour and animal intelligence suffered a kind of censorship, whereby animals were seen as passive entities subject to the external (environmental) factors of natural selection. Starting with Thomas Hunt Morgan's (1866–1945) studies on the fruit fly (*Drosophila melanogaster*), which began in 1910 at Columbia University in New York, the focus shifted on recognising gene carriers in chromosomes, demonstrating their ability to give rise to mutations and laying the genetic foundations of heredity. This led to the development of the Modern Evolutionary Synthesis, which closely linked Darwinian theory, based on natural selection, with Gregor Johann Mendel's (1822–1884) theory of heredity, in the belief that biological variety is due to random genetic mutations. In addition to Hunt Morgan, the authors who contributed to this transformation of evolutionary theory—defined by Romanes as neo-Darwinism, in deference to the considerations of August Weismann (1834–1914) and Alfred Russel

Wallace, who rejected the heredity of acquired traits—include Ronald Fisher (1890–1962), Theodosius Dobzansky (1900–1975) and John Haldane (1892–1964). It is clear that the Synthesis was a turning point, focusing entirely on genetics and mathematical models, leaving little room for behavioural anecdotes.

It is only since the 1970s, thanks to the pioneering work of scholars such as Richard Lewontin, Stephen J. Gould and Niles Eldredge, that this monolithic vision of evolutionism has been called into question, with the emergence of historical considerations that focus on animal populations as active entities and not just as simple gene containers. Animals become once more the protagonists of this process, as had already been clear to Charles Darwin himself: the latter had not been afraid to devote the final part of his life to observing and experimenting with the flexible behaviour of earthworms. On this basis, in the last decades of the twentieth century, we witnessed a slow transition that would lead to recognising the combined effects of the environment on developmental processes, epigenetic inheritance, the symbiotic nature of living beings and niche theory, in a global vision now defined as the Extended Evolutionary Synthesis, which puts the protagonism of organisms back at the centre.

The Extended Evolutionary Synthesis

The ability of a given individual to originate different phenotypes has been defined as phenotypic plasticity and can be considered as an organism's ability to be susceptible, during the developmental phase, to particular environmental releasers that can produce morphological and behavioural changes. We know, for example, that the development of larvae in social hymenoptera is regulated by the dietary factor, which will determine their fate as sterile or fertile individuals. We should not consider these elements as accidents or deviations from the normal autopoietic pattern of development, which is usually assumed to be controlled exclusively by the genome. On the contrary, we must accept the fact that the organism's own developmental plan already includes the cofactorial intervention of the environment. Obviously, this word means something different from the traditional conception of the individual's reception in the space-environment or of the stimulating factors. The cofactorial environment in the developmental process is not something external, but part of the formative ingredients of the individual morphotype. Phenotypic plasticity—that is, the property of giving rise to differently shaped organisms, even from the same genome, according to environmental variables—is certainly a major adaptive advantage.

The possibility of giving rise to standard and distinct phenotypes within a species is called *polyphenism*. An example of this is the dual morphotype of the orthoptera *Schistocerca gregaria*, which has a solitary morphism with short wings and green colour and a gregarious morphism with long wings and intense colouration. Already in 1963, in his essay *Animal Species and Evolution*, Ernst Mayr differentiated between morphological differences due to genes and those produced by the environment. We can identify different forms of polyphenism. It can be (1) seasonal, that is linked to the time of development: for example, the caterpillars of *Nemoria arizonaria* that hatch in spring have a shape reminiscent of oak flowers, while those that hatch in summer resemble twigs (Greene 1989). Polyphenism can also be (2) based on exposure temperature in the embryonic process, as in the sexual definition of some reptiles, for example turtles and crocodiles, or (3) nutritional, that is related to how the larval organism is fed: royal jelly, through the increase of neotenin, increases the body size of bee larvae and the development of fertile females with functional ovaries. Polyphenism can be (4) defensive, that is predator-induced: for example, tadpoles of the treefrog *Hyla versicolor* are able to change the colour of their tails and body structure if molecules from their predator are present in the water (McCollum and Van Buskirk 1996). Finally, it can be (5) induced by conspecifics: the fish *Thalassum bifasciatum*, for example, develops as male or female depending on whether the opposite sex is prevalent in the area. Indeed, countless factors determine the phenotype, including the microbial environment and the symbionts accepted by the organism during development, as well as certain physical parameters such as gravity, pressure and light exposure.

Evo-Devo theory acknowledges the importance of the developmental process as a key moment in the definition of the phenotype and, consequently, in the evolutionary processes. Phenotypic plasticity affects the processes of adaptation and sexual selection, since preference in partner choice is based on the traits displayed. Today we are discovering that phenotypic induction, triggered by the environment and made possible by the developmental plasticity of the genotype, can be channelled and stabilised through selection, as Waddington had already intuited with the concept of genetic assimilation. But there is more! The developmental steering carried out by the environment on a particular character is never limited to that character, but leads to relevant systemic effects, affecting the organism as a whole, through the process of *phenotypic accommodation*. West-Eberhard (2003) defines this phenomenon as the mutual adjustment of parts during development: a new regulation of the developmental process can have significant effects on the entire anatomy, as no individual part can change independently. This can be observed in the different breeds of dogs, which, despite differences concerning—say—their splanchnocranium, such as prognathism and enagnathism, have their muscles, cartilage and bones coordinated with each other in such a way as to enable functional optimality.

When a phenotype has advantageous adaptive traits, the phenomenon of genetic accommodation may occur, whereby if there is genetic variability within that population, selection may favour those alleles or genetic combinations capable of stabilising, integrating or even perfecting the expression of that phenotype. In these cases we speak of *cryptic genetic variation*, as it is kept hidden by the uniformity of environmental induction. Diversity can suddenly arise due to variations in environmental situations or in the lifestyle of a particular population. Thus, the phenotypic uniformity of a population is not merely an expression of genomic uniformity, but a containment mechanism on the part of environmental factors, for example through *niche homeostasis*. Variability, on the other hand, can be brought to light by environmental or style variation and then subjected to new forms of natural selection. The results of this dialogue between genetic characteristics and environmental factors in the development process are numerous: for example, they may favour a particular morphotypic declination or, on the contrary, maintain a certain gradient of phenotypic plasticity. What is certain is that the behavioural factor, which lies at the heart of the relationship between animal and environment, plays a fundamental role in both the micro- and the macroevolution of the species. Moreover, the threshold systems of phenotypic and genetic accommodation processes help us to better understand the phenomenon of stasis and punctuation in palaeontological finds.

The great paradigmatic shift produced by authors such as Werner Callebaut, Gerard Müller and Stuart Newman in *The Organismic Systems Approach: Evo-Devo and the Streamlining of the Naturalistic Agenda* (2007, 25) is to consider ontogeny as the engine of phylogeny, reversing the Haeckelian concept of recapitulation. The issue of the generation of new forms during geological eras, which can also be defined as macroevolution, is revealed as the construction of new body plans (*baumplan*) and as the latter's rearrangement through speciation. It is evident that in both cases what produces morphopoiesis is achieved by altering developmental programmes. Let us consider the concept of threshold in morphogenesis as an emergent (we could say quantum) effect, whereby there is no passage through intermediate forms, but only morphotypic orbitals that are characterised by a condition of equilibrium and correlation of individual parts. As noted by Gerd B. Müller in *Le origini della novità morfologica* (2013), the idea of threshold effects is based on the fact that most morphological changes are associated with alterations in the size and proportions of body parts in organisms. There can be processes of heterochrony or chronological displacement of developmental processes with hypertrophies and restraints given by the reciprocal action exerted by tissues. Müller points out that the biomechanics of embryonic development affect morphogenesis by virtue of correlative or tendency processes: for example, in whiptails (of the *Teiidae* family) the progressive elongation of the body is combined with a reduction in limb length.

But is it possible to link animal creativity and its consequences for niche construction with the changes in the developmental process noted by the Evo-Devo theory? My answer is definitely affirmative: behaviour is the most important factor in niche construction and the true developmental environment of organisms belonging to a certain species is precisely the niche. In *Endless Forms Most Beautiful* (2005), Sean B. Carroll recalls how the modification of the gene coding for the development of the temporalis muscle (MYH16) in humans means that the fibres of this muscle in our species are reduced to 1/8 compared to other primates like the macaque. This resulted in a reduction in the mandibular musculature, which had significant developmental effects by reducing tension in the skull. According to Carroll, this allowed the skull to become thinner and larger, leading to an expansion of the encephalon in hominid evolution (260–265). I propose, then, a hypothesis that brings together Evo-Devo and niche construction theory: (1) the use of fire and early lithic tools, that is a new niche, lowered the selective pressure on temporalis muscle performance, enabling the survival of subjects with a mutation in the MYH16 gene and the deactivation of the protein responsible for the development of temporalis muscles; (2) the lesser development of this muscle had an effect on embryonic development, reducing tension on the cranial bones, promoting encephalic growth.

Another fundamental chapter in the formation of the phenotype is the concept of *epigenetics*, coined in 1968 by Waddington to define how the regulation of gene expression induced by environmental factors is able to influence morphogenesis. With the rise of Evo-Devo and the discovery of gene modulation mechanisms, the term epigenetics came to be understood as the set of mechanisms involved in the regulation of gene expression during development. We now know, for example, that certain genome marking mechanisms, such as the methylation of certain gene regions or the modification of histone proteins, alter their expression, that is, their transcription into messenger RNA and thus the production of the protein. We can therefore consider epigenetics as the set of mechanisms that produce phenotypic variability without modifying the nucleotide sequence of the genome. Epigenetic regulation or modulation can be caused by factors that directly affect transcription by acting on the DNA, such as methylation or regulation of the proteins that package genes, either favouring or not favouring transcription by RNA polymerase. The regulation of transcription by hormonal action or by environmental interference, while acting on a gene, can have significant consequences due to pleiotropic effects or when involving particular areas such as homeoboxes. The effects of environmental agents, for example temperature or moisture gradients, can give rise to marked phenotypic differences: this is the case of the butterfly *Bicyclus anynana*, which has a morphotype for the dry season and one for the wet season.

We speak of *epiallelic inheritance* and variations in the form of *epimutations*: crustaceans of the *Artemia* genus can show environment-induced changes in the chromatin configuration that increase heat resistance, and this character is passed on to subsequent generations even in the absence of the corresponding environmental stimulus; mice can transmit olfactory experience over several generations through the hypomethylation of certain regions of the relative gene in spermatozoa; likewise, a metabolic alteration due to a low-protein diet in rats (*Rattus norvegicus*) can be inherited over several generations. A *transgenerational epigenetic inheritance* has also been found in humans, relating to cardiovascular problems, susceptibility to diabetes or certain neoplasms.

The term *epigenetic inheritance* indicates transmissible morphotypes that are not encoded at genome level. As Scott Gilbert and David Epel report in *Ecological Developmental Biology: The Environmental Regulation of Development, Health and Evolution* (2015, 70–73), in order to document the heritability of environmentally induced epigenetic changes in mammals, it is necessary to refer to the F3 generation. If a female (F0) is exposed to the environmental induction factor, the foetuses in her womb will also be exposed (F1) and so will the foetuses' germ cells (F2). For example, in the case of epigenetic expression of the *Agouti* gene in rodents, the ancestors' diet is able to influence the phenotype of the F3 generations (Morgan et al. 1999), even if they are not themselves subjected to that diet. Obviously, if it is the male who has undergone induction, living in a different environment from the female, or if conception occurs by in vitro fertilisation, we can analyse inheritance from the earliest generations.

The role of behaviour is also central to epigenetic inheritance because, as we have seen, it is a form of inheritance of acquired traits, not relating to the structure of the genome but to its expression during the development process. The term *acquired traits* refers to the individual's experience in terms of (1) type of diet, especially poor or unbalanced; (2) conditions of prolonged stress or fear, for example associated with certain smells; (3) exposure to particular molecules, such as endocrine disruptors or chlorine compounds, or irradiation; (4) infection with certain viral or bacterial agents; (5) alterations in metabolism or hormonal homeostasis, as a result of excessive competition, crowding or domestication. One particular form of epigenetic inheritance involves the transfer of small non-coding RNAs that are transmitted through the germ line and can modulate gene expression. In the nematode *Caenorhabditis elegans*, starvation can produce small RNA fragments capable of silencing particular genes, and this process is passed on over several generations. In some ways, epigenetic inheritance seems to be reminiscent of the inheritance of acquired characters proposed by Lamarck in his *Philosophie Zoologique* (1809), although it is clear that epiallelic action takes place through mechanisms other than the simple use or disuse of organs. On the other hand, it must be said that both phenotypic modulation of the developmental process and epigenetic inheritance involve a third factor: the ecological relationships between living beings and their symbiotic or interdependent dimension.

As we know, the concepts of individual and autopoiesis are slowly giving way to other definitions due to the growing discoveries in terms of developmental symbiosis, which reveal the organism more as an ecosystem than a genetically pure entity: suffice it to think that a human being is composed of 90% prokaryotic cells and that about 35% of the metabolites present in blood derive from bacteria (McFall-Ngai et al. 2013). The ubiquitous presence of symbiotic mechanisms, not only in the environment or on an organism's external or internal surfaces, but even within cells, was highlighted by Lynn Margulis in her essay *Symbiotic Planet* (1998). This is also true for developmental processes, so we can say that the morphotype is deeply influenced by the microorganisms hosted by the animal species. The forms of symbiosis within an organism are divided into aspects of mutualism and commensalism, favouring functional processes, such as cellulose metabolism in the foregut of ruminants, wood digestion in some termite species through the symbiont *Mixotricha paradoxa* and coral colouration due to zooxanthellae algae providing nutrients and oxygen. The molecular techniques developed in the twentieth century showed that no organism can properly be called an individual, but is rather a *holobiont*: an entity that develops and functions on the basis of complex processes of multi-species relationships within the organism itself.

Natural selection therefore cannot operate exclusively on presumed gene pools, but must also rely on functional consortia or adaptive teams. However, what is most surprising is not so much the symbiotic performativity of organisms, but the way in which symbiosis enters into the morphogenetic process itself, which is why we speak of *symbiopoiesis*. Hence the concept of *endosymbiosis*, whereby microorganisms hosted within the cell itself are able to influence gene modulation through the already mentioned mechanism of selective transcription. As noted by Scott Gilbert and David Epel (2015, 92–105), in the nematode *Brugia malayi*, the zygote relies on bacteria of the genus *Wolbachia* to regulate the cell divisions that give rise to the antero-posterior axis. In particular, a crucial role played by bacteria of the genus *Wolbachia* in the developmental stages has also been found in the parasitoid wasp *Asobara tabida* and in many other invertebrates, where there is also evidence of a vertical transmission process: these bacteria are permanent residents in the cytoplasm of oocytes. In mammals too, it has been shown that bacteria play a key role in the development of particular organs through the production of releasers that influence gene regulation. For example, some symbiotic bacteria have been found to influence brain development in mice (Wang and Kasper 2014).

The importance of cooperative evolution, in some ways mirroring the classical competitive evolution, has been proposed by Ashley Montague (1955), Lynn Margulis and Dorion Sagan (2002), as well as Donna Haraway in her essay *Staying with the Trouble: Making Kin in the Chthulucene* (2016), through the concept of symbiopoiesis. The same notion can be derived from my own concept of *heteronomy of the living*, whereby a being foresees in itself the effect of the external referent. In the essay *At Home in the Universe* (1995), Stuart Kauffmann pointed out that all evolution is co-evolution: I think this statement sums up what has been said so far, considering organisms and their evolutionary trajectory not only as entities caught in a continuous competitive dance with each other, but as beings that express a multiplicity of interactive modes, sometimes even through cooperation. The idea of the selfish gene, akin to computer viruses that simply replicate using hardware as containers, thus gives way to a much more articulate and pluralistic vision. As Massimo Pigliucci and Gerd B. Müller point out in *Evolution, the Extended Synthesis* (2010, 14), "Far from denying the importance of genes in organismal evolution, the extended theory gives less overall weight to genetic variation as a generative force". Many authors of the Extended Synthesis theory believe that genetic variation follows the evolutionary process, initially driven by phenotypic plasticity in response to changing environmental conditions. Genetic variation would basically guarantee the consolidation of this developmental direction, the fidelity of its inheritance and the progressive fixation of the phenotypic trait.

Creativity and Niche Construction

As Scott Gilbert and David Epel suggest in their aforementioned work, the environment does not merely select the morphological variability of living things, but is capable of engineering it: environmentally induced changes in phenotype, if they allow for better adaptation over a sufficiently long time period, can become the rule. However, while it is true that the environment has *objective inducing factors*—such as temperature, the presence of particular symbionts or predators, and the intervention of certain chemical molecules—it is equally certain that animals are not passive in the face of external conditioning factors. Animals can respond to environmental fluctuations in many ways: (1) by moving in search of more favourable conditions, (2) by changing their habits or (3) by inventing new adaptive strategies. Modification of the environment through niche specialisation is one of the most frequent ways of doing this, but at this point we need to clarify the term *niche*, because it is clear that the ecological regime of a species is in itself a niche positioning. Foraging habits, styles of interaction with other species, territoriality and reproductive patterns undoubtedly define the niche of a species, even if this does not constitute a true environmental modification. In other situations, for example in the case of beaver dams or termite mounds, on the contrary, we find the creation of a microenvironment capable of maintaining some parameters constant, such as temperature, water availability, and protection from disturbing factors. In these cases the niche construction factor is more evident.

On the other hand, in both, the behaviour of a species represents a kind of tradition that has a relevant impact on the environment, influencing its active variables and thus acting on the factors of selective pressure. If a population develops an expressive style that produces fitness in terms of reproductive success, the ability to perform that particular behaviour will also become a character subject to selection. The same can be said for the parameters of aesthetic preference or recognisability in the choice and acceptance of a partner or in competitions relating to sexual affirmation. According to the traditional ecological vision, the occupation of an ecosystemic space is characterised by objective parameters imposed from the outside, in terms of a precise position in the trophic chain. This view, here, gives way to the awareness that, in a more or less explicit manner, all animals actively contribute to constructing their niche, organising a sort of microenvironment projecting their needs, almost a sort of *amnion* capable of maintaining environmental values within a certain range of homeostasis. In this sense, it is wrong to speak of the environment exclusively in terms of something external, because each species reworks this parameter and *makes it its own* through ethological expression.

As John Odling-Smee recalls in *Niche Inheritance* (2010, 175–176), "Half a century ago a developmental geneticist, C.H. Waddington (1959) proposed the concept of an *exploitive system* in which animals choose and modify their environments, and by doing so, change some of the natural selection pressures they and their descendants confront. Later, a population geneticist, R.C. Lewontin (1983) proposed a similar idea: organisms do not adapt to their environments, they construct them out of the bits and pieces of the external world". According to the twentieth-century Modern Synthesis paradigm, organisms are subject to selective pressures independent of animal populations. Natural selection was seen as a kind of objective filter that rewarded some individuals over others, generating the processes of adaptation. However, it is clear that organisms are not passive entities suffering the effects of the environment: they move around various territories and modify, through their action, the very characteristics of the environment, and consequently the selective pressures at play. We must, therefore, question the concept of autonomous selection because, while it is true that various environments provide characteristics that can direct adaptive correlation processes, it is also true that the organisms themselves contribute to defining parameters that act as factors of selective pressure. This, moreover, shows that there is a vertical transmission not only of a genetic and epigenetic type but also in terms of *ecological inheritance*.

Through niche construction, animal behaviour produces what in humans we call an artificial environment or the domestication of environmental parameters: a set of new factors that interact with external ones, giving rise to a kind of *adaptive Umwelt*. As Salvatore Tedesco points out in *Estetica e scienze della vita* (2013), edited together with Andrea Pinotti, niche construction theory marks the transition from an externalist perspective, entirely focused on the pressure of natural selection and the adaptive response of organisms, to a conception that can be defined as internalist, whereby it is possible to theorise reciprocity in the co-evolution of organisms and their environments. Niche evolution can result from several behavioural events, like (1) the colonisation of a new environment through migration; (2) the modification of a lifestyle, for example in foraging or in sexual choice canons; (3) the invention of a tool that enables new performances; (4) the advent of a symbiotic relationship with another organism; (5) the co-optation of a behavioural expression into other domains by *ontogenetic exaptation*; (6) collateral or epiphenomenal development of a character entrained by others, producing transformations in style; (7) modification of developmental processes or gestation parameters; and (8) the establishment of a cultural tradition.

The importance of animal protagonism and creativity is evident in niche construction, precisely because variations in behaviour represent the key factor in this event. If the environment influences the individual's phenotypic expression with its ecological parameters, it is clear that the colonisation of a new territory, the modification of a dietary style, the advent of a symbiotic relationship and the like have a significant impact both on the conversion processes of developmental plasticity and on the selection of allelic variants, modifying their adaptive parameters. Creativity may express itself in the updating or slight modification of the expressive strategy, while paving the way for a new space of adaptation. In certain cases, it may involve adjustments, adaptations, innovations or inventions of a more significant nature: then the modification leap of the selective matrix is more substantial because the environmental parameters change.

Animals modify their behaviour for an infinite number of reasons, often giving rise to real discontinuities. The repercussions of a small deviation are usually unpredictable, because they can have amplifying effects, resulting in real metamorphic cascades, or they can be more contained. Sometimes the action is given by the degree of evolvability of that organism, other times it remains quiescent until other boundary factors change. Some transformations stem from the direct action of an innovation, others from side or recursive effects. The modification of a niche, as mentioned, can be the result of co-optation or an epiphenomenon driven by other expressive evolutions. In this regard, I believe, for example, that the domestication of animals by humans is a side effect of the development of an increased epimeletic sensitivity to pedomorphic signals in response to the neonatal immaturity of our children. The adoption of pups from other species, which is the basis of the domestication process, is thus an epiphenomenon of the need for parental care in human infants. At other times, innovation may result from imitation of another species' behaviour or from the discovery that said behaviour can serve as a useful clue to achieve a certain objective. Some interspecific collaborations, such as that between the honey badger (*Mellivora capensis*) and the greater honeyguide (*Indicator indicator*), may have arisen in this way.

What are the effects of the niche on the evolution of the morphotype or on possible speciation processes? They can be of different types. They can (1) limit environmental fluctuations and therefore reinforce the morphotype's stasis mechanisms, acting as a buffer or stabilised microenvironment over particular ranges. They can be (2) cryptic with respect to selective pressures or environmental releasers that would otherwise have a relevant role on that population. They can be (3) emergent, that is inducing new selective or guiding parameters in the development process and therefore typifying it. Other types of effects are (4) pluralistic, that is capable of maintaining a great variability in the morphotype, in terms of both genetic variability and polyphenism; (5) generalist, that is capable of lowering the volume of correlative specialisation and thus ensuring the permanence of evolvability contents; or (6) expressive, that is strongly oriented towards specialisation and reduction of plurality. The niche also intervenes (1) on the development process, because as we have seen, this is affected by environmental modulation factors, and therefore on phenotypic expression; (2) on epigenetic inheritance processes, that is on the vertical transmission of gene regulation and therefore on phenotypic fixation and (3) on selective pressures, and therefore also on the allelic variability present in the population. If we consider the phenomena of genetic assimilation from phenotypic plasticity, epiallelic inheritance from vertical transmission of epigenetic mechanisms and the predisposition of selective pressures in the population, we realise the importance of expressive protagonism in evolution.

We can distinguish between two kinds of niches. The first is a *living niche*, which can be defined as a sort of home for the animal that serves as a place of refuge or ambush, such as burrows or cobwebs, or as a place to stay, such as beaver dams or the hives of social hymenoptera. The second is a *developmental niche*, such as the paedotrophic nests of social wasps, the litter dens of carnivores and the nests of birds. In both cases, animals have to perform innumerable tasks of supervision beyond the mere application of innate endowments. The niche, therefore, not only influences the morphofunctional characteristics of the organism and the connotations of innate behaviour—in terms of (1) project molarity and (2) operational molecularity—but also exerts a selective action on other characters that are more related to situations of operational superintendence, such as the type of intelligence, the creative predisposition, the experiential database, the learning modes, and the tendency towards social and cultural transmission. The niche is given by particular interactions with ecosystem counterparts, applicative capacities in the performance of one's functions, dependence on certain factors or parameters, activities aimed at preparing the environment and choice of the appropriate site for one's purposes.

With regard to the latter aspects alone, it is easy to recognise the following cognitive needs: (1) identification of the place where one's site is to be located, based on a careful assessment of opportunities and risks; (2) learning the spatial references for orientation around the site, since set-up operations often require moving away to retrieve material and returning to it; (3) analysis of the adaptation needs that characterise that particular site and monitoring whether the results of the actions taken correspond to them, correcting any inconsistencies. Before choosing a site, an animal does a lot of research and in some cases surveys several points to compare possibilities. A particular example is the choice of swarming sites for bees, a task to which the scout workers devote four days. They visit competing sites and, on returning to the colony, dance around to make their proposal, trying to recruit other scouts to visit the site. The final decision is made through a dialogue between the different engagement groups which, much like a kind of parliament, eventually lead to a count to determine the most appropriate site, resulting in a collective decision. Once the chosen location has been determined, its location must be remembered and this applies to all animals. Beewolves (*Philanthus triangulum*), also known as bee-killer wasps because they mostly prey on *Apis mellifera*, must remember the precise location of their paedotrophic nest, as they return there many times during construction and supply trips.

Building operations involve expressive protagonism, as well as molar-molecular innate competences, and in any case are based on motor coordination and sensorimotor adaptation processes that require adaptive superintendence. But even if we stick to the expression of the innate schema of competence, it can still be traced back to the concept of tool-usability rather than associative automatism. As James and Carol Gould point out in *Animal Architects* (2007, 30–39), the classical stimulus-response mechanism is unsuitable to satisfy the complex motor articulations of the niche set-up, because each receptor would have to be linked to a specific programme, with the result of increasing the required circuitry in a very cumbersome way. It is far more parsimonious to imagine sense-motor maps capable of integrating the input data and reflecting them onto precise action schemes.

On the other hand, in order to develop a niche, an animal must evolve not only sense-motor patterns but also a real *personal space*. Constructing a niche, in fact, means moving in a spatial dimension that relates one's own body to the surrounding environment, in a continuous exchange capable of locating the points of convergence between the animal and what surrounds it, looking for suitable arrangements and often implementing shortcuts. When we observe some of these complex constructions, such as, for example, those of Trichoptera larvae capable of intercepting the flow of the current, we realise that, beyond a preordained architectural plan and equally predefined operational molecularities, the larvae must pay constant attention to the feedback received during the set-up. It is interesting to note that the way to make adjustments and, in certain situations, repair damage varies from subject to subject. These differences have been noted in much of animal architecture, and are sometimes striking: for example, trapdoor spiders of the suborder *Mygalomorphae* are able to hide their burrow mouths and ensure waterproofing in different ways depending on the characteristics of the environment.

Indeed, the world of spiders offers a large number of examples of niche construction and of superintendence in how a design plan is carried out, adapting it to the environmental situations and using the characteristics present. Hammock webs, constructed by certain spiders such as *Linyphia triangularis*, involve very different capture techniques, each correlated to the characteristics of the prey. Orbicular webs also require considerable cognitive activity, both in their construction and in their use. There is substantial agreement among scientists that the choice of anchor points and the number of spokes varies to the point that each web is a tailor-made product. In certain situations we find a kind of convergence between the concept of niche and that of tool, or at least we find cases where the line between the two becomes extremely blurred. One example is the net-casting spider, *Deinopis subrufa*, which makes a small web and then throws it at its prey to catch it. The complex work involved in preparing the web in this case meets the spider's need for manageability—and therefore for reference to its own body—as well as for effective predatory behaviour. The use of somatic space as a reference model for construction is also evident in the way wasps construct their cells: for example, wasps of the *Polistes gallicus* species utilise their heads as moulds by using their antennae, one held inside the wall and the other outside. A similar situation can be found in the weaver ants of the genus *Oecophylla*, which are able to build real bridges with their bodies to bend the leaf margin in order to make their final structure, which requires the larvae to produce silk filaments to glue the walls together.

We can therefore speak of *behavioural niche* as an identity dimension just like the morphological structure or the general physiological characteristics. There is no doubt about the presence of an expressive niche, which makes a species recognisable in various aspects, such as foraging patterns, dispersion over the territory, social dimension, courtship choreography, reproduction times and modes, communicative semiotics and *Umwelt* structure. Even if we were to exchange the morphology of a dog with that of a cat and vice versa, we would still be able to recognise the two species from their respective behavioural patterns, so the behavioural repertoire is a phenotypic feature like any other. The species, therefore, is an entity related to a certain niche and this should be understood primarily in terms of style, where it is difficult to discern the external aspects from the internal ones. It is a *modus vivendi* that translates into a singular interpretation of the generic usability or virtuality permitted by a certain environment. The action of the population on niche construction recalls a process of actualisation and therefore of interpretation, that is, what I defined with the metaphor of the statue brought out of the marble slab. That's it: the niche is the statue, one of the possible forms that the environment can take, the one chosen by the species-sculptor.

On the other hand, the image of the species, in its features of somatic presentation and style, does not only concern the way of interpreting a fragment of the world, transforming it into a sort of *amnion* in its own likeness. The niche is the way in which the species looks at itself and invents its own *code of segregation* from all that is external: an intimate and coherent-faithful dimension that projects itself outside as a comfort zone. The niche allows for a presentation that is always a belonging, a manifestation that does not end in the body of the individual, but extends into environmental space and intergenerational time. The niche is the most obvious form of recognition signature—based on specific mate recognition systems (SMRS), to echo Paterson's proposal. The recognisability allowed by the niche as a style of belonging has a central function in sexual selection, precisely to ensure fertile mating outcomes. This homologation—as well as the recognition apparatuses expressed by Adolf Portmann (1897–1982) in his phanerology and by Amotz Zahavi (1928–2017) in the concept of the handicap principle, but remarked upon even earlier by Darwin himself in his 1871 essay—suggests that the species is first and foremost a segregative entity, that is a way of constructing a separate reproductive unit that at a certain moment interrupts a more general gene flow, carving out its own identity.

Behaviour is the expression of a body that interacts with the environment, a somatic dimension that has been defined precisely by the interruption of gene flow and *panmixia*, enabling the disjunction or production of biodiversity that seems to govern the entire organisation of living things as a primary principle. More than a space, however set up and interpreted by the creativity of a species, the niche is a style that allows habits and customs to evolve within a certain population. It would be wrong, however, to consider this ethographic repertoire as a direct expression of the allelic heritage. As with every other aspect of the phenotype, the expressive identity of a species is the result of a developmental process in which the niche is primarily responsible for igniting and configuring a genetic expression to obtain the final result. A niche is still a *space of possibilities*, capable of defining limits and time frames even with regard to the becoming of form, its expressions of *heterochrony*, *heterotopy* and *allometry* of development. The punctuations shown by Eldredge and Gould are expressions of the niche, of its tendency to reconfigure the regulators of development. Epigenetics has shown how environmental factors are fundamental in gene regulation processes during development—think of Hox genes, the pleiotropic effects of certain genes, or marking factors—and Evo-Devo biology has highlighted the constraints, we might say the quantum nature, of possible variations. Not all forms are possible, and at the same time the configuration of an organ is capable of determining a general restructuring. The niche must therefore be regarded as a developmental environment that can influence the organisation of the body.

In general, we are inclined to consider the anatomical structure of the body as an entity that presents certain flexibility with respect to environmental variables—take, for example, height in human populations as it changes depending on the diet—but which nevertheless carries a broadly defined conformation. The niche provides particular selective pressures—for example, the ability of an argiope to build its web, of a paper wasp to create its hive, of a mantis to correctly position its ootheca—so that a species' identity is kept under control by the same expressions related to the niche. Constructing a niche does not simply mean lowering the level of selective pressure, but moving from an environment-referred type of selection to a niche-referred type, that is personalising selection. The niche performs a selective action on the morphotype, including the physiological characteristics that underlie behaviour. At the same time, by influencing a certain organisation of the body, through the modulation of developmental dynamics, it not only influences the cognitive functions, but also the possible experiences that become available. In the human being, hands are a somatic instrument that enables certain prehensile functions, but they are also an organ of knowledge of the world. The same can be said for wings or fins: the body is a flywheel and also a channeller of cognitive functions.

Behaviour is central to how a niche is defined, not only because it is an expression of the individual-world interaction that leads to environmental modification, but also because it is itself a predicate of the niche, given that the latter is a *modus vivendi* and not an occupied space. Now, a behavioural niche refers to a large number of variable factors, such as (1) the environmental elements with which the individual interacts; (2) the type of food and the current way of obtaining it; (3) the perceptual demands to which it is subjected, in terms of specific research or essential monitoring; (4) the athletic activities to which it is called; (5) the type of interaction with conspecifics, for example, the social dimension. Imprinting in birds and socialisation processes in mammals show that full identity of belonging in these species is linked to proper exposure to parental relationship in order to develop key aspects of reproductive recognition, for example singing in birds and proper presentation choreography in mammals. Well-established research on sexual imprinting illustrates that in many ways this is a partially open system that uses the parental relationship as an externalisation of information. This also demonstrates the co-evolutionary aspect of the relationship between gene and niche.

I do not believe it is possible to understand niche construction without the creative factor, or in any case without the subjective interpretation that animals display in the fulfilment of their species needs and in the expression of the most common behavioural patterns. The phenotypic plasticity of a species also manifests itself in behaviour, so the species repertoire presents itself as a sort of *expressive range*, within which individuals perform ordinary tasks. This variability oscillates as if subjected to a *strange attractor* which, while keeping faith with the *niche project*—such as the typical hive structure—and the *niche operations* that guide the individual's expressive master plan, relies on subjective flair to get the job done.

In the construction of termite mounds, insect activity proves to be anything but based on simple automatisms, not least because of the great variability in the way each subject intervenes in the construction of walls and arches. With regard to the work of termites, James and Carol Gould (170–178) point out that only columns with a certain radius can be successfully connected to each other, but the height and curvature depend on the initial distance between them, which is always variable. In the case of termites, the niche must ensure not only protection but also air circulation, a certain humidity gradient and a constant temperature, also in order to support the symbionts. Many animals use niche specialisations to keep the temperature constant. Crocodiles, for example, dig into the ground and build mounds of vegetation to incubate their eggs, as the fermentation process allows the eggs to warm up. The megapod birds of Australia and Indonesia also build similar nests, using the incubator method, that is, piles of compost from which they obtain the necessary heat for their eggs through fermentation. It should be noted that these birds must continually monitor the state of fermentation and heat produced, adding vegetation if the temperature is low and making ventilation holes if it is too hot.

Nest-building in birds can be particularly costly, not only in terms of finding the best site for the nest but also in terms of finding the building material: the cliff swallow (*Petrochelidon pyrrhonota*) can fly over 200 km to collect the pellets for its nest, while the weaver bird (*Ploceus cucullatus*) covers a distance of over 300 km. In preparing their nest, birds must adapt the deposit of material to the characteristics of the site, choosing the most suitable kind as the work progresses: for example, coarser material at the beginning to make the supporting structure, and then thinner and softer material to cover the nest. Some birds, such as hornbills (family *Bucerotidae*) almost close off the entrance to the hole, leaving enough space for their beaks; others such as grebes (*Podiceps cristatus*) create floating nests out of stacked water plants, whereas bitterns (*Botaurus stellaris*) trample the vegetation to produce a platform. The nest of the ovenbird (*Seiurus aurocapillus*) is spectacular: it builds a cup-shaped nest out of a mixture of mud and plant material, drying it in the sun to create a brick-like structure. When one observes the composition of the most common bird nests, for example those of blackbirds (*Turdus merula*), goldfinches (*Carduelis carduelis*) and wrens (*Troglodytes troglodytes*), one notices that each individual uses the material it finds in a very flexible and creative way, very often using material left by humans. In conclusion, the niche is always a work based on creativity and cannot be reduced to mere automatism.

Works Cited

Carroll, Sean B. 2005. *Endless Forms Most Beautiful.* New York: W.W. Norton & Company.

Darwin, Erasmus. 1794. *Zoonomia or The Laws of Organic Life.* Dublin: P. Byrne, and W. Jones.

Darwin, Charles. 1871. *The Descent of Man and Selection in Relation to Sex.* New York: Merrill and Baker.

———. 1881. *The Formation of Vegetables Mould Through the Action of Worms, with Observations on Their Habitats.* London: John Murray.

Dawkins, Richard. 1982. *The Extended Phenotype.* Oxford: Oxford University Press.

Forestiero, Saverio. 2009. *Ambiente, adattamento e costruzione della nicchia.* Padova: Cleup.

Gould, Stephen Jay. 2007. *Punctuated Equilibrium.* Cambridge, MA: Belknap Press.

Gould, James L., and Carol Gould. 2007. *Animal Architects. Building and the Evolution of Intelligence.* New York: PublicAffairs.

Gould, Stephen Jay, and Elisabeth Vrba. 1982. Exaptation-A Missing Term in the Science of Form. *Paleobiology* 8 (1): 4–15.

Greene, Erik. 1989. A Diet-Induced Developmental Polymorphism in a Caterpillar. *Science* 3 (243): 643–646.

Grinnell, Joseph. 1914. An Account of the Mammal and Birds of the Lower Colorado Valley. *University of California Publications in Zoology* 12: 51–294.

———. 1917. The Niche-Relationships of the California Trasher. *Auk* 34: 427–433.

Hall, Brian Keith. 1992. *Evolutionary Developmental Biology.* London: Chapman & Hall.

Haraway, Donna J. 2016. *Staying with the Trouble, Making Kin in Chthulucene.* Durham: Duke University Press.

Hubert, François. 1814. *Nouvelles Observations sur les Abeilles.* Paris: J.J. Paschoud.

Kauffman, Stuart A. 1995. *At Home in the Universe. The Search for Laws of Self-Organization and Complexity*. Oxford: Oxford University Press.
Kirby, William, and William Spence. 1828. *An Introduction to Entomology: or Elements of the Natural History of Insect*. London: Longman, Rees, Orme, Brown and Green.
Knight, Thomas Andrew. 1837. On the Hereditary Instinctive Propensities of Animals. *Philosophical Transactions of the Royal Society of London* 127: 365–369.
Kupiec, Jean-Jacques. 2019. *Et si le vivant était anarchique*. Paris: Les liens Qui Libérent.
Laland, Kevin N., John Odling-Smee, and Marcus Feldman. 1996. The Evolutionary Consequences of Niche Construction: A Theoretical Investigation Using Two-Locus Theory. *Journal of Evolutionary Biology* 9: 293–316.
Lamarck, Jean-Baptiste. 1801. *Système des animaux sans vertèbres*. Paris: L'Auteur-Deterville.
———. 1809. *Philosophie zoologique*. Paris: L'Auteur-Dentu.
———. 1815. *Histoire Naturelle des animaux sans vertèbres*. 7 voll. ed. Paris: Déterville.
Le Roy, Charles-Georges. 1782. Istinct. In *Encyclopédie, ou dictionnaire raisonné des sciences, des arts et des métiers, par une société de gens de lettres*, ed. André Le Breton, vol. 8, 795–799. Neuchâtel: S. Fauche.
———. 1802. *Lettres sur l'intelligence et la perfectibilité des animaux, avec quelques lettres sur l'homme*. Paris: Imprimerie de Valade.
Lewes, George. 1874. *Problems of Life and Mind*. London: Trübner & Co.
Lewontin, Richard. 1983. The Organism as the Subject and Object of Evolution. *Scientia* 118: 63–82.
———. 2000. *The Triple Helix*. Cambridge, MA: Harvard University Press.
Lyell, Charles. 1830. *Principles of Geology*. London: John Murray.
Malthus, Thomas R. 1798. *An Essay on the Principle of Population, as it. Affects the Future Improvement of Society*. London: J. Johnson.
Margulis, Lynn, and Dorion Sagan. 2002. *Acquiring Genomes. A Theory of the Origins of Species*. New York: Basic Books.
Mayr, Ernst. 1958. Behavior and Systematics. In *Behavior and Evolution*, ed. Anne Roe and Simpson George, 341–362. New Haven: Yale University Press.
———. 1963. *Animal Species and Evolution*. Cambridge, MA: Belknap Press.
McCollum, Andy S., and Josh Van Buskirk. 1996. Costs and Benefits of a Predator-Induced Polyphenism in the Gray Treefrog Hyla Chrysoscelis. *Evolution* 50 (2): 583–593.
McFall-Ngai, Margaret, et al. 2013. Animals in a Bacterial World, a New Imperative for the Life Sciences. *Proceedings of the National Academy of Sciences of the United States of America* 110 (9): 3229–3236.
Morgan, Hugh D. et al. 1999. Epigenetic Inheritance at the Agouti Locus in the Mouse. *Nature Genet.* 23: 314–318.

Müller, Gerd B. 2013. Le origini della novità morfologica. In *Estetica e Scienze della Vita. Morfologia, biologia teoretica, evo-devo*, ed. Andrea Pinotti and Salvatore Tedesco. Milano: Raffaello Cortina.

Newman, Stuart A. 2007. The Organismic Systems Approach: EvoDevo and the Streamlining of the Naturalistic Agenda. In *Integrating Evolution and Development. From Theory to Practice*, ed. Sansom Roger and Brandon Robert. Cambridge: MIT Press.

Odling-Smee, John. 1988. Niche-Constructing Phenotypes. In *The Role of Behavior in Evolution*, ed. Henry Plotkin, 73–132. Cambridge, MA: The MIT Press.

———. 2010. Niche Inheritance. In *Evolution: The Extended Synthesis*, ed. Massimo Pigliucci and Gerald B. Müller, 175–207. Cambridge, MA: MIT Press.

Paley, William. 1802. *Paley's Natural Theology*. London: R. Faulder.

Paterson, Hugh E. 1980. Comment on "Mate Recognition Systems". *Evolution* 34 (2): 330–331.

Piaget, Jean. 1976. *Le Comportment, moteur de l'évolution*. Paris: Gallimard.

Pigliucci, Massimo, and Gerd Müller. 2010. *Evolution, the Extended Synthesis*. Cambridge: MIT Press.

Pinotti, Andrea, and Salvatore Tedesco. 2013. *Estetica e Scienze della Vita. Morfologia, biologia teoretica, evo-devo*. Milano: Raffaello Cortina.

Réaumur's, De, and René-Antoine. 1734. *Mémoires pour servir à l'histoire des insects*. Paris: Impr. Royale.

Romanes, George. 1883. *Mental Evolution in Animals: With a Posthumous Essay on Instinct by Charles Darwin*. London: Kegan Paul Trench & Co.

Scott, Gilbert, and David Epel. 2015. *Ecological Developmental Biology: The Environmental Regulation of Development, Health and Evolution*. Oxford: Oxford University Press.

Sebright, John. 1836. *Observation upon the Instinct of Animals*. London: Gossling & Egley.

Smellie, William. 1791. *The Philosophy of Natural History*. Boston: Hilliard, Gray: Little and Wilkins.

Spencer, Herbert. 1855. *The Principles of Psychology*. London: Longman, Brown, Green, and Longmans.

Waddington, Conrad H. 1959. Evolutionary Adaptation. In *Evolution After Darwin*, ed. S. Tax, vol. 1, 381–402. Chicago: University Chicago Press.

Wallace, Alfred Russel. 1864. The Origin of Human Races and the Antiquity of Man Deduced from the Theory of Natural Selection. *Journal of the Anthropological Society of London* 2: 158–170.

Wang, Yan, and Lloyd H. Kasper. 2014. The Role of Microbiome in Central Nervous System Disorders. *Brain Behav. Immun.* 38: 1–12.

West-Eberhard, Mary Jane. 2003. *Developmental Plasticity and Evolution*. Oxford: Oxford University Press.

CHAPTER 11

Plural Intelligence

Premise

We have seen how the ability of various species to perform at a high level—as observed, for example, in social hymenoptera or in weaver birds—does not necessarily indicate the possession of high cognitive skills on the part of the individual, because these skills can be attained through predefined operational schemes developed by phylogeny. On the other hand, it should be clear by now that an innate or learned skill, however well instilled down to the smallest details—think of the skills of a craftsman, refined through apprenticeship and experience—does not oppose or replace the need to manage the work, moment by moment, in its operational course. A species may well be equipped by phylogeny, both in terms of the blueprint to be produced and the sequenced actions to carry it out, but more often than not the individual has to reconcile these tasks with the specific circumstances. The tendency of social wasps to build a nest of a certain shape by means of very standardised operations does not dispense the insects from having to constantly supervise their worksite with actions of adjustment and correction.

Having particular skills means having to apply them, but conditions are ever-changing, so it is not enough to have knowledge, that is a useful tool, but it is essential to be able to use it with the necessary flexibility. It is clear, therefore, that the more complex the equipment, the greater the need for precise adjustments. We mistakenly believe that innate competences are

R. Marchesini, *The Creative Animal*,
https://doi.org/10.1007/978-3-031-07414-1_11

autonomous properties, which lie beyond the individual's control and are self-sufficient, so that they do not require application checks on the subject's part. We thus set them against intellectual qualities, sometimes expressed through resourcefulness and the ability to solve unusual situations, which focus on individuality and its presence in the singularity of the situation. But these are two big mistakes: (1) the first is believing that competences are operationally self-sufficient and (2) the second is thinking that intelligence is based on nothing and is best expressed in the absence of competences. Innate endowments—but I would say, more generally, all knowledge—require intelligence: they are not two separate domains, let alone inversely proportional to each other, but they are linked by a factor of direct proportionality. An animal that has a very articulated and complex social dimension, as is the case with many primates or lycaons, not only will be endowed with high-profile innate and learned skills to navigate within complex group dynamics, but also will have an equally substantial social intelligence.

If we understand this profound relationship between carrying instructions and intelligence, the meaning of creativity, as the continuous construction and management of new knowledge, becomes clear. Creativity is one of the characteristics of intelligence and it is therefore important to shed light on this expression of the animal mind, which we can define as the subject's ability to reason on impacting situations, rummaging beyond appearances to bring out possibilities that are not immediately perceivable or already expressed. Intelligence is the ability to identify a new solution to a problem, to express a different configuration in the repertoire of psychomotricity, to identify an instrumental possibility in an object, to discover a new order of things, to imagine relationships between events, to read the social dynamics within a group in a systematic way. Intelligence therefore means reading between the lines of the world to bring out new meanings within the virtuality of reality.

Intellectual Diversity

I have said that the evolution of a species does not depend exclusively on objective shaping by environmental forces, but also on the construction of a *modus vivendi* situated in an eco-ethological space of personalisation and reproductive segregation, which can be defined by the term *niche*. In *The Evolution of Beauty* (2017), Richard O. Prum points out that the development of ornaments, through the mechanism of subjective aesthetic experience, has played a key role in the differentiation between species on the basis of processes of recognition and choice. As I remarked in the last two chapters, the concept of subjective experience and segregative innovation—that is the ability to construct new canons of belonging through the mechanism of species-specific signatures—is fundamental to understand the evolution of species, in an extended conception of phylogenetic processes that go beyond the algorithmic claim of adaptationism. When we observe the emergence of a new criterion of aesthetic preference, as well as when we are faced with the development of a cultural tradition or use of a tool in a population, we must refer to this ability of intelligence to produce innovations through constructivist work on the unexplored possibilities of the real.

Diversification as well as belonging, the complex interplay between stasis and punctuation in the morphogenetic dialectic, can be attributed to this quantum character of the phenotype, which buffers mutations up to a certain threshold and then expands them beyond that gradient. In fact, the life of a population, and more generally of a species, is subject to continuous mutations, not only genetic but also epigenetic, environmental and due to creative events operated by the individual. Many of these transgressions of the canon are washed away, in a process of continuous dissipation, linked to the continuous reshuffling of heritages; others accumulate until they reach the critical gradient capable of making the morphotype leap onto a new morphogenetic orbital. Intelligence is the great engine of evolution in animal species, which shows how absurd it is to keep considering the individual a totally passive entity.

On the other hand, intelligence cannot be still considered as a universal function, but must be viewed in terms of plural expressions or forms of intelligence, as we learn from Jeremy Narby's essay *Intelligence in Nature: An Inquiry into Knowledge* (2005). In fact, intelligence is a property that takes on different forms in different species, just like the other features of the morphotype. The intelligence of a species, just like its innate endowments of competence, is the result of the progressive adaptation of the species to its niche, that is to that set of behaviours of environmental adaptation and recognisability that characterise a given taxonomic entity. In fact, it manifests itself not only through the novelty or particularity of the morphotype and not only by virtue of innately predefined habits, but also—and this is the most interesting point as regards creativity—in a very specific way of understanding the here and now on the part of the individual. It is clear that each species has a certain sensory access to the outside world, that it is electively receptive to certain stimuli, that it presents predefined patterns of expression that are fairly uniform across individuals, in short, that it has a certain behavioural as well as morphological identity. The point here is to understand that each species, in addition to this, has its own intelligence, that is a certain way of metaphorically looking at reality, applying and adapting its skills to the particular situations faced by individuals.

In other words, we could say that each species has a certain way of reasoning—and therefore of reflecting on the hidden sides of phenomena and on the possibilities that can be explored—that depends on the joint work of some of its specific *meta-operational* features, such as working memory, the prevalence of certain logical functions, dispositional components, the type of attention or metacognition. The appearance of a species or niche—from my point of view the two terms are basically equivalent—brings qualities with it that are shaped in mutual coherence: one of these is always the type of intelligence, that is the particular way of looking into the situations that arise. Intelligence is, therefore, the ability to *intus legere*, to delve into and go beyond appearances or the every day, not stopping at the already experienced, but trying to bring out possibilities, working on the data offered by contingency instead of mechanically applying one's skills. Intelligence, therefore, challenges the situation; it does not stop at the epidermis of appearance but rummages through the mesh of hidden possibilities: it raises problems for the world rather than solving supposedly objective problems posed by reality. An animal with an emphatic social attitude, such as a dog for example, is not only endowed with great communication, engagement and mediation skills, and in general with a set of competences that facilitate its relationships, but also considers social relations as a problematic dimension to be interpreted between the lines in order to bring out new opportunities.

As Charles Darwin already pointed out, there is no opposition between instinct and intelligence, and the reason is very simple: the greater the expressive endowments in a certain field, the more complex the related activity of adaptation and reflection will be. But then it is clear that we cannot speak of intelligence in the singular, precisely because each species has particular cognitive vocations, that is ways of interpreting its own behaviour. Both the endowments of knowledge (innate and learned) and intelligence are called upon to carry out a certain *modus vivendi* in their relationship with the world, seeking to make the most of the possibilities offered by a fluctuating environment characterised by singular situations.

An animal is peripatetic, in continuous movement, not only from the point of view of somatic mobility in the world but also from the point of view of cognitive mobility. An animal does not wait for an opportunity to present itself, but literally goes in search of it and does everything possible to make it happen. Adaptive specialisation is not simply a matter of bending to environmental conditions, but rather a way of constructing a certain segregative dimension by exploiting the available environmental potential to build a certain niche of recognisability. It follows, then, that intelligence is both the creator and the product of this process of morphotypic personalisation of the species, so it cannot be spoken of in the singular.

If we observe solitary animals, as in the case of most felids, mustelids, but also some canids like foxes, we will discover a predominantly solution-oriented intelligence, designed to understand the specific requirements of the problems faced and endowed with a strong curiosity about the particularities of situations. The solution-oriented intelligence of these animals appears to be highly innovative precisely because it is inclined to construct *ad hoc* solutions based on personal ingenuity. On the contrary, a social animal addresses the problem as a matter that does not concern itself individually, but the whole group. Its way of addressing the problem is not, therefore, one of direct engagement: the subject will try to construct a specific social systemic, a cooperative play scheme useful for addressing that particular situation. In the same way, we can identify species that are particularly predisposed to using external objects as extensions of their bodies, whereby individuals are inclined to use tools, favouring certain practices over overall body movements and, conversely, other species that are more predisposed to athletic skills. We would be wrong to consider problem-solving as a more developed form of intelligence than social intelligence, nor should we believe that the use of tools makes that subject more intelligent than another who shows great capacity for psychomotor organisation and kinaesthetic creativity.

As Howard Gardner suggested with the theory of multiple intelligences, we must consider intelligence not as a unitary faculty, but rather as an integrative capacity of multiple cognitive functions that takes shape in a plurality of predispositions. For example, among human beings there are individuals with a greater capacity for logic and mathematics than for visual-spatial imagination, others who excel in cataloguing, narrative or somato-kinaesthetic skills. If all this is true, it does not make much sense to ask who was more intelligent between Einstein, Dostoevskij or Picasso, because they all had different intellectual propensities. For the same reason, we cannot put the different species on a sort of hierarchical scale in terms of intelligence, because the specialisations at play are even more substantial than what Gardner refers to with respect to the human being. Just as it is absurd to ask whether the sensory properties of one species are superior to those of another—but the same reasoning could be applied to endocrine, immune, enteric or locomotor faculties—so it is unreasonable to think of an absolute intelligence, comparable in different animals, because intelligence is a function and as such responds to the logic of niche specialisation like all other species endowments. A solitary animal does not need great social capacities, either in terms of skills or intellectual faculty; in the same way, some species present such a psychomotor complexity that they have no need to use tools, so this lack is not at all detrimental to their cognitive capacities.

Just as the different bodily organs are consistent with each other—think of the changes in the entire musculoskeletal system due to human bipedalism, or the entire body structure in birds to enable them to fly—so there is a close relationship between the lifestyle and niche of a species on the one hand and its innate skills and type of intelligence on the other. Intelligence includes some basic cognitive performances or activities—such as problem-solving, memory, prefiguring, simulating, orienting, reflecting, self-referencing—but these are differently developed in different animals to form specific types of intelligence, which cannot be placed on hierarchical levels. The fact that a species makes frequent and variable use of objects as tools does not *ipso facto* make it more intelligent than another species that never uses them. If anything, we can identify some species that present a greater variability of cognitive performance, for example in the communicative, social and technopoietic spheres, as is the case with corvidae and psittacidae, which leads us to think of a more significant development of the cognitive apparatus. In other words, one can say that one species has greater working memory and another has better episodic or spatial one, that a particular animal is more communicative and another is more self-referential, or more able to build tools, but one cannot say that a species is more intelligent than another.

In 2008, I published an essay entitled *Intelligenze Plurime*, which took up and expanded on some of my previous publications on animal intelligence. In it I presented some types of intelligence, which I would like to revisit and update in this context. Observing animal behaviour, I note the following intellectual macrocategories, each of which is the prerogative of certain species:

- *Social intelligence*, the ability to move freely and innovatively within complex systems, knowing how to construct mediations and negotiations, but also cooperative or confrontational structures, remembering the dynamics in the group
- *Problem-solving intelligence*, the ability to tackle situations not through standardised expressions but by virtue of the propensity to look inside a problem, trying to identify its structural requirements
- *Aesthetically oriented intelligence*, the propensity to identify clues or signs to transform them into canons of choice or orientation maps
- *Technical and operational intelligence*, the ability to identify new strategies or tools to achieve one's objectives, through a pragmatic-functional outlook in the relationship with the world
- *Kinaesthetic intelligence*, the ability to organise one's own psychomotricity by adapting it to situations and exploring all the possibilities of movement of one's own body
- *Referential intelligence*, the tendency to learn from external referents, both conspecific and heterospecific, using others as models and facilitating the development of cultural traditions
- *Communicative intelligence*, the ability to use abstract symbols to refer to particular entities to be transmitted to a referent, but also having metalinguistic skills, that is using signs that refer to other signs
- *Empathic and self-reflective intelligence*, the ability to understand the conditions of others, to act on the basis of this interpretation and also to know one's own dispositions and characteristics
- *Abstract intelligence*, the ability to construct concepts, such as same-different, bigger-smaller, above-below, known-new or categorical prototypes referring to entities, or, again, mathematical abilities
- *Inferential intelligence*, the ability to extract rules from the outcome of situations, a hypothetical-deductive propensity in the face of events

The theme of animal intelligence is fundamental to this long discussion of creativity, because it shows that an individual's behaviour cannot be traced back to its previous endowments; on the contrary, the very presence of these endowments requires a capacity for application and innovation on the part of the subject. In other words, behaviour is always defined in the here and now, thanks to the complex action of reasoning, which uses endowments as tools that must nevertheless be adapted, revised and corrected time after time. But the issue of intelligence leads us to view the animal world as a pluriverse of possibilities, so that just as we have spoken of an immersive *Umwelt*, we must also recognise a cognitive *Umwelt*, whereby every animal expression is, within a perimeter of possibilities, a creation of worlds. Sometimes, however, even that perimeter is stretched and an individual or population finds itself embarking on new existential wanderings, thanks precisely to the C-factor of intelligence. Speciation is not only the result of a genetic drift, but also a modification of the belonging systems in force within the species, including aesthetic experience. Two events show more than any other how creativity affects the phylogenetic trajectory: the emergence of a cultural tradition and the use of tools.

Culture in the Animal World

In *The Evolution of Culture in Animals* (1980), John Bonner suggests an interpretative model for identifying the signs that lead us to recognise a cultural condition in the animal world, or, at least, a first approximation that applies to all situations where a tradition is present: culture is defined as a passage of information, through teaching and learning processes. Although every categorisation has blurred boundaries, I would like to point out that, when using such wide-ranging definitions, the confines of a phenomenon become so vague as to render the concept itself evanescent. If we assume that we are faced with a cultural act every time there is an exchange of information—incidentally, identifying precise teaching situations in a set of parental activities is far from easy—we risk that the concept may come to include communication, modelling, target orientation and facilitation. We know, in fact, that parental care in mammals and birds, while aimed on the one hand at protecting, looking after and feeding the young, also represents a kind of apprenticeship that leads to the individual's refinement of the species-specific dimension. Parental teaching encompasses behaviours of simple interaction or transmission of patterns up to full-fledged teaching activities.

Animals living in complex societies, or performing their main survival functions through cooperative actions, must necessarily develop communication codes and articulated social experiences that begin with the parental relationship. The latter often provides the basic social regulators that allow the individual to move on to a relationship with the peer group in order to achieve social maturity as an adult. As we have seen, one of the most accepted theories of intelligence development takes into account precisely the demands of community life. For Joseph J. Campbell (1904–1987), culture is the set of behavioural patterns of a social group that are transmitted from generation to generation through learning. This definition, taken from the essay *Human Evolution* (1967), already involves a first narrowing of the semantic field, given that the transmission of information is circumscribed within an intergenerational relationship, whereby culture seems to take on a vertical dimension, based on traditions that are maintained over time. In my opinion, however, this definition still fails to distinguish between a genuine cultural transmission and other types of intergenerational passage that do not display strictly cultural characteristics. This is a difficulty we also find in Wilfrid Le Gross Clark (1895–1971) when, in his essay *Man-apes or Ape-men?* (1967), he speaks of pattern transmission through imitation or teaching.

As I mentioned, parental care always entails, albeit implicitly, a transfer of information and models from parent to offspring, so if we were to follow these interpretations we would be unable to differentiate culture from maternal learning. Should we, for example, consider a mother cat teaching her kittens to hunt as cultural behaviour? In my view, no, because that behaviour is strongly linked to the cat's innate characteristics: it is generally an intrinsic part of the cat's parental patterns, without distinction between one population and another. A similar situation has been found by Alex Thorton and Katherine McAuliffe of Cambridge University (2006) in meerkats (*Suricata suricatta*), where adults bring the young a live scorpion, albeit without its stinging tail, to train them. In the case of both cats and meerkats we cannot speak of culture because those behaviours are widespread in the species. A similar situation applies to call-food behaviour in adults of thrushes and other nidifugous birds, as well as to complex behaviour such as cooperation between honey badgers and indicator birds.

A cultural tradition (from the Latin *tradere*, meaning to transmit) can be defined as a particular condition of intergenerational passage, which occurs when two populations of the same species develop behavioural repertoires that differ in certain aspects, and when these traditions are transmitted to subsequent generations. A cultural event may be the consequence of a particular situation experienced by that population, but it may also derive from an individual's creative behaviour, followed by a particularly successful dissemination within the group. In this case, it is not situational specificity that constitutes the segregating factor—which differentiates the two populations—but the emergence of a new behavioural style that favours the detachment of that population from its basin of origin. If, on the contrary, a certain habit or expression is part of the species' repertoire, so that the translated information does not differentiate a group from the canonical or ethographic repertoire, we are not dealing with a cultural event, but with social learning. Often, in fact, some innate behaviours involve completion through social relations—see Spalding's concept of imperfect instinct—whereby the *dimension of belonging* is linked to social interaction: namely, it is due to the externalisation of certain information within the parental care. A culture, on the contrary, does not produce homologative events within the species, but within a particular community, whereby that tradition is typical of that group. So, I think we can speak of a cultural condition when a certain expression has wide variability in the species and great uniformity within each population.

Clearly then, it is not enough to define culture as the transmission of information or as intergenerational passages, because if the teaching behaviour by the social counterpart and the content of the intergenerational transmission are already foreseen at the ethogram level, the release from phenotypic expression is only apparent. The species-specific repertoire, which makes a certain species recognisable from the point of view of behavioural identity, may be hereditary, and therefore manifest itself innately, or may result from the relationship with the parent. We know, in fact, that maternal deprivation at an early stage in endothermic vertebrates alters the individual's behavioural identity. I have emphasised parental care because, from imprinting in birds to the process of attachment in mammals, the assiduous relationship with the parent and the extreme plasticity of the young illustrate an extremely complex passage of information, determining a real formative process that will influence the individual in all respects: from social interaction with its conspecifics to sexual orientation. Culture is undoubtedly something that is learned in parental and social relationships, but it represents a special case of vertical transmission of a certain behaviour. In order to be able to speak of culture, there must be not only an intergenerational passage but also (1) variability in that behaviour within the species, such that it cannot be included in the repertoire; (2) uniformity of that habit, style or model in a specific population and (3) non-attributability of that particular expression to an environmentally determined factor.

If the behaviour detected in a certain population turns out to be necessitated or at least directly produced by the environmental conditions to which that group is specifically subjected, one might still argue that it is not culture, but a variation of the behavioural phenotype due to the particularities of the paratype. A cultural phenomenon, on the other hand, is indisputable if two populations of the same species, living in similar environments, showcase two different styles of expression. Hal Whitehead, in his book *Sperm Whales: Social Evolution in the Ocean* (2003), has offered valuable information on the development of specific communicative repertoires, veritable dialects, in different groups of sperm whales (*Physeter macrocephalus*), where all individuals within a group use vocalisations of the same frequencies. As Michelangelo Bisconti points out in *Le culture degli animali* (2008), "the different sound peculiarities of these groups cannot be related to particular ecological conditions of the habitat, because several groups were observed sharing the same stretch of sea, and it was demonstrated that each group carried its own vocal dialect" (144–145). A similar condition, albeit referring mainly to dietary characteristics, was found by John K.B. Ford et coll. in *Killer Whales* (2000), a study that subsequently led to highlighting marked differences also from the standpoint of vocal communication: a good example is the article by Volker Deeke et al. entitled *Dialectic Change in Resident Killer Whales: Implications for Vocal Learning and Cultural Transmission* (2000, 60, 629–638).

We can therefore imagine the emergence of a cultural element as a case of contingency, neither necessarily accidental nor determined by the environment or attributable to particular ecological conditions. It is safe to assume that the environment may favour, influence or enable a certain cultural emergence, but the latter is still an unpredictable event. Cultural evolution always has a historical character: it's an event that Stephen J. Gould would define as not necessarily repeatable. On the other hand, there is no doubt that a cultural phenomenon must include elements that are easily transmissible, because it does not rest on innate patterns to be completed, to take up Douglas Spalding's conception, but exploits contingency factors, as in the ontogenetic exaptation mentioned above.

In this regard, it is interesting to mention Richard Dawkins' concept of *meme*, introduced in his 1976 book *The Selfish Gene* to indicate an information unit that uses neurobiological circuitry to spread. Like a gene, though with a different replication mode, a meme is an entity that resides in the brain and propagates through culture. Although this transmission is profoundly different from the one linking the genotype to the phenotype—and this gives us a further reason to differentiate culture from parental teachings—it is evident that cultural phenomena also have a bearing on evolutionary processes. The first reason is the one usually put forward in group selection processes, reported for example by Richard

Lewontin in his essay *Biology as Ideology* (1991), whereby competition does not only exist between individuals, in order to secure the best reproductive chances, but also between different groups within the same species. In this case, any factor that strengthens that particular group compared to others will increase the assertion of its characteristics in subsequent generations. Moreover, if a habit influences the survival chances of the population that carries it, it will favour the replication of the meme itself.

The link with niche theory is evident, since a new habit can bring about a significant transformation in the *modus vivendi* of that particular population, often through a concatenation of factors with a snowball effect. A niche, in fact, is a metastable structure that maintains its homeodynamic equilibrium through threshold systems, so that sometimes a small modification is sufficient to give rise to high-impact catalytic effects, until the system quantitatively finds a new state of equilibrium. By modifying the environment, the animal also produces significant variations in its own lineage for two main reasons: (1) because the phenotype is the true field of selection, but we know that the phenotype is simply a manifestation of the genotype through the influence of the paratype, especially the environment; and (2) because each environmental modification changes the selective pressures and hence the resulting predicates. We must imagine the combination of these two factors—the first concerning the effect on the developmental process, the second concerning the phylogenetic affirmation of a morphotype—as an action that does not reshape but opens up new morphogenetic and behavioural possibilities.

Finally, let us not forget that the introduction of a cultural element can modify the mate recognition system highlighted by Hugh Paterson (1980, 34, 330–331) for its clearly segregative effects. We have seen, in fact, that cladogenesis can be ascribable to a geographical interruption of the gene flow, as in the case of peripatry and allopatry, but can also be due to a modification of the recognition signature. The creativity factor is therefore of fundamental importance not only in the life of the individual, but also in terms of the consequences it may have for the overall evolution of a population. As we shall see when discussing the use of tools, a cultural acquisition can have two apparently opposite effects: (1) increase the chance of conquering new lifestyles and colonising new territories, inaugurating new evolutionary events in that population; (2) reinforce niche specialisation, helping to buffer any environmental fluctuations and thus becoming a stasis factor in the morphotype. Suppose, then, that within the same species, one population acquires a cultural contribution that facilitates effect 1 of evolutionary diversification, while the second population is subjected to effect 2 of niche reinforcement: we could explain the long-term phenomenon of punctuated equilibrium noted by Gould and Eldredge. Of course, the emergence of a culture represents a hiatus in the phylogenetic trajectory of a species, capable of generating implementation effects that go beyond the specific significance of the acquisition.

Cultural phenomena are also sometimes referred to as mechanisms typical of large organisms, due to the fact that generations live longer than environmental fluctuations, which, as we know, is always a setback for adaptation processes. While fluctuations in bacteria are absorbed through rapid replication and the production of many variants, this is not the case in complex organisms, which have long development times and a low regime of genetic variants or exchanges. We can certainly say that the phenotypic modifications produced by learning a certain culture can influence the selection of particular genotypes, as reported by Feldman, Cavalli Sforza (1976). It is also undeniable that, by acting on the phenotype, culture can provoke physical modifications, like the transformation of the muscular-skeletal system caused by the use of a particular instrument, or the metamorphosis of metabolic regimes and intestinal microfauna determined by the consumption of certain foods. Fagen (1978) suggests that the inventiveness of a given social group may underlie positive selection and that cultural changes brought about by inventions may in turn affect genetic evolution.

There are obviously some morphological characteristics that favour the development of cultures: (1) complex social behaviour, (2) parental care, (3) imitative capacity (allomimesis and allelomimesis), (4) complex communication, (5) living in an environment with strong fluctuations, (6) mnemonic skills, (7) length of developmental age and (8) brain development time. With respect to the characteristics of the brain, it is clear that a greater number of neurons—it is not a question of overall size, as in birds neurons are more densely packed—determine the database enabling the ability to retain information. Among insects, for example, hymenoptera are greatly skilled in constructing cognitive maps even of very tortuous paths; in 1979, Randolf Menzel demonstrated that bees have both short-term and long-term memory. The prevalent type of communication can also influence cultural evolution. In *The Evolution of Culture in Animals* (1980), Bonner points out that when signals are based on gestures or the emission of sounds, the basic structures involved do not have to undergo major genetic changes because a vast repertoire of signals can be invented or learned. These signals are beneficial to an organism as long as it is flexible in its behaviour.

Learning abilities also have a bearing on the development of cultures. In both true and estrildid finches, it is not uncommon to observe song patterns characterising a certain population. Each chick learns from its father, especially during hatching, thus producing the cultural phenomenon known as "dialect". J.L. Brown suggests that learning is essential to recognise the local population. Consequently, I believe that the evolution of an aesthetic canon can also arise as a cultural event, which subsequently gives rise to a phenomenon of cladogenesis. It is therefore evident that strong social learning capacities are required for the evolution of cultures, starting above all with the relationship with the parent. As W.A. Mason points out, when a parent is teaching its young it sometimes reprimands them if their imitation is poor. The parent monitors the level of fidelity and appropriateness of the imitation and prompts the offspring to perform better. These teaching behaviours are very frequent in anthropomorphs. In captivity these behaviours are even more intense, because the mother, no longer busy searching for food, draws more stimulation from her young.

It is not easy to establish whether a cultural acquisition should be considered as a response to an environmental stimulus, for example a problem with the usual resources or the opening up of new possibilities, or whether the transformation of niche variables is due to cultural evolution itself. In all likelihood, this phenomenon is favoured by the convergence of these two factors: on the one hand, a changing environment with untapped opportunities or dwindling traditional supplies, and on the other, a certain expressive plasticity in the species and the presence of individuals endowed with a degree of creativity. Ethologists have been able to document, for example, that a tradition generally arises, thanks to the inventiveness of a particular individual, who suddenly starts a tradition because the new behaviour is productive, that is able to satisfy a certain need, and because it is reproducible and replicated by other individuals in the group. One aspect that may favour the evolution of culture is the presence of behavioural polymorphism in the species, or the possibility of making multiple choices on how to achieve a certain adaptive outcome. In many species, flexibility is a fundamental endowment, because it allows them to live in highly fluctuating contexts, to occupy previously foreclosed environments with greater versatility, to resist resource constraints and to access new forms of foraging.

I believe that creativity is always the most important factor in cultural evolution, even in the face of fortuitous eventualities, that is, occasions of acquisitive serendipity. In fact, to be able to translate an event into a habit, the subject must be able to understand it and construct an application model that can be used beyond the occasion on which it occurred. Intelligence, as the ability to make an expression fluid, freeing it from a canon and problematising it, therefore plays a central role in cultural evolution. A culture can arise when a subject makes an important discovery, when it suddenly finds the solution to a problem, or when it invents a way to reach a much desired but inaccessible resource. A cultural tradition can develop if that behaviour becomes part of the ordinary customs of a certain population and if it is the focus of transmissive dynamics, such as the mother-young relationship or social positioning. On the other hand, one should not underestimate the presence of *superstitious behaviour* in some cultural expressions, whereby a correlation is mistaken for a causal mechanism and all activities that serve to reproduce the desired situation are ritualised.

In the above-mentioned essay, Bisconti identifies three essential elements for cultural evolution: memory, communication and creativity: "When in a group of social animals an individual develops a new behaviour, perhaps in order to solve a certain problem, this can be observed by the other members of the group who may have to deal with the same problem". Of course, the consequences can be countless, but I want to mention just four: (1) the other members of the group do not realise the importance of that behaviour, which therefore remains confined to the individual; (2) the other members of the group are not able to reproduce that behaviour, because it is strongly correlated to the characteristics of the individual who developed it; (3) the behaviour is repeated only by some members of the group, so that with time the habit is lost and (4) that behaviour has strong transmissibility, so that the majority of the individuals in the group adopt it, quickly establishing a tradition. As can be seen, only in case 4 does the individual's creativity give rise to a culture.

In addition to this, the success of a certain cultural tradition is closely linked to the affirmation of the population that developed it. It should be noted that ethnozoology, that is the presence of cultural differences within the various populations of a species, has only recently begun to be studied methodically, just as is happening with the phenomenon of animal personality. In the essay *L'animale culturale* (1974), Danilo Mainardi stresses the importance of a continuist vision, whereby the human being is not considered as something different from the dimension of animality. This approach reveals how culture is something extremely widespread in the animal world. Cultural expressions—as I have defined them in terms of variability of traditions within the same species and uniformity within a population—may concern (1) the phenomenon of the aforementioned different dialects and thus the modification of the signature of recognisability in sexual choice. They also include (2) the tendency to use certain tools and the way of doing so; (3) the assumption of a new food habit, both in the type of food and in the way of treating it or (4) a particular innovation in niche construction. Finally, cultural expressions are also (5) the definition of new migratory routes; (6) the use of medicinal plants and psychotropic substances; (7) the evolution of aesthetic preferences.

With regard to the presence of dialects, we have seen that in birds, thanks to the process of filial imprinting and antiphonal singing, melodic singularities are created that characterise the individual or the pair, but which have affinities within the parental lineage and which often evolve into dialects specific to a certain geographical area. The phenomenon of dialects is widely present even among cetaceans, so it is always difficult to include a foreign individual in a group. The use of tools, in terms of what objects are used and how they are used, is another important cultural parameter. Chimpanzees, for example, have profound technical differences between one population and another. Adriaan Kortland (1965, 327–332), Jane Goodall and Van Lawick (1971, 212–213) found that forest chimpanzees used sticks in a much more rudimentary way than savannah chimpanzees: while forest chimpanzees swung or threw the stick, savannah chimpanzees were known to strike the enemy (a stuffed leopard).

As regards the evolution of a new eating habit, two cases are still paradigmatic. The first (1) is the Japanese macaque (*Macaca fuscata*) from the island of Koshima: in particular, the female Imo dipped a potato in water to clean it, giving it a new taste, a habit that spread throughout the group (Itani 1959; Kawamura 1963). The second (2) is the blue tit (*Cyanistes caeruleus*), which acquired the habit of piercing the aluminium cap of milk bottles to get the cream, as observed in England (Hinde and Fisher 1951). As we have seen, animals modify their environment through their canonical behaviour or through inventions that often become cultural traditions enabling a species to establish itself in a particular territory. Niche construction can therefore be seen as a process by which a custom gradually becomes part of the species canon. But this tendency can bring about the emergence of different traditions. The construction of a shelter can take on different forms depending on the traditions at play in different populations. This can be seen, for example, in beavers, as Donald Griffin reports in *Animal Minds* (1992).

The study of migratory routes of Canada geese has shown that they are undertaken within very stable family groups that live in close contact for at least a year. Returning to the same nesting sites every year has been proven to be an outcome of cultural transmission. Lorenz suggested that geese base their orientation on geographic familiarisation processes that are preserved and spread by tradition. The relationship between culture and geographical orientation is a topic that is gaining increasing interest among scholars. This introduces the concept of *social knowledge archives* in the animal world, illustrating that in some species it is essential to protect older individuals, such as matriarchs in elephants, precisely because they are the bearers of knowledge that is fundamental to the survival of the group. Once again, culture here is a local element, a depository, capable of reinforcing specific adaptive capacities in a certain context. Cultural segregation is a very important issue, creating profoundly different life strategies within the same species. In some species, such as sperm whales (*Physeter macrocephalus*), there are culturally distinct groups that can form social barriers between each other. When this occurs we have the emergence of different styles that correspond to a different ability to adapt to change (e.g. climate change).

According to recent ISPRA research, it is essential to protect this cultural capital. In some species, it should be possible to characterise animal populations based on their cultural profile rather than the more traditional approaches based on genetic diversity or geographical isolation (Ciucci et al. 2019). This intergenerationally transmitted knowledge can reveal fundamental aspects for the conservation of these species. The 2020 conference (CMS-COP 13) also considered culture in relation to the preservation of different species. For example, the culture that allows chimpanzees to crack nuts in the westernmost parts of Africa enables these populations to survive during the dry seasons. It was emphasised that human activities that may disrupt these cultural traditions can seriously jeopardise the survival of these populations.

The Use of Tools Among Animals

Humans consider themselves to be the technological animal *par excellence*, and there is no doubt that technology in our species has reached an apex and a pervasiveness not found in others. The traditional approach to anthropology sees this eminently Promethean tendency of the human being as evidence of poor or incomplete equipment from a phylogenetic point of view and, consequently, a weak predisposition in terms of innate behaviour. This is a mistake even just from a purely descriptive point of view, since humans have significant morphological specialisations directing them towards a particular lifestyle—think of the conformation of the spinal column, the evolution of the foot, the lowering of the larynx, the structure of the skull, the increase in the gluteal muscles, the development of the neocortex, the no longer manifested oestrus, the degeneration of the vomero-nasal organ, the loss of the penic bone, to name but a few examples. However, this is a fallacy above all with respect to the very principles of selective dynamics, because it is impossible to imagine a species not shaped by natural selection. Darwin himself emphasised the inconsistency of the inverse proportionality between instinct and intelligence, and he noted a strong innate equipment in human beings.

Once again we must point out the extraordinary relevance of Darwinian thought, unfortunately distorted by the rise of Wallace and Weismann's neo-Darwinism in the twentieth century. In a logic of continuity, the English naturalist pointed out that complex instincts were precisely what required greater intelligence in terms of application and adaptation. In fact, if we study the use of tools in the animal world, we discover that certain orders—such as primates, corvidae, hymenoptera and cephalopods—that are extremely complex in their behavioural repertoire, are the ones that use tools most easily and in the most articulated way. There is no doubt that there is a very close relationship between the phylogeny of a species and its tendency to use certain tools, where the innate-intelligence flywheel plays an important role alongside the species' practical tendencies, interaction with objects in the environment, curiosity and exploratory aptitude, mnemonic skills and expressive flexibility. There is a biunivocal or recursive relationship, whereby the innate (morphological, functional or behavioural) characteristics of a species enable or facilitate the evolution of tool use, and at the same time the use of tools is an important selective steering factor.

Creativity undoubtedly plays a central role in technopoetic evolution, and the tendency to play, as a training ground for hybridising behaviour, exploring new possibilities, probing and developing imaginative attempts, is often the key factor in the propensity to invent tools or exploit existing objects for instrumental use. Not all individuals in a population have the same predisposition or readiness to identify a possible use for a tool, and the parameters are manifold—such as the habit of interacting with objects, the curiosity and exuberance of that subject, the sense of security. In short, there is such a thing as a *technopoietic personality*. It is interesting to note that it is generally the young who invent new tools, but adults are the ones who use them best. The complexity or incongruity of the living environment also act to promote technology: for example, in captivity some animals tend to show a greater predisposition to using tools. With animals endowed with strong allelomimetic tendencies—such as corvidae, psittacidae and primates—continuous contact with humans is also thought to have a strong influence on promoting tool use.

In nature we find the canonical use of certain tools as part of the species repertoire, like sea otters using stones to open molluscs, but also in less established situations. Moreover, there is a close relationship between the canonical use of an instrument and the cultural tradition, since most of the time we notice the use of different instruments or different ways of using the same tool by two or more populations of the same species. In this sense, technopoietic evolution has an effect on the phylogenetic trajectory of the species or the development of a cultural tradition. A tool can act in different ways on the selective pressures that characterise a species or a population within that species. For example, (1) it acts paratypically on the phenotype by changing selection parameters; (2) it encrypts certain selective pressures and thus lowers selection on certain genotypes; (3) it ushers in or emphasises other selective pressures, raising selection on certain genotypes; (4) it gives rise to new organ performance contexts through exaptation; (5) it binds particular organs of the body in a new way by establishing new somatic correlations.

It is not easy to classify instruments homogeneously, but generally speaking there are two types of effects: (1) the externalisation of functions, so that the organ that was previously assigned to them modifies its performative repertoire; (2) the somatisation of the external element, so that the conjugation apparatuses, both muscular and of motor coordination, tend to acquire hybrid performative coherence. It is important to differentiate the instrumental use of an object, for example the use of a stone found in the environment, from the preparation of an instrument, for example clearing a stick of side twigs or bending its end to make a sort of hook. On the other hand, there are nuanced situations where it is not always easy to make clear distinctions, such as choosing stones suitable for the purpose or taking them from a different place from where they will be used, demonstrating a precise plan of action. Ordering the various types of tool use according to their function is more convenient than other classifications. As Benjamin Beck writes in *Animal Tool Behavior*, in most cases when an animal uses a tool, it does so to extend its ability to reach a distant object, to intensify the mechanical force it can exert on its surroundings, to increase the effectiveness of its display behaviour or to improve its control over liquids (1980, 127).

It is worth recalling Wolfgang Kohler's findings on chimpanzees' use of crates to reach bananas placed high up in the cage. In fact, a frequent use of a tool is precisely to reach something placed out of the animal's reach: (1) in spatial terms, for example high up, by stacking objects to reach it; (2) in terms of foreclosure, if the target is placed in an inaccessible crevice. Often the target is located in a cavity or hole and the animal uses a stick to reach it or a lure to entice it, if the target is an animal. Sometimes it may also be necessary to use a stick to pull in or knock down something placed high up. In these cases the tool becomes a way of bridging a distance between the animal's body and the target, acting as an extension of the body, so that its use requires a psychomotor somatisation process that is not easy. This is why a young animal needs a long apprenticeship in order to become comfortable with a tool and is at an advantage if, in the early stages of development, it had the opportunity to observe adults handling it. Birds that make extensive use of an instrument usually hold it with their beaks, while mammals hold it in their forelimbs.

Other times, the use of a tool is linked to the objective of intensifying mechanical force. This includes the use of stones and mortars to crush and break the shell of nuts, the use of levers to lift something or sharp objects to dig or apply pressure, and the use of sticks or stones to obstruct or close a passage. The use of a tool can contribute to increasing the mass and the force of impact, decreasing the elasticity of the prehensile structures, or it can be linked to competition, conflict, adversarial exhibition, to the aim of frightening, challenging and hurting the opponent. The use of tools for aggressive purposes is also of primary importance for humans, so much so that it has been the subject of reflection in a number of essays by Konrad Lorenz, such as *Die acht Todsünden der zivilisierten Menschheit* (1973). In the animal world, the most frequent ways of doing this are (1) dropping or throwing sticks, stones or other objects from above; (2) throwing an object directly at an opponent; (3) using a stick to strike; (4) brandishing and waving an object as a threatening gesture; (5) dragging or rolling something towards an opponent; (6) striking an object in order to make noise. In all these activities, the object may be a weapon or more simply a demonstrative tool.

Another kind of tool use may concern activities that have to do with the control of liquids. Anthropomorphs frequently use leaves as sponges to collect liquids or to contain them. Other times, containers are used to hold water or, conversely, leaves are used as a sort of umbrella to avoid getting wet when it rains. We also find the use of tools for dabbing or cleansing, and these activities are related to other similar self-referred activities, such as using a stick to scratch oneself. In some cases the instrument has a plurality of purposes, for example status or play: the same object can allow the animal to play or to perform particular practical activities. Many social animals use objects to engage their mates, displaying the object as a flag and then running away to attract attention and be chased. This behaviour is common among wolves, for example, and particularly among pups, and can also be seen very easily among dogs. Some breeds in particular very frequently use objects as gifts (e.g. retrievers), as a means of challenge (e.g. molossers) or as a status-possession symbol (terriers).

It is not always easy to distinguish the proper use of a tool from other niche-modifying behaviours, based on the construction of useful structures for particular activities, such as spiderwebs or bird nests. We have seen that animal behaviour is not made up exclusively of body expressions, but also of more or less complex constructions, such as dens or termite mounds. According to most authors, these are not to be considered tools in the proper sense of the term, so that nests, beehives, spiderwebs and the like are excluded from the list of tools, but it is clear that we are dealing with borderline situations or, as mentioned, with blurred boundaries. For example, the arbours built by bowerbirds are not considered a tool, whereas the stick that some species use to paint their arbour is. The hard surfaces that some animals use as anvils are also of uncertain placement, for example when seagulls drop molluscs from above to cause them to open. The same can be observed in the white-winged chough (*Corcorax melanorhamphos*) or the Egyptian vulture (*Neophron percnopterus*), which drops pelican eggs onto hard surfaces.

This behaviour is common among corvidae; in particular, carrion crows (*Corvus corone*) have been found to scatter palm fruits on roads and wait for cars to break them as they pass over them. Throwing objects at hard surfaces is also observed in mustelids and viverrids. Likewise, shrikes of the genus *Lamius* stab their prey insects into the thorns of trees, immobilising them and preventing them from escaping while they consume them. The nuptial gift found among spiders of the genus *Pisaura*, but also in some dolphins or Galapagos cormorants (*Phalacrocorax harrisi*), also has an uncertain attribution. What about using ingested objects to perform particular functions? Some crustaceans ingest sand to aid their posture, while birds do so to facilitate the function of the gizzard or muscular stomach and mechanically grind their food. Other behaviours that are difficult to categorise include anting, placing oneself on top of an anthill to be sprayed with formic acid for disinfestation purposes, or catching ants with one's beak and depositing them on one's body, a behaviour found in many birds.

How do these behaviours arise? Initially, they might simply be discoveries, but over time they become part of the cultural tradition of a given species, subsequently modifying the structure of the niche and, with it, the selective pressures confirming all the characters that increase that behaviour's integration into the species' repertoire. Speaking of epigenetic effects, as shown in Conrad Hal Waddington's (1905–1975) concept of genetic assimilation, lifestyle and in particular responses to niche modification are seen to influence gene regulation, transforming what the Edinburgh scholar called the epigenetic landscape. As mentioned above, the genome is not only a recipe book of characters, but also regulatory genes that make morphogenesis very sensitive to environmental changes in a logic of gene interaction. Waddington stressed that individual behaviour is central to evolutionary processes, precisely because of the tendency to modify the developmental niche. Creativity, therefore, does not merely switch on the usual light bulb in the individual, enabling it to achieve its goals, but can sometimes give rise to a tradition which, in the long term, through its epigenetic and niche effects, produces substantial changes in the whole species.

The use of a tool, when it becomes a tradition, is certainly one of the most important processes of niche modification. The emergence of a technical expression is strongly linked to the type of intelligence, so certain animals are more prone to engage in technopoiesis events, not because they are more intelligent than other species, but simply because they have a vocation for this type of interpretation of the world. There are also other factors, such as the presence of morphological determinants, like the predisposition of prehensile organs and good coordination with a sense organ (e.g. oculo-manual coordination in primates). Another important factor is the ability to learn and in particular the tendency to make use of experience. Exploratory motivation, curiosity, the tendency to interact with objects, investigative manipulation, the tendency to play with objects or to use an object as a lure for playful engagement all play a part. The general habit of playing with objects has also been shown to give rise to a kind of predisposition or apprenticeship that later facilitates the use of tools. As reported in a study by H. Birch (1945), chimpanzees that were allowed to play with sticks when young proved to be much more capable of using them as an extension of the arm when they reached adulthood. It is believed, therefore, that the increased ability of adults to use tools is attributable to sense-motor development.

The greater capacity of adults to use tools, indeed, can be traced back to two convergent developmental processes: (1) the development of psychomotricity and experience with manipulating objects; (2) the possibility of spending the developmental years in an enriched environment that stimulates adaptation strategies or intervention in external situations. Factors that facilitate the use of tools include expressive versatility, the tendency to diversify one's behaviour and modify it, the ability to innovate, experiment, correct, without becoming discouraged but persisting with obstinacy. The researcher Sue Savage-Rumbaugh points out that there is a sensitive or critical period, quite early according to her, during which experiences with objects have the greatest effect on future abilities to use tools or think through them. The use of a tool, in fact, involves a precise motor organisation and the sensorimotor correlates develop in the juvenile period. Peeling back a plant to find insects is properly an act of feeding, but coming into contact with pieces of bark is fundamental for their subsequent use as tools. Some behaviours, such as throwing branches from trees to ward off intruders, probably derive from the habit of moving about in trees in an excited manner, shaking branches as a display of strength. According to some authors, such as John Alcock in *The Evolution of the Use of Tools by Feeding Animals* (1972), certain tool-use behaviours may have originated as frustration responses. Sometimes there are random events, for example a reinforcing effect following from touching an object, other times there are more complex processes, as is the case with the insight observed by Kohler.

There are varying opinions on this, from the idea that the use of a tool can arise as a casual response to frustration, to the view that it derives from joyful and playful behaviour (Lawick-Goodall 1972). In any case, the use of a tool is the result of the articulation or refinement of a simpler behaviour. We have seen this when talking about primates throwing branches, or in the case of the Egyptian vulture that uses a stone to crack the shell of ostrich eggs. As Beck points out, since the evolutionary source of tool use lies in the remotest past, our understanding of original behaviours and their evolution rests entirely on conjecture (1980, 188). In my opinion, the origin of a technique does not follow a universal law but may vary greatly: what's certain is that the use of a tool does not rest on anything, but is based on behavioural patterns already partly present in the species. Sometimes these patterns favour the occupation of new environments or new resources, so the presence of tool use might enhance colonisation. Alcock (1972), for example, hypothesised that the main animal species that use tools have invaded niches that are not characteristic of their phylogenetic groups. William Hamilton (1936–2000) described these species as scavenge-hunters (1973, 128), that is not very specialised in predation, but undoubtedly flexible in behaviour and highly creative. There is no doubt that the effects of technopoiesis can be very diverse: technology can have a compensatory, subsidiary or vicarious action, but is always innovative.

One aspect that should not be underestimated is the segregating effect of technical traditions in different populations. West African chimpanzees open nuts by crushing them with stones, while Gombe chimpanzees crack similar nuts by beating them against trees: the different ways of opening nuts by using external entities have given rise to two different cultural traditions. A similar situation is found in the behaviour of Californian and Aleutian sea otters. As mentioned above, different chimpanzee populations use sticks differently, and different gorilla populations resort to different uses of medicinal plants. Carel Van Schaik of Duke University has shown that orangutan populations in Sumatra's Suaq Balimbing Forest use sticks to extract insects from trees, whereas this behaviour is absent in populations in the rest of Sumatra and Borneo.

Sometimes we can hypothesise an environmental influence in the induction to use a certain tool and its subsequent incorporation within a culture. The use of sticks to catch termites or stones to open nuts, eggs or molluscs has shown that there is a very close relationship between the tendency to use tools and the habit of foraging for food inside a shell. The tool is very often an extractive device and is the only means of accessing a particular food source in the absence of other food alternatives. The transformation of the sporadic use of a tool into an intergenerational tradition depends on the fact that some uses require an apprenticeship. When using stones to crack open molluscs or nuts, young individuals usually fail in their attempts and may even injure themselves, but despite this, and despite the behaviourist concept of positive punishment, they keep trying. Mr and Mrs Boesch (1983), studying the behaviour of chimpanzees in the Tal forest in Côte d'Ivoire, were able to record cases in which the mother had actively intervened to teach her offspring the correct way to operate, underlining how mothers encourage their young to practise this skill by giving them nuts and stones.

The search for food by extraction lies somewhere between using the tool to expand one's reach and using it to increase mechanical force. Certainly, the use of a tool in a repetitive way and in the same context must be differentiated from the use of the tool in different contexts and with equally specific techniques: this is an *intelligent use of the tool*, which can be observed not only in primates but also in insects. Once again, it is clear that the early manipulation of objects has a fundamental influence on the versatility of use of the instrument itself. This of course brings us back to the argument of so-called environmental availability, since tool use is more likely to develop in an arid area than in a forest or swamp. But we should not think that technical behaviour is just a reaction to some environmental determinants. Observing termites fishing or seagulls throwing molluscs, we detect a persistence that shows an undoubted finality of behaviour. Moreover, the way in which such behaviour is carried out is always related to specific difficulties or characteristics. Seagulls, for example, never throw at random but according to the compactness of the shell and the place of impact: in fact, they drop molluscs on rocks, not on sand. Chimpanzees are able to carry their tools from the place where they are found to the place where they will be used, which are often far apart. Gombe chimpanzees fish for ants in a different way than for termites, because of the different dangers involved: ants are fished from a higher place to avoid being bitten.

In the world of invertebrates, the use of tools is far from uncommon. For example: antlion larvae set traps for passing ants and throw sand at their victim to make it fall; weaver ants use larvae to build and repair their nests; Sphecidae close their nests with pebbles; hermit crabs occupy empty shells; crabs place sea anemones on their backs to disguise themselves;

octopuses use stones to hold open the shells of large bivalves; the ant *Dorymyrmex bicolor* uses pebbles to block the exit hole of enemy ant-hills. There is also evidence of the use of tools in reptiles: for instance, crocodiles use sticks to attract birds. In the realm of birds there are many uses for tools: bowerbirds use pieces of bark to paint their arbour; American robins (*Tordus migratorius*) use twigs held in their beaks to move dry leaves and look for insects; Egyptian vultures and black-breasted buzzards (*Hamirostra melanosternon*) use stones to crack large eggs; white-winged choughs use pieces of mussel shell to open other bivalves; woodpecker finches (*Camarhynchus pallidus*) insert sticks in tree holes; parrots of the genus *Cacatua* use sticks to scratch themselves and many species of herons use straws as bait.

Also, primates are not the only mammals equipped with tools. For example, California ground squirrels (*Otospermophilus beecheyi*) throw sand at snakes; African and Asian elephants use twigs to scratch themselves or ward off flies; some deer such as the greater kudu (*Tragelaphus strepsiceros*) and *Elaphurus* tend to adorn their antlers with hay and mud; spectacled bears (*Tremarctos ornatus*) use sticks; sea otters (*Enhydra lutris*) use stones; polar bears throw large blocks of ice or stones at seals and walruses; some mustelids and viverrids use stones. There is no doubt, however, that the presence of tools is particularly frequent in primates. For example, howler monkeys of the genus *Alouatta* throw objects to drive away intruders; Colombian white-faced capuchins (*Cebus capucinus*) use sticks to strike rivals or snakes; white-fronted capuchins (*Cebus albifrons*) use objects to reach for food; grey langurs of the genus *Semnopithecus* throw stones to scare away intruders; sooty mangabey monkeys (*Cercocebus atys*) use sticks to dig into the ground and widen the entrance to underground nests; rhesus macaques (*Macaca mulatta*) roll stones from above to ward off intruders; Barbary macaques use competitive buffering (*Macaca sylvanus*); baboons of the genus *Papio* throw sand into their enemies' eyes.

There is a great deal of research on anthropoid primates. I have already mentioned some findings, for example how chimpanzees fish for termites by not only using a stick as a tool but adapting it to their needs, removing side twigs. Tool-using behaviour seems scarce in hylobatids (gibbons and siamangs) apart from the occasional throwing of objects at intruders. However, there have been a few cases of individuals in captivity soaking pieces of cloth in water and sucking on them, a behaviour that would suggest a similar practice with leaves in the wild. The use of tools is widely documented in orangutans, mainly in a destructive sense when in captivity.

Object throwing is also present among gorillas, in addition to the use of liquid containers. In conclusion, we can say that the use of tools is just one of the many ways in which animal behaviour can manifest versatility and creativity, with the individual's action as the driving force behind expressive evolution. Having said this, I also think it is incorrect to think that only animals that use tools demonstrate a mind in action, because in reality technical intelligence is just one of the many forms of intelligence present in the animal world.

Works Cited

Alcock, John. 1972. The Evolution of the Use of Tools by Feeding Animals. *Evolution* 26 (3): 464–473.

Beck, Benjamin. 1980. *Animal Tool Behavior*. New York: Garland STPM Pub.

Birch, Herbert. 1945. The Relation of Previous Experience to Insightful Problem Solving. *Journal of Comparative Psychology* 38: 367–383.

Bisconti, Michelangelo. 2008. *Le culture degli altri animali*. Bologna: Zanichelli.

Bonner, John Tyler. 1980. *The Evolution of Culture in Animals*. Princeton: Princeton University Press.

Cavalli-Sforza, Luigi L., and Marcus Feldman. 1976. Evolution of Continuous Variation: Direct Approach Through Joint Distribution of Genotypes and Phenotypes. *Proceedings of the National Academy of Sciences of the United States of America* 73 (5): 1689–1692.

Ciucci, Paolo, et al. 2019. Animal Cultures Matter for Conservation. *Science* 363 (6431): 1032–1034.

Deeke, Volker, John K.B. Ford, and Paul Spong. 2000. Dialectic Change in Resident Killer Whales: Implications for Vocal Learning and Cultural Transmission. *Animal Behaviour* 60 (5): 629–638.

Fagen, Robert. 1978. Evolutionary Biological Models of Animal Play Behavior. In *The Development of Behavior: Comparative and Evolutionary Aspects*, ed. Gordon M. Burghardt and Marc Bekoff, 385–404. New York: Garland.

Ford, John K.B., Ellis M. Graeme, and Kenneth C. Balcomb. 2000. *Killer Whales: The Natural History and Genealogy of Orcinus Orca in British Columbia and Washington State*. Seattle: University of Washington Press.

Goodall, Jane, and Baron Hugo van Lawick. 1971. *Innocent Killers*. Boston: Houghton Mifflin.

Griffin, Donald. 1992. *Animal Minds. Beyond Cognition to Consciousness*. Chicago: The University of Chicago Press.

Hamilton, William. 1973. *Life's Color Code*. New York: McGraw-Hill.

Hinde, Robert Aubrey, and J. Fisher. 1951. Further Observations on the Opening of Milk Bottles by Birds. *British Birds* 44: 393–6.

Itani, Junichirô. 1959. Paternal Care in Wild Japanese Monkeys, Macaca fuscata fuscata. *Primates* 2: 61–93.
Kawamura, Syunzo. 1963. Subcultural Propagation Among Japanese Macaques. In *Primate Social Behavior*, ed. Charles A. Southwick. Toronto: Van Nostrand.
Kortlandt, Adriaan. 1965. How do Chimpanzees Use Weapons When Fighting Leopards? *Yearbook American Philosophical Society* 1965: 327–332.
Le Gros, Clark Wilfrid E. 1967. *Man-Apes or Ape-men?* New York: Holt, Rinehart and Winston.
Lewontin, Richard. 1991. *Biology as Ideology*. New York: Harper Perennial.
Lorenz, Konrad. 1973. *Die acht Todsünden der zivilisierten Menschheit*. Berlin: Piper Verlag.
Mainardi, Danilo. 1974. *L'animale culturale*. Milano: Rizzoli.
Marchesini, Roberto. 2008. *Intelligenze plurime*. Bologna: Oasi Alberto Perdisa.
Menzel, Randolf. 1979. Behavioral Access to Short-Term Memory: The Honey Bee as a Model System. *Naturwissenschaften* 70: 504–511.
Narby, Jeremy. 2005. *Intelligence in Nature*. New York: Tarcher Perigee.
Paterson, Hugh E. 1980. A Comment on "Mate Recognition System". *Evolution* 34 (2): 330–331.
Prum, Richard O. 2017. *The Evolutionary of Beauty*. New York: Doubleday.
Thornton, Alex, and Katherine McAuliffe. 2006. Teaching in Wild Meerkats. *Science* 313 (5784): 227–229.
Van Lawick-Goodall, Jane. 1972. A Preliminary Report on Expressive Movements and Communication in the Gombe Stream Chimpanzees. In *Primate Patterns*, ed. P. Dolhinow. New York: Holt, Rinehart, and Winston.
Whitehead, Hal. 2003. *Sperm Whales. Social Evolution in the Ocean*. Chicago: University Chicago Press.

CHAPTER 12

The Conditions of Creativity

Premise

My long reflection on animal creativity is based on an inescapable assumption: the acknowledgement that a state of subjectivity should be extended to other species and indeed characterises animality as a whole. I have defined this subjective condition as articulated in four levels. The first (i) is the sensory level, which leads an animal to differentiate between sensibility (what is present out there in the world) and agency (the changes it produces in the world), giving rise to a first archetypal definition of *Self*. Then there is (ii) the affective level, indicating the subject's inherent interests, which can be referred to the hedonic index and is linked to the variables of physiological homeostasis, aesthetics or preferences, as well as the animal's dispositional nature, capable of producing a *for-Self* or copulative condition for the subject. The third level is (iii) the cognitive one, based on the subject's endowments of knowledge, considered as tools and not as automatisms, such that the individual is the one who uses them, adapting them to its actual needs, whereby the subject is the bearer of contents or *in-Self*. Finally, there is (iv) the mental or integrative level, which establishes an integration between the components (emotions, representations, etc.): these produce a psychological state (fearing that something dangerous may happen) which in turn determines a propositionality (I believe that if I run away I will avoid the risk) that translates into a behaviour (running away).

R. Marchesini, *The Creative Animal*,
https://doi.org/10.1007/978-3-031-07414-1_12

Without subjectivity, every reference to intelligence, the ability to evaluate situations, the consequent flexibility of expression up to the most blatant peaks of creativity, not to mention the varying levels of awareness, will inevitably evaporate. Subjectivity is presence, that is, the ability to *be present* in the moment and in the situation through a diachronic dimension of *Dasein* that brings out a present by inaugurating a dialectic between past and future. An object or a machine evidently lacks this condition, despite having functional or performative connotations and sometimes even a singularity of state. Certainly, speaking of consciousness—a possible fifth plane of subjectivity—without clarifying the *principle of subjectivity* means, in fact, not addressing the heart of the problem, leaving an unsolved philosophical issue to the clutches of explanatory circularity. An entity can only express creativity if it is in a position to emerge from the here and now by projecting itself elsewhere; to do this, consciousness is not indispensable, but it is certainly necessary to be able to reconfigure the past in a future projection and vice versa, that is, not to be captivated in the moment.

The concept of *being present in the moment and in the situation* therefore refers to a state that may seem paradoxical when related to the concept of the present, because it implies not being entirely or totally comprised in the moment and in the given circumstance. One could say that being present means being *somewhere else* at every moment of experience. If we can and must speak of animal *Dasein*, the term must translate the sense of not being confined right here and right now. The subject cannot be fixed in a point: it is constantly experiencing the perspective towards something that isn't there because it has already passed or is yet to come. Subjectivity is a diachronic condition of being, not because it is endowed with an evolutionary or functional unfolding—all phenomena are—but because it constantly oscillates between different temporal dimensions. Creativity is an expression of this magmatic nature of the animal condition, so recognising its presuppositions is the only way to avoid the mere description of the *mirabilia* of the zoological universe: these accounts are interesting from a kaleidoscopic point of view or as material to be examined, but certainly not exhaustive for understanding subjectivity. After all, what is creativity, if not the way in which an animal constantly reinvents itself? Where does it lie, if not in the subject's incessant motion that leads it to go beyond itself, to become something else at the very moment it expresses what it is?

Subjectivity is essentially a reversal of objectivity, so any mathematisation of the entity—by positing process isochrony and developmental determinism—clashes with the subjective attribution, which is, conversely, anecdotal, arbitrary, aleatory and above all free in its singularity. Subjectivity is an intrinsically non-repeatable phenomenon, which therefore jars with the very methodology by which one claims to translate it into an explanatory model based on such coordinates. Biomechanics can explain many physiological mechanisms that characterise living beings, I have no doubt about this (even if in order to make ends meet we often pass off stochastic events as noise, which is a serious mistake). But it is misleading when it comes to subjectivity, which is always an expression of freedom in its manifestations of mental states and behaviours. One of the most relevant characteristics of this animal protagonism is the *condition of ownership*, which we detect from two ontological predicates: the presence of inherent interests and the availability of the apparatus of competence. The fact of being able to use one's endowments as a tool, in fact, produces a condition of ownership, which means that the subject is not determined from the outside but follows its own inclinations and is directed towards its own inalienable ends.

Here, too, I would like to clarify something. These ends are not inalienable because they cannot be alienated for some reason, but because to do so would mean stifling the qualities of subjectivity itself. Self-ownership is the ordinary state of the animal and the indicators of coherence in this respect are guided by the hedonic system: this is the reason why I see the joy of living as the true causal source of creativity, in its various reconfigurations of reality. A subject is one who *can experience pleasure*: this, in my opinion, is the reversal we should make with respect to Bentham's expression. Now, in order to understand the causal meaning of pleasure and not fall into the error of considering it a mere consequence or marker—as in the behaviourist concepts of reinforcement and punishment—it is essential to envisage the hedonic system as a *flywheel of desires* that drive the subject to explore possibilities. Exploratory curiosity, imaginative expectation, motivational appetite, restlessness, exuberance, to name but a few, are forms of expression of the joy of living, able to lead the individual to perform an undefined activity which therefore necessarily involves creativity.

One hundred and fifty years or so after Darwin's 1871 essay and his analysis of preference systems, we are paying the price of an entire revisionist century, with few commendable exceptions, which has completely erased subjective orientation from the predicates of animality. But perhaps the hindrance is far deeper and more pervasive than the reductionist counter-revolution of the past century: it concerns the anthropocentrism of Western culture as a whole. In our collective imagination we are still caught up in traditions that undermine the concept of animal subjectivity. And I am not just referring to the mechanistic view of the Cartesian tradition, back in vogue in twentieth-century determinism, because the history of thought has accumulated many prejudices regarding the concept of animality. Undoubtedly, the idea that animal expression is the result of automatisms, whatever their denomination or the explanatory model adopted, still plays a far from minor role. However, I believe that the conceptual obstacles to the philosophical analysis of animality are more articulated and require a wide-ranging critical reflection that cannot be limited to the variously revised models of Cartesian biomechanics.

The Emancipation of Animality

A problem that poses great difficulties for *philosophical ethology,* called upon today to reflect on animality as a metapredicative condition, concerns the very fundamentals of *Western thought.* It is evident, in fact, that we can only speak of animality as inclusive of *Homo sapiens,* since any comparison, when dealing with predicative aspects, should be made with respect to another species and not to some imaginary "animal" category. Unfortunately, ontological reflection is flawed: it is built on a bogus categorical antinomy. This is not, therefore, a small detail or a specific criteriology, but the very foundation called upon to denote the human dimension through the animal counterpart, in a dichotomous game of oppositional qualities that is ontologically incorrect. The not-too-veiled objective is to show the uniqueness but, above all, the special nature of man compared to the jumble of brutes who, although different from each other, would be united in lacking the faculties of *humanitas.*

Throughout the history of philosophical thought, there have been several variations on the theme of human speciality: at times referring to Epimetheic incompleteness or lack of rank, at others to the faculty of reason or membership in the civil and political community or, again, to the possession of language and symbolic thought, right up to the Socratic concept of *imago dei* as well as Abrahamic heritage. Thus, different predicates have been used time after time to mark the distinctive counterterm against which to base the uniqueness of man, brought out of the *mare magnum* of animal indistinction. Ultimately, beyond the dialectical particularity of individual authors, it is easy to find a common thread running through these traditions: the claim to derive the human dimension by opposition to the animal one. Clearly, if this archetypal antinomic construction were to fall apart, there would be a risk of a sort of ontological implosion, especially for visions that need clear boundaries and foundational self-referentiality, that is, the claim to understand humanity *iuxta propria principia*.

This explains why the human versus animal dichotomy has undergone different variations over time, even though the dialectical mechanism—what Giorgio Agamben called the anthropological machine in *The Open* (2004)—has remained unchanged. The metamorphoses of oppositional pairs—variations on the same theme—certainly reveal the philosophical approach of individual authors, but never deny the anthropocentric background that links together a large part of Western philosophy and often does not spare even those authors who would have liked to distance themselves from the anthropocentric tradition. For example, in *The Animal That Therefore I Am* (2008), despite his intention to criticise and dismantle the anthropocentric conception, Jacques Derrida does not go beyond the oppositional and categorical use of the animal and inevitably falls into the usual discriminatory problem, whereby the animal is incapable of autobiography while the human being is lacking in self and exposed to nudity. In other philosophers the discontinuity is even stronger, starting with Aristotle who considered the human being *zoon logon echon*, a property that does not remain confined to the descriptive aspect, but produces an ontological elevation. The desubjectification of the animal passed through various attributions in several philosophers—such as Immanuel Kant's conception of a necessitated entity and Martin Heidegger's notion of poverty in world—without ever addressing the problem of its basic correctness.

As we can see, the two terms nevertheless maintained a relationship of specularity, through the play of contralateral attributions which make the human being not a kind of animal but the non-animal *par excellence*. However, I would like to clarify one thing. It is not a question of not recognising the exceptional qualities or, if you like, the particularities of the human being, because that would be an equally short-sighted and dogmatic attitude; nor can we agree with the idea that the only aspect that characterises our species is language or the autobiographical capacity. My criticism is against the claim to define the animal as an ontological category *opposed* to the human being, an epistemological obstacle that leads to mistakes in the conception of both animality and the human condition. Indeed, because of this claim we fall into the following errors. We (i) believe that we can compare predicative aspects, that is, typical of our species, with metapredicative aspects, which do not concern a particular animal species but a whole category. We (ii) consider animality as something extraneous to the human dimension or at most spurious (regressive, ancestral, profound) from which we ought to move away in order to assume truly human predicates. Finally, we (iii) think that there is a common denominator shared by all other species and that this predicate is not present in the human being, so that a chimpanzee belongs to the same category as a jellyfish, as opposed to man.

If we can speak of animality—and I personally believe so—this term must ontologically and not only biologically include the human being. Animality must be a metacondition that is (i) definable through metapredicates, such as subjectivity; and can be (ii) conjugated in a predicative sense in the different *expressions of being-a-subject*. The latter are developed on various taxonomic levels, such as vertebrate, tetrapod, homeotherm, mammal, down to the dimensions of species, ethnic group, family, individual and the latter's here and now. The concept of animality must be reformed from its foundations, though its emancipation from the captivity of fictitious attributions—the kind that is only functional to our need to find a counterpart to humanity. Philosophical ethology, therefore, has a task that, while remaining in close relation with the descriptive ethology of animal behaviour, mainly concerns the ontological aspect of animality, developing the necessary critique in order to free animality from this deceptive positionality. The instrumental precept of viewing animality as humanity's dark mirror has made it impossible to construct a coherent metapredicative ontology from which to derive the predicates of the various species, including our own. Emancipating animality, including our own animal condition—at this point it should be clear that it is not a question of freeing the brute in us or giving free rein to feral irrationality, as the animal was traditionally described—would not only help us to better understand other species, free of prejudice, but also help us to have a better image of ourselves.

Indeed, more than a simple bias against other species, we must speak of a condition of captivity that has enchained animality as a whole: this has prevented any ontologically objective reflection, undermined by the constant need to draw an unbridgeable gap between us and them. In this sense, I think there is an urgent need to emancipate animality, that is, to free it from the constraints to which we have subjected it, also to recover the ontological dimension of humanity while avoiding the trivialisation of ancestrality or instinctiveness—the animal that is in me—or the need to construct the human through some imaginary extra-biological dimension. The emancipation of animality is also human enfranchisement, liberation of the animal *that therefore I am*, to borrow Jacques Derrida's phrase. On the other hand, to recognise ourselves fully as animals, by considering humanity as a conjugation of animality, does not mean to fall within the traditional parameters of the latter, but rather to redefine the *metapredicative condition of animality*, getting out of the fetters of oppositionality. And when reference is made to a metapredicative condition, this does not mean highlighting the characters that define a species or a particular type of animal—such as ferality or aggression—but rather connotations that go beyond one's species.

The construction of animality as the counterpart to the human dimension has been based on a sort of cherry-picking of the characteristics that can be defined as negative, often with the addition of emphases or mythologies to increase their oppositional character with respect to the human being's presumed positive nature. Thus, the wolf has become the symbol of animal brutality, the sheep of lack of personality, the bear of sexual intemperance and the chicken of stupidity. But beyond this gallery of symbolic definitions, which do not always present the animal in a negative way—think of the industriousness of bees, the strength of bulls or the courage of lions—the stereotypical use of the predicate-animal pair has led to the alchemical extraction of certain traits to be attributed to the animal condition, in order to bring out a profile that can aspire to represent all non-human species. In other words, we passed from predicative extraction to metapredicative attribution: a macroscopic ontological error. While the portrait of Dorian Gray reminded the protagonist what his real image was, animal specularity is supposed to remind man of what he is not or should not be—a view that unites, despite the differences, thinkers who cover the entire span of the modern age, from Pico della Mirandola to Schopenhauer. Dismantling this anthropogenetic device, which transforms animality into a counterpart to allow the emergence of human protagonism, is not easy. This is an insidious disease and the relapses of the anthropocentric disorder are frequent.

Therefore, the great philosophical laboratory of human elevation through animal opposition did not only create the machine-like animal.

First and foremost, it painted the picture of a brute, devoid of reason and sentiment, completely encompassed by base instincts, uncontainable in its appetitive irrationality: the beast characterised by dark and gloomy hues as a primordial menace and as a monstrosity to be vanquished or subdued, to be transformed into a trophy or a scapegoat to reaffirm the cosmopolis. There is also the theriomorphic demon, the expression of contamination and wickedness, the tempter capable of misleading man from the right path, the fall into ancestrality of William Golding's *Lord of the Flies* (1959) or of Lombrosian physiognomy, the horrific metamorphosis of Gregory Samsa or the regressive one of Pinocchio, that is zoonosis. Western thought has therefore achieved human emancipation from animality through the denigration of the latter. The animal is the gaze that does not rise to the sky, the poverty in world of the tick stirred by butyric acid, the dialectic between *bios* and *zoe*, the ape incapable of learning a language, the replicative fury of mating. The animal lacks something, even in the completeness of its rank, and only the human being is capable-of.... Hence the claim to transform the species predicate into an ontological metapredicate. Yet even that predicate turns out to be wobbly, because of some odd researcher who finds another species capable of doing the same thing as us humans. So the bar must be raised, and so on and so forth.

It seems very difficult today to get away from this urge to affirm the *non-animality* of the human being: it's as if it were at odds with a deep-seated need, which I am not sure is attributable exclusively to a specific culture. What is certain is that in the Western philosophical tradition, the dichotomy between human and animal is so deeply rooted that it has even affected Darwinian thought, whereby the non-human is reduced to an ancestor, in order to bury our animal condition in the mists of remote ages and in the depths of the unconscious. The aim is always the same: to create an insurmountable barrier between us and animality. And when one criticises the oppositional stereotype and the reductionist conception of the heterospecific, one is immediately accused of anthropomorphising animals: a condemnation without the possibility of appeal. On the contrary, any form of simplified and mechanistic explanation of animal behaviour is credited because of a misreading of Lloyd Morgan's canon. In fact, there is no explanatory simplicity in the farrago of atomic stimulus-response mechanisms: the use of three models of explanation (associative, psycho-energetic, cognitive) for the same phenomenon is anything but simple, not to mention the claim to place an unbridgeable gap between humans and anthropoids, against Darwinian continuism.

But in what situations can we speak of true anthropomorphisation of other species? First of all, I would like to say that in general, it is more correct to use humans as a term of comparison rather than machines. The reasons for this are not only related to the complexity of animals, admitted by Descartes himself, but are based on the theoretical consequences of evolutionism. That said, it is clear that sometimes attributing human behaviour to another species is an anthropocentric way of not confronting diversity. When we believe that a dog or cat shares our perception of the world or the same communication code, is interested in what we are passionate about or has the same social dimensions, we are caught up in trivial forms of *humanisation*. However, when we observe play or parental care in mammals or emotional reactions in other classes as well, if we find similarities with our own behaviour, so that we interpret it based on overlapping criteria, we are not wrong. Above all, in such cases, we are not anthropomorphising, because we are not attributing human qualities to the other species, but we recognise shared traits, homologies inherited from a common ancestor. If we refer to certain laws of learning—such as that of effect or exercise—or of projection into a purpose, just to give a few examples, interpreting behaviour according to the parameter of similarity does not mean anthropomorphising. Obviously, one must always be careful to assess the gradient of taxonomic proximity and thus of expressive homology.

We have seen how each species has its own dimensional bubble of perceptive and operational immersion in the world, as we find in Jakob von Uexküll and Georg Kriszat's essay *Streifzüge durch Umwelten von Tieren und Menschen* (1934). However, it is a mistake to consider these dimensions as detached monads because, in the logic of common ancestry and therefore of the inevitable presence of shared characters, the *Umwelten* partially overlap. Asserting the unknowability of the heterospecific as Thomas Nagel does in his essay *What is it Like to Be a Bat?* (1974) is only apparently an *epoché* because, in fact, the continuism and ontological consubstantiality of animality is indirectly denied. There is no doubt that the greater the taxonomic proximity (space of homology) and the greater the adaptive convergence (space of analogy), the greater the overlap. However, recognising similarities of belonging does not mean anthropomorphising, because we do not attribute human predicates to other species: we recognise that those predicates are shared. In such cases, it is a mistake to consider that particular predicate as human, because it actually predates the origin of our species: therefore, it must be evaluated as a non-human inheritance. Having said that, on some occasions it is also useful to rely on a form of projection or, better, to use the method of critical anthropomorphism, introduced by Gordon Burghardt in *Cognitive Ethology and Critical Anthropomorphism* (1991), which starts from comparison as a hypothesis to be tested.

Using the parameter of absolute difference, instead, is most often a stretch, especially when we are talking about the behaviour of species close to our own and with overlapping neurobiological structures, as is the case with mammals. It is no coincidence that mice and rats are used in neurophysiological experiments to draw conclusions about humans with regard to the functions of neuromodulators or the topography of specialisation of the neocortex. On the other hand, I believe that the comparative model is also useful for understanding the expressive diversity of different species. I call this methodology *reversible anthropomorphism* (meaning that it can return to the starting conditions), because it is based on the recognition of needs—such as eating, reproducing and learning—that unite the human being to other species, albeit distant from a taxonomic point of view. Starting from these basic needs, which do not characterise certain species but the condition of animality as a whole, helps to understand predicative evolution, that is, how each species has constructed its own identity by expressing those needs in a specific way.

Going Back to the Subject

When we say that an animal is a subject, and not just an object, we are making an important statement that, however, requires explanation, since it is not easy to pinpoint what is actually meant by the term subjectivity. If we avoid the convenient tautology of consciousness as an explanatory factor of subjectivity, recognising its nature of *petitio principii* rather than of true explanation, we discover that subjectivity is an attribute that is intuitively easy to understand, but very elusive when it comes to clarifying it. From an etymological point of view, we know that the word *subiectus* means "that which is placed underneath", that is to say, which lies at the base and configures the predicates of the entity. According to ancient thought, the subject, for example the Hellenic *hypokeimenon*, is what lies at the foundation, the unchanging substance of being: it is constitutive but at the same time hidden, evading any definition. For Aristotle, the subject is characterised as an intrinsic and immutable principle, the result of form and substance, a *synolon*, that is, the indissoluble union of matter and form.

Subjectivity, however, soon assumed a character of disjunction from the phenomenology of nature: an operation that led it to a sort of ontological sublimation. Man differs from other species precisely because he is freed from nature: he is a subject insofar as he is free and self-determined. In Plato's *Protagoras*, the tale of the two Titans describes animals as endowed with natural predicates, and therefore predetermined, as opposed to human beings who, neglected by Epimetheus, can enjoy art and ingenuity thanks to Promethean reparation. This founding myth is anthropocentric, as it describes the human condition as (i) special compared to other animals because it is the result of a different genealogy; (ii) lacking natural predeterminations, seen as limitations and (iii) self-defining and autopoietic in its ability to choose its own ontological path. What should draw our attention most is the attempt to ground subjectivity in absolute freedom and autonomy, linking these faculties to the lack of belonging to the animal world. This myth established an ontological difference between human and animal based on predicative genealogy—Promethean versus Epimethean—and led to the desubjectification of other species, uniformly described as brutes.

This vision would find its most complete expression in the humanists of the fifteenth century and in particular in the *Oratio de hominis dignitate* (1942) by Giovanni Pico della Mirandola (1463–1494). Following the Promethean mirage, the (human) subject was referred to as inwardness, freedom of choice and, consequently, as self-determination, elevation and non-specification. During the Renaissance, while animals were regarded as specialised to express a certain *habitus* and therefore confined to a natural rank, the human being was believed to lack predetermination and thus able to conjugate with the universal: think of Marsilio Ficino's concept of *anima mundi*. The verticalisation of the human found in these authors would manifest itself in the paradigm of *Homo mensura mundi*, which overturned the Sophist principle of relativism into a universal condition of humanity. The humanist parable emphasised this sense of human uniqueness, which distances the subject from the telluric and from nature. The cultural climate was still a long way from the mechanism of the seventeenth-century scientific revolution, but the seeds of this antinomy were already present. The ontological question of animality thus reached the modern age, after having been debated by Aristotle, the Stoics and Augustine himself, calling for a clear distinction based precisely on the subject-object dichotomy that imposes the loss of a subjective dimension on non-human phenomenology.

In this sense, Descartes (1596–1650) identified the subject with *res cogitans*, designating that which is able to think, to doubt, to take a stand with respect to necessitated unfolding. The subject is what *cannot be thought except in function of thought itself*, unlike the object, which exists independently. But for this very reason, with the French philosopher, subjectivity seemed to lose its concreteness and was washed away into broad meanings that take leave of matter and nature altogether. The doubting subject cannot be mathematised and hovers above all necessary gravitation, the same gravitation that instead governs somatic mechanics and the animal condition. The *res cogitans* summarised the predicates of *inextension, freedom and awareness*, sanctioning the dualism between spirit and body and thus placing subjectivity in the exclusively human faculties of thought and language. On the other hand, the convergence of subject and interiority led to considering the subject as an entity independent of others: Descartes' *cogito* was undoubtedly a process of solipsistic revelation, but detachment was also found in Gottfried Leibniz (1646–1716), who saw the individual as a monad without openings. Disembodying subjectivity also meant introducing a dangerous precedent with respect to the risk of *biomechanism*—Michel de Montaigne (1533–1592) had well foreseen this—with the related anthropocentric isolation.

A diametrically opposed position can be found in Montaigne's writings, which drew mainly on Plutarch. In *Apologie de Raimond Sebond* (1580–1588), Montaigne emphasised the ability of animals to build complex social relationships through cooperative performance, to solve challenging problems using logic and to display great communication skills. In the *Méditations métaphysiques* (1970), on the contrary, Descartes claimed that animals lack a real language and have no internal representations of the world, therefore their behaviours are mere reactions to stimuli instructed on the basis of instinct and due to their nature: their subjectivity is an optical illusion. An important response to Cartesian reflections came from the Epicurean philosopher Pierre Gassendi (1592–1655) who, in his *Fifth Objections* to Descartes, underlined the presence of *phantasia* in animals, that is (i) the capacity for immediate apprehension, that is the capacity for imagination, (ii) naming, that is the capacity to refer or attribute signs to past entities, (iii) inference, that is the capacity for association to construct general ideas and make deductions. For Gassendi, *phantasia* is first and foremost a capacity for imagination, so that animals, like humans, are able to learn by experience and reasoning.

The risks inherent in Cartesian biomechanics were also noted by the philosopher Pierre Bayle (1647–1706) and were taken up by many eighteenth-century authors: for Théophile de Bordeau (1722–1776), the vision of animals as machines could lead to include humans in a mechanistic interpretation as well; Jean Meslier (1664–1729) considered the opposition between humans and animals as unrealistic and David Renaud Boullier (1699–1759), author of the *Essai philosophique sur l'âme des bêtes* (1937), thought animals should not be compared to machines, but to human beings. Criticism of Descartes also came from Leibniz, who regarded animal subjectivity as endowed with a vital energy (*appetitus*) that underlies its freedom of expression and purpose. The subject is an active individual and is the primary source of its own orientation, according to a logic of maintenance and expansion. An animal is never passive and, as posited by Baruch Spinoza (1632–1677) especially in the concept of *conatus*, finds its adequate cause in its affective conditions, that is to say in feelings and passions. In the third part of *Ethics*, indeed, Spinoza (1951) states: "in no case do we strive for, wish for, long for, or desire anything, because we deem it to be good, but on the other hand we deem a thing to be good, because we strive for it, wish for it, long for it, or desire it" (E III, 9s). The subject is the real protagonist of action, as an affective entity which, as such, is involved in the interface with the outside world: its being in the world always means taking a direction, choosing, preferring, acting, that is, imprinting its own will or following its dictation on the stage of events.

Appetite (understood in the broadest sense of the desiring condition), sensitivity (as the ability to react to stimuli and choose one's own direction) and intelligence (generically denoted as the property of learning and adapting to particular situations) seem to bring subjectivity back to the animal condition. Subjectivity therefore indicates a way of behaving driven by orientation and electivity, following rules of partiality or non-objectivity, as opposed to the spontaneous movements of objects. In this sense, while an object simply follows the course of the current, a fish goes against the current, an antelope can jump over an obstacle, a bird can fly and not fall like a stone on the ground. The character of electivity, as well as opposing that of impartiality, gets rid of the passivity which, according to Spinoza, characterises actions that have their adequate causes in the external world. Subjectivity can be illuminated by consciousness but does not necessarily imply it, since the turbulent motions of the unconscious are also part of subjectivity. One is the protagonist of one's own progress in the world because (i) one is the bearer of intrinsic goals; (ii) one is the owner of one's own existential endowments. Subjectivity, in this sense, cannot be reduced to biomechanics, since the action expressed is not simply the unfolding of a pre-established causal cascade.

An important author in the transition between Cartesian thought and the study of animal behaviour that was to characterise the nineteenth century was undoubtedly Étienne Bonnot de Condillac (1714–1780). He believed that the theory of the machine-like animal was completely unfounded, and regarded animals as *sentient entities* capable of learning by experience and of making comparisons and judgements. In *Traité des animaux* (1755), Condillac maintains that there is no substantial difference between instinct and reason, since they are part of the same system originating in sensation. Condillac, however, undoubtedly retained a difference between humans and other species precisely in the capacity for general reflection and language. An author that was much more inclined to consider animals as endowed with reflective thought, albeit through simple and concrete ideas, and intentional behaviour, was Charles Bonnet (1720–1793): in his *Essai analytique sur les facultés de l'âme* (1760), he asserts the importance of pragmatic communication and survival behaviour. However, the need to place the human dimension at the centre of the Enlightenment facilitated Cartesian relapses, returning animality to a desubjectivised and repetitive condition, as absent, stunned, a prisoner of its own heritage and closed-off within the perimeter of its rank.

As said, the discussion between continuism and divergence would characterise the eighteenth century, as part of the rationalistic and encyclopaedic project, undoubtedly a feature of the cultural climate of the Age of Enlightenment. In many ways, it provided fertile ground for later reflections on evolutionism, in a comparative conception that was to form the basis of continuist thought. The champion of the idea of a single structural plan in animal forms, even in species that are very distant from each other, was Etienne Geoffroy Saint-Hilaire (1772–1844), who believed that the differences between *embranchements*, established by George Cuvier (1769–1832) in his essay *Le régne animal* (1836), were not absolute and therefore called for comparative criteria to help better understand the relationships between species. Interestingly, Geoffroy's intuitions would find new confirmation in the Evo-Devo theory. As Alessandro Minelli reports in *Forme del Divenire* (2007) "Time would, to a large extent, prove

Geoffroy right, both about the general principle of the unitary structural plane common to all animals, and about several more specific issues, such as the equivalence between the dorsal side of vertebrates and the ventral side of arthropods and other animals" (10).

The fathers of evolutionism were certainly sceptical of mechanistic models, adopting a parameter of continuity between humans and other species and only allowing for a difference in degree. In fact, it was only from the twentieth century onwards that the biomechanical conception was definitively affirmed through neo-Darwinism, which resumed and in many ways achieved a scientific modelling of the machine-like animal. In this period, behaviourist reductionism and the theory of tropisms took hold, in spite of all the previous remarks on animal intelligence made by evolutionists and natural theologians. The reductionism that followed Lloyd Morgan's formulation of his canon seemed to mark the end of the dualism of instinct and intelligence, whereby animal expression was seen as a convergence of competence and adaptive capacity. On the other hand, the claim to explain animal behaviour exclusively through the model of automatisms failed within a few decades, when the insights of Kölher, Tolman, Lorenz, Piaget and other authors were picked up by Donald Griffin, triggering the cognitive revolution also in ethology. Once again, however, the theme of subjectivity, while becoming the explanatory core of behaviour, remained in a sort of limbo masked by the concept of awareness.

Let us return, then, to subjectivity as a pre-existing condition that *precedes* the illumination of consciousness. My proposal of the four levels of subjectivity, in this sense, seeks to define precisely this *subiectus*: the foundation that lies beneath and supports the expressions of subjectivity, including consciousness. I am not ignoring the importance of reflection on consciousness; on the contrary, I believe that this faculty is actually more common and articulated than we usually assume. Authors such as Donald Griffin, James Gould, Marc Bekoff and Frans de Waal have done well to show how much awareness there is in animal behaviour. On the other hand, I believe that using higher levels of intentionality as a starting point might suggest that where there is no consciousness we can justifiably speak of automatism or that subjectivity is founded on consciousness when, conversely, the opposite is true. Subjectivity is first of all the appearance of a Self, even if in a rudimentary form—the simple ability to differentiate sensory feedback—which is then necessarily enriched with affective qualities of orientation and knowledge content, until it reaches forms of integration capable of transforming the here and now into a propositional attitude. Subjectivity, therefore, is first and foremost an expression of self-ownership and therefore of a constant invention of the future, which focuses precisely on creativity as a way of affirming one's presence.

The Subject as Presence

If we step back from the usual evaluative parameters, we realise that creativity is the salient characteristic of animal orientation and of the way the individual constructs its here and now. The animal's action is *performed* at the very moment it is *expressed* precisely because it is conducted on the basis neither of retrospective causality nor of an attractive teleology. It is not possible to ignore the dialogical—and therefore open and innovative—becoming in which the animal lives its present. The way in which an animal deals with the present is not based on static instantaneousness divorced from everything else, but on a diachronic oscillation: the past is reinterpreted by virtue of its specific projection into the future and vice versa. Eschewing the causal determinism of external or internal stimuli and, equally, the finalism determined by a deficiency or a project result that is desirable in itself, the subject creates its own condition by relating these possible influencing factors and going beyond them with the imagination.

As Henri Bergson points out in *L'évolution creatrice* (1907), circumstances, however similar they may be, never act in the same way on the same subject, since they always catch it at a new moment in its history (15). We are faced with fluctuating variables—the changing characteristics of the world and the physiological and psychological transformations of the subject—so the multiple intersections of these differences always produce singular conditions. However, it is not just that. When we speak of presence we are not referring to a simple singular composition of variables, but to (i) a subjective inclination to work on the influencing factors and (ii) the capacity to add something, precisely because of the emergent effect of the mind. Subjectivity goes beyond its components, because it is capable of producing a superintendence. While not necessarily conscious in this creative expression, the subject is the helmsman of its orientation insofar as it has ownership over the processes that affect it.

For Jeremy Bentham (1748–1832), this inherence, which is reflected in the definition of what can be called a moral patient, is the capacity to suffer, while in the evolutionary epistemology of Donald Campbell and

Konrad Lorenz subjectivity is based on the possession of a-prioris and therefore on the protagonism of knowing, which is never reduced to mere exposure to the world. Both the capacity to feel and the capacity to desire, as well as the specificity of perceiving or knowing, point to the electivity of the subject, which is partial in the intersection with reality not because it is lacking, but precisely because it is founded on such metapredicates. The animal is a subject because it constructs, or literally extracts, a *forma mundi*: its being present on the scene does not end with registering what is there, but with realising *what could be there*. The individual is not limited to the already given, whether apparent or to be discovered, but actualises new states *ex novo*, while respecting the constraints of reality. The presence of the animal, therefore, can only be based on creative expression.

On the other hand, it is necessary to be very clear about what I mean by creativity: as I said, it is never *ex nihilo* nor is it free from the limits imposed by the world. Animal creativity is conditioned: it is subject to strict conditions that cannot be avoided. First of all, the subject's entire past, in every inheritance content, establishes a well-defined plane of access to the world: for this reason, presence, which is always also experience, can never be unlimited. Moreover, presence is intrinsically a state function, so the subject is never neutral and complete with respect to all possible physiological and psychological conditions. Indeed, (i) it is always present in a certain way, and (ii) it can only activate certain functions at a given time, so it is never expressed in its entirety. The creativity that the subject manifests in its here and now is therefore correlated to and limited by these intrinsic variables. On the other hand, the most relevant aspect here lies in the very concept of the creative process which, in the current mindset, is thought of as an event that produces something out of nothing and is realised in a solipsistic way by a creator. On the contrary, by creativity I mean (i) a dialogue between subject and world, whereby if creation is a novelty, it is also unexpected, because both subject and world develop something unprecedented, becoming co-factors in the process; (ii) an extraction of a possible form (one that is conceivable even if not yet present), from the virtual condition of reality through a new organisation of things.

The animal, even when motionless in the face of events, is always creative, because it is caught up in itself, in its feelings, emotions and desiring condition, which drive it to produce a surplus. The subject is the protagonist of its life, because it is not content with what is already available on the world's plate, but wants to set its own table through the extraction of possibilities. The subject is the opposite of a closed monad: living off heteronomy, as a dynamic process of participation and introjection of the world, it continuously creates new existential contingencies. Protagonism is expressed in creative openness, that is, in the animal's capacity to act in accordance with a future condition, which still resides in the limbo of virtuality. For this reason, a subject can never be compressed into the present, because its dimension is always found in the creative reconfiguration of the flow of what has been and what is not yet. To quote Henry Bergson, again in the abovementioned essay, "Wherever anything lives, there is, open somewhere, a register in which time is being inscribed" (1907, 24–25). The animal's protagonism lies on the one hand in being part of this flow, and on the other in emerging from it, in being elsewhere.

I wanted to emphasise this positional multiplicity—between phylogenetic and ontogenetic, near or distant pasts, immediate occasions or possible futures, and above all various kinds of elsewhere—in order to affirm a sort of overlap between subjectivity and positional unspeakability, which can be translated into a *quasi-absence*. This means that even agency, the most striking aspect of animality, comes as a surprise through the unpredictability of its expression: it is an instant decision. One is a subject because one's action takes place at the very moment in which it is accomplished, as it is neither included in the factors acting at a certain moment nor due to predetermined causal sequences. The subject is a protagonist because it reinterprets the past according to the score of an *occasional* projection towards the future: every memory or acquisition is not the repetition of a fixed pattern, but the innovative reconfiguration of dynamic models. In fact, both reminiscence and imagination are influenced by a fleeting and unambiguous present in the individual. Moreover, this clash of temporal dimensions does not follow a direction of flow, let alone respect a causal linearity, so the subject's time is always one condition among many possible ones. Hence a "being in time" that cannot be confined to the simple parameter of becoming.

The animal invents its own *chronodimension*, creating it in the moment, reconfiguring what it has received or acquired from the past and projecting itself into an open future from which it can extract a possibility by acting on the virtuality of reality. The animal's protagonism lies precisely in its ability to interpret the conditions it has acquired in order to construct an open future. The continuous comparison between heritage and projection means that subjectivity is endowed with the present but not entirely contained by it. The present, in fact, is precisely an invention of the temporal scale produced by the individual. In this sense, taking up the great insights of Martin Heidegger in *Sein und Zeit* (1927) but extending them to subjectivity in a broader sense, we can say that *Dasein* is situated but at the same time transcends positionality. The subject constructs the coordinates of its *peripathesis* in the world, but this is not simply a definition of unfolding. In fact, there is not only an *operative agency* or expressive electivity, but also a *diachronic agency* or temporal electivity, by virtue of which the subject can never be assigned to a precise position. Presence, therefore, also lies in being able to invent one's own time, which can no longer be considered in terms of the duration of a process, but in terms of the subjective construction of becoming.

Paradoxically, it is just because of this positional multiplicity that we can speak of the subject's presence in the here and now, able to attend the situation through diachronic oscillation precisely because it is not completely absorbed in a single temporal dimension. Subjectivity and presence are superimposable: this is one of the reasons why I do not base subjectivity on consciousness. Presence, in fact, can also be unconscious or preconscious; it exists in awakeness as well as in the oneiric condition, because it arises from the systemic and continuous work of the mind. Presence is the fruit of the ability to create a unitary condition producing a continuum of psychological states in the flow of time through retrieval and projection into states to come. Presence makes the set of *influential elements*—such as phylogenetic a-prioris, experiential a-posterioris, physiological conditions and the contingencies of the world—not determinants but influences, a sort of map within which to trace one's own path. The influential elements are never zeroed out—protagonism is not absolute freedom—but they become useful references to build a route that is enriched by the individual's creativity. Presence therefore lies in the ability to put these elements together, imposing an order on them that can be functional to the subject's tendency to look beyond.

When I speak of presence I mean being faced with all these influential elements, but at the same time being emancipated from the determinism of each as well as the combined action of their intervention. Animal behaviour always involves going beyond these factors, while taking them into account: taking the initiative and using the variables at play, as a helmsman manoeuvres his vessel to reach his destination. I refer to the animal's capacity to *be present in its here and now* without being totally at its mercy, inventing a proposal—deployable in an orientation, a choice, a decision, a foreshadowing or a projection—which is always exposed to indeterminacy.

The animal is not totally immersed in the moment of convergence of the factors involved but makes them its own by going beyond them: it is not captive of its here and now but owns it. The various influences can inspire orientation but not determine it, not even in co-factorial terms.

An animal is always present when it operates in the world, even in the most minute of occupations, even when it engages in behaviour with a slim margin of distance from its inherited skills. This presence cannot be ignored, downgrading it to mere appearance, because it is a central factor in explaining an animal's behaviour, its adaptive capacity in the flow of circumstances that always present margins of novelty and its pursuit of its own course. Indeed, protagonism lies in being innovative (unpredictable) and not simply deducible from the factors involved, because presence has room for that *something more* which is the expression of the mind's productive arbitrariness, that which goes beyond the situation itself. Here it is a question of understanding the relationship between that "something more" and the consistency of the intrinsic and extrinsic already given. We must avoid falling into polarities which, while describing expression as a recapitulation on the one hand, on the other aim at eradication, as if expression were a kite that, having broken its thread, wanders in the sky.

The concept of presence, the animal *Dasein*, risks being interpreted as a distance from reality, in line with what has been said about emancipation from all forms of imposed conditions. But from my point of view, things are exactly the opposite. Precisely because an individual is interested, it never approaches the world according to the sole directives of the here and now, but invests it with memories, searches it for opportunities or hidden risks, projects itself into a future condition, plans actions to come. In other words, the here and now cannot suffice for those who bear interests, so presence never yields to repetition, but neither is it detachment, in the sense of taking leave. Rather, it is the capacity to embrace a longer time than the immediacy of phenomenal development. Presence cannot be separated from the interested being that always brings its initiatives into a closer relationship with reality instead of the disjunction of creative solipsism. Therein lies the fascinating reason for the creative motive: if it cannot be subjugated through the enumeration of the variables present, neither can it be considered detached from the influencing factors—in a word, from reality.

A subject is present insofar as it is interested, so presence implies a close connection and, conversely, non-deductibility with respect to events. Presence is a deep and extended contact with reality, which increases the space for expressive freedom. It is never an act of mere glossing over phenomenal appearances, but rather an insight not only into what is clearly there but also into hidden opportunities and the dimensions of the possible. Presence is therefore the ability to look at the factors of influence with new eyes, widening the mesh of circumscriptions and, in a way, ***maximising the virtual power*** of characteristics that at first glance appeared simply constraining. For a very skilled player, a football pitch is immense, the rules of football are like words that lend themselves to composing the most diverse phrases and the body seems to lose gravity. Freedom is therefore the ability to interpret one's own here and now, not to transcend it.

Spaces of Creative Freedom

The individual is never completely free, but neither is it totally subject to influencing factors: this, in my opinion, should be the starting point for understanding the *conditions of creativity* that allow for the subject's protagonism. The phylogenetic heritage, past experiences, physiological and psychological conditions of the individual define certain propensities and copula skills—what to extract and how to interpret presences—which influence the relationship between the subject and contingencies. This is already true in terms of the orientation or interpretation of agency attributed to the environment: we have seen this by comparing the way a cat and a child interpret the accessibility of a meadow. Innate and learned endowments define ways of intersecting with the world and, consequently, establish coordinates of creativity. If for a cat being in a meadow most often means wondering what is moving and how to chase it, for a human it means asking what stands out and how to collect it. These are copula prevalences, which however do not limit the subject's options to predetermined expressions, but open up spaces of creativity. Being movement or icon-oriented and proactive in chasing or collecting opens up the individual's protagonism, it doesn't limit it. Although equipped with seeking images and standard operants, the subject must accept the challenge of interpreting the situation and facing the unknown.

After all, the environment also establishes its own range of usability, which in this case is objective: it is undeniable, in fact, that one cannot collect shells in a meadow. External reality contains the individual's protagonism in terms of a *field of possibilities*, that is, opportunities, risks, constraints and practicability. However, even if we knew in detail the characteristics of this field and the innate and learned copulative endowments, we still could not apply the claims of the *Laplacian demon* to the behavioural phenomenon: the physiological variable proves this approach untenable, because the body's condition is ever-changing. And I am not just referring to phasic, metabolic or biorhythmic oscillations, but to the fact that every physiological moment of the individual is singular and non-repetitive. On top of this there is the psychological dimension, the integrative positionality I have defined as the fourth level of subjectivity, which constantly produces new propositional attitudes. Moreover, the environment too is a dynamic entity that constantly changes both as such and at the very moment in which the subject intersects it. In fact, even the environment-meadow cannot be addressed in a repetitive way, because each moment in the same meadow produces singularities of state.

Therefore, I do not wish to claim that an animal has complete copulative freedom in relation to the world and that, likewise, it is free to disregard what each environment allows for or even prescribes; I am only saying that, within these perimeter factors, protagonism is never absent, but, on the contrary, it is a necessary condition precisely for dealing with specificities, that is, the particular constraints encountered. From my point of view, protagonism does not indicate the subject's absolute freedom, but rather its capacity for productive superintendence over the situation. Consequently, creativity also moves through a complex fabric of influencing factors, makes use of them and does not contradict them to create new possibilities. Ultimately, it is a question of seeing the final expressive product neither as an *objective figuration*, already present and simply discovered by the individual, nor as an arbitrary, totally unrelated creation. The creative result is the construction of a new organisation of state, an invention in its own right, resulting from the dialogue between the individual's predicates of subjectivity and the potential of reality.

When I speak of a range of arbitrariness in the state assumed by the subject in the here and now, which sometimes becomes explicit in a behaviour (but not necessarily), I wish to underline the unpredictability of the mind in formulating proposals that sometimes go beyond the mere management of the situation. Of course, when faced with the need to eat or drink, an animal must primarily respond through highly referred behaviour, often adopting standard strategies that have already been accredited. But an animal's life is not a whirlwind of occupations and actually contemplates a lot of free time, which can be spent resting, playing, exploring and—why not?—lost in the imagination. The mind is creative by definition, because it is capable of envisioning something that is not present in the situation, perhaps by mixing together the mnestic threads that come from different experiences. Individual action obviously takes place within canonical ranges of expression: a dolphin cannot behave like a sea lion, just as one cannot pick up daisies on the seashore. However, the way an animal behaves is always the result of an addition, that is, an innovative proposal, which depends on its mental turmoil.

That is why I speak about spaces of expressive freedom, areas well circumscribed by influential variables where, however, the creative inspiration of the mind is made possible. As we have seen, a great deal of creativity can be expressed on a football pitch, albeit within a pre-established spatial perimeter, on a ground that, as such, defines the possible types of activity, in accordance with very precise sporting rules and by virtue of the equally binding physical conformation of the players. Similarly, we must not think that influential factors, that is the innate, past experience, momentary physiological conditions or environmental constraints, limit creativity and protagonism. In reality, these factors do no more than circumscribe a *space of expression* for a subject's creativity, while also giving it a dimension in which it can be released. Creativity needs to be circumscribed. While it is true that projection towards unrealised conditions and the plurality of integrative states are fundamental qualities of the mind, it is equally undeniable that both protagonism and creativity rely on a dimension in order to be realised: creative freedom is by no means at odds with the definition of expressive spaces.

Creativity is never absolute, because at any given moment one finds oneself within a range of possibilities given by the influencing factors. This is why I speak of a space of expressive freedom both in terms of the implementation of behaviours of competence (i.e. already experienced as effective) and of the creativity that goes into it. I think that every behaviour is made of competence and creativity. Both competence and creativity are related to the space of expressive freedom that the situation (i.e. the influencing factors) provides. The space of expression is like the oft-mentioned marble slab, from which one can extract a statue that is possible but not yet actualised. When I say that creativity is a deeper and wider embrace with reality, I mean that it is the capacity to bring out from reality also the forms that still lie in the limbo of virtuality. Creating an instrument from a piece of wood essentially means unveiling a potential use in that object, organising its characteristics on the basis of hidden performative properties and functional needs. The presence of the subject is expressed in the ability to lift this veil, arranging a new organisation of reality and bringing together different aspects located in equally different domains, such as the physical laws that regulate performative mechanics, the functional potentialities of the object, the relationship with the animal's practical knowledge, the techniques for constructing the instrument, as well as the implicit characteristics of the object itself.

In other words, presence is not total immersion in the apparent condition, but the capacity to reveal hidden properties, possible outcomes, new correlations, that is, the virtual richness of a situation—including possible setbacks and deviations from previous experiences. Precisely for this reason, presence establishes a much stronger conjugation compared to simply reacting to a stimulus on the basis of a preset mechanism. Unveiling unexplored possibilities means, in fact, building an even closer relationship with the world, inaugurating new figurations enables by the expressive space. Undoubtedly, it is precisely this deep conjugation that makes adaptive protagonism possible, otherwise an animal could do nothing more than stick to what is already there, repeating actions that have already been predetermined or implementing random responses, demonstrating a distance from the evenientialpotential, as if it were looking at the situation with the eyes of the past or in total darkness.

If the animal were merely to take note of what is given, it would inevitably have no real presence but only an eternal present. As can be seen, to deny creativity is to disconnect the subject from a dialogue with the world that can bring out not only adaptation to situational specificity, but also the invention of new practices. It would mean imprisoning it in an isochrony that does not allow it to *be present in the here and now.* Indeed, the explanation usually given for animal behaviour always oscillates between these two boundary conditions: (i) the triggering of a pre-established automatism, either through instinct or by virtue of conditioning; (ii) random attempts when the subject is in a condition of absolute novelty. If we deny animal protagonism, we inevitably arrive at the explanatory dichotomy that alternates between random and repetitive behaviour. I do not mean to say that these two conditions cannot be regarded as borderline or extreme occurrences in the animal's expressive landscape, but for this very reason they do not represent the everyday life of an individual who, conversely, must use its skills to adapt or innovate.

Protagonism is never absolute, but contemplates degrees of creativity, depending on the subject's need to manage a greater or lesser margin of novelty and to consequently fine-tune its action of adaptation or innovation. These are two very common situations, which place animal protagonism not in the Empyrean of high cognition, but in the banality of ordinary behaviour. In fact, competence is unarmed if it cannot be subjected to functional reconfigurations with respect to what the situation requires. The subject, therefore, must know how to use it as a flexible tool allowing for small or big adaptations. On top of this, there is what I regard as the most interesting aspect of creativity: the constructive expression of novel conditions, which is not limited to adaptation, but literally produces new existential dimensions. Adaptation is important, because in most situations the animal must be able to find solutions, however small, to the flood of variations in the world. But if we understood creativity only in terms of adaptation, we would miss the most exquisite part of animal protagonism: the creation of new conditions.

In fact, operating within a space of expression does not only mean finding a way to achieve the usual objectives by applying the usual strategies in a context that is constantly changing the rules of the game. The animal does not only find new forms of adaptation to the environment, adjusting its equipment to small situational changes, because new goals and strategies are often inaugurated in creative projection, changing the order of factors and constructing new conditions. It is wrong to discard the principle of adaptation, but it is also misleading to think of it in terms of passivity, that is of simple exposure to the selective filter. Likewise, it is wrong not to realise that, alongside adaptation, there is the subject's creativity, capable of extracting *new existential dimensions* from the field factors. The denial of animal creativity, therefore, lies in removing expressive protagonism—being a regulatory and innovative entity within one's own space of expressive freedom—from ordinary behaviour. In repetition, as well as in the randomness of non-oriented attempts, there is no regulating subject, no intersection with the singularity of reality, but only subjection to it. But this is not what we commonly observe.

Animals always use their endowments creatively, not only to adapt them to new specific situations, but also to create new conditions of reality and invent new existential dimensions. Creativity has been banished from animal behaviour because of an epistemological—rather than anthropocentric—flaw. After all, in order to bring out a quality, in this case creativity, we must not turn it into noise in relation to codified categories. The need to mathematise animal phenomenology has inspired a dichotomous explanatory logic, with attribution by approximation, as in the case of the *automatism vs. random attempt* dualism. If, on the other hand, we observe an animal, noticing its active presence in the interaction with different situations, we discover that adaptation and innovation, leading to different lifestyles or niche constructions, are indispensable operations in ordinary life. Creativity is a basic component of living. For this reason, it is not to be sought in attitudes that are unusual, complex, reminiscent of human expressions or surprising for their extraordinariness. As soon as we understand the animal's protagonism in ordinary situations, we have a good basis for dealing with the higher levels of cognition.

Creativity takes place without betraying the expressive spaces defined by influential and determining factors. A highly talented football player does not violate the rules of his sport, does not change the perimeter within which he operates and, at the same time, is aware of his physical performance, yet he is able to think of something that goes beyond the simple summation of the factors involved. And he does so precisely because he does not act by mere repetition or blindly, but is firmly conjugated-interested in what unfolds before him: he knows well the space in which he is operating. He can bring out a hidden opportunity, he can see the differences with respect to past experiences, he is capable of finding imperceptible similarities between apparently distant conditions, he can find correlations even far apart in time or notice attributions that are unprecedented at first sight. Thus, precisely by attending to the field factors, he is able to change the spatio-temporal order of the situation, bringing out possibilities that were not previously known. The space of expressive freedom is, therefore, always present, but it does not limit the subject's protagonism, both in ordinary situations requiring adherence, and in less frequent ones calling for innovation, that is, probing the unexpressed potential of the situation.

My intention is not, therefore, to deny the possibility that animal behaviour may, in certain circumstances, fall into situations that resemble the mere repetition of an acquired skill or the expression of random attempts, but rather to emphasise the exceptional condition of both repetition and random attempts. Both have the same probability as revolutionary behaviour, that is, as those moments of vivid creativity that leave us astonished. Ordinary behaviour, however, is an entirely different matter! In most behaviour we witness: (i) the phenomenon of adaptation, when there is a strong overlap with previous experience and a small to medium margin of novelty; (ii) the phenomenon of innovation, when the situation is more fluid, hinting at untapped possibilities. Both these forms of creativity have different degrees and connotations, since adjustment can be made by co-opting other strategies or by discovering possibilities, while innovation ranges from simple updating to outright invention that radically changes the field of expression.

There are, however, common aspects in various creative expressions that show two things: first (i) that the subject is the one who uses its endowments instead of being driven by them; and second (ii) that without a touch of creativity there can be no form of adherence or experimentation. The interpretation of the space of expressive freedom, that is the usual protagonism of the subject, is realised in these two conjugations of ordinary creativity. So imagine placing the different behaviours on a segment that marks their degree of creativity. The graduated indicator starts with a response reminiscent of randomness, where creativity is zero: there, each expression is equally likely and the animal appears to be absent from the situation. This is followed by a large area of the segment, with increasing creativity, where the animal tries to invent new conditions through a wide margin of expressive freedom. The peak of creativity is reached at a point in the segment where innovative attempts and adaptation of experience reach a certain balance: in this condition, knowledge becomes fluid. From that point onwards, creativity decreases: trials increasingly take the form of co-opting previous equipment, limiting the gradient of variation. This leads to the other end of the segment, back to zero creativity, but this time because behaviour is repetitive.

I believe that situations involving polar conditions are very few in an animal's life, but I do not feel I can rule them out. Likewise, the individual may sometimes be able to go beyond adaptation or updating, by performing behaviour with a more significant gradient of novelty. For example, creativity is accentuated when a behaviour is recycled, that is co-opted from a totally different context, or when two behaviours belonging to different expressive patterns are combined in a single action. Generally, however, the individual, even in innovation, only makes small changes to its niche style. If these changes become part of the tradition of a certain population, over time, they can give rise to very articulate constructions. On the other hand, it is not necessarily the case that a small invention may produce far-reaching effects or that the subject brings about such a productive invention that it marks a watershed between before and after. Typically, creativity lies in the two broad segments of adapting an endowment and using it in a new, albeit similar context to make a targeted attempt. In the most complex situations there is a median or balance point where both the freedom of the attempt and the experience of co-optation converge in the subject's strong protagonism: this is where the most surprising creative behaviour takes place.

The Joy of Living

Creativity is, therefore, the main ingredient of the subject's protagonism, that *quid* that brings the animal out from the depths of necessitated phenomena, inaugurating conditions that are not presupposed by the field factors. In other words, thanks to creativity, at every moment the animal adds something that was not there before, a surplus that makes it impossible to explain away the behavioural phenomenon by limiting it to retrospective variables. This is why I find the machine-like animal model inadequate: not because I do not recognise the presence of physiological mechanisms in animal phenomenology and certainly not because I question the causal explanation, but simply because the psychological dimension, emerging from a plurality of influential variables, creates the very conditions for the protagonism of presence.

Of course, in order to admit this protagonism, we need to change how we interpret the animal condition, because creativity obviously requires preconditions, such as the acknowledgement of a systemic-emergential entity capable of supervising expression, avoiding the autonomy of individual automatisms. Expressive protagonism is the subject's ability to (i) grasp the particularity of the situation encountered along one's path, trying to understand the level of similarity to previous conditions; (ii) identify spaces of possibility through experimentation, targeted attempts, probing and imagination. In other words, it means identifying (i) the margins of novelty that call for adaptation of the skills possessed; (ii) clues in unknown conditions or, in any case, useful references to find possible solution keys; (iii) risks and opportunities that are not obvious, but virtually possible; (iv) new ways or conditions to realise one's interests. In order to grasp the specificity of a situation, it is necessary to know how to look inside it, penetrating the surface of apparent coincidence with already experienced situations. But in order to do this, it is necessary to place all the factors that can be drawn from the situation within a systemic complex, capable of relating with the heritage and the prevailing interests and, undoubtedly, of penetrating or going into depth, questioning appearances and the status quo. Big or small, creative work is part of an animal's existential habits, not an exception.

We would be wrong, however, if we considered creativity as a solipsistic event, aimed at transcending the real without taking into account the specific conditions in place. When I say that protagonism lies in going beyond the influential factors, essentially all the variables converging on the subject, I do not mean that one should not weigh them out. Creativity is a dialogue with reality and, *a fortiori*, with the situation, both within the animal itself and in the world. It means bringing out a novelty precisely thanks to continuous evaluation, capable of opening the space of possibilities in a Socratic way and allowing the birth of something not already given, precisely because by paying attention to the already given. In order to understand creativity, a fundamental prerequisite is therefore the concept of dialogical experience, understood as active presence in the intersection with the world. It is not possible to consider the subject as a passive entity that is simply impressed by what is present, capturing the givenness of reality based on sensory and cognitive endowments and, finally, simply adapting to it. Protagonism indicates the possibility of giving a certain direction to the experiential event, bringing out opportunities that are not simply given. Experiencing recalls the act of experimenting, a process that leads to a condition of competence—from the Latin *peritus*—whereby one becomes an expert, but at the same time one has given birth to a certain happening through experience, bringing it to light.

The word *experience* also includes the meaning of "testing", that is attempting and undertaking. Hence the concept of experimenting, entering into a deeper relationship with reality or checking a possibility. It means coming into contact with a phenomenon and activating a flywheel of metamorphoses, made up of questions and answers, which will have repercussions on the biographical structure of the subject: each step in this dialogue modifies what follows. In dialogical experience you know the starting points but you cannot predict the points of arrival, because each interchange opens up dimensions of virtuality that did not precede the exchange undertaken. This encounter transforms *the way* an animal experiences itself, probing not only the virtuality of situations external to it, but also its own virtuality, its possible aims and interests. If we imagine an individual's experiential endowments as a sort of bubble that allows one to look out onto a certain horizon, that is, a *proximal plane of experience*, each experiential process modifies this plane and consequently the horizon that the individual looks out onto. It modifies it by the mere fact of expanding the bubble, although the most relevant aspect is that it gives rise to a lack of homogeneity in the plane itself, which becomes more detailed in some areas and more generalised in others. In this sense, experience defines not only the *experienced* (the contents acquired by the process) but also the *experienceable*, the desirable experiences.

Experience, in fact, is solicited by the individual's projection towards the world, in a way that can be compared to a sort of active search: it is the individual who searches reality in order to bring out an experience which, therefore, is always linked to the type of search undertaken. Even when the animal witnesses events that may appear as objectively stimulating-interesting, there is still a marked individual variability at play. Experience arises from tension, both proper and implicit in the subject; it's a projection that conjugates it to the outside world, sustained by the animal's need to find aims and interests—in short, the *joy of living*. Experience, therefore, breaks down the phenomenon into its intersecting features: weighing its importance, linking it to particular inherencies, understanding the factors that make it possible, learning to relate it to other events, assessing its circumstantial significance in relation to other phenomena and identifying the correct actions to deal with it. Clearly, this is not a simple induction, but rather a search in which the individual carries out investigative actions of various kinds. In the end, experience produces an interpretative legacy of the event, a scheme of connections or, if you like, a functional representation, which does not correspond to an objective and passive analysis of the facts, but rather to the creation of new configurations. Experience, therefore, produces a subjective translation of the phenomenon.

We have seen that pleasure arises from a tension that finds its constraints, negations, risks, occasions, confirmations or opportunities—in a word, its problems—in the world. Experience is the attempt to release this tension which, incidentally, is the source of the problems themselves: if the animal were not projected towards the world to satisfy its hunger for joy, it would have no problems. The subject is constantly faced with these problems and can solve them because it possesses *inferential meta-rules* that it applies according to the particular situation. For example, an animal will use deductive inference if what it needs is a useful model to extract a result, while it will choose inductive inference when it wants to generalise or abductive inference if it wants to get to the bottom of a circumstance. Experience, therefore, is never mere exposure to the world, but means actively working out of possibilities with respect to the events that the individual encounters, challenging oneself on the basis of probabilistic or Bayesian parameters. We should not consider knowledge as an end in itself—knowing for the sake of knowing—because what the animal seeks is a space of affirmation.

Knowledge is a creative act in the sense of probing the space of possibilities, seeking to enlarge one's territory, to find new opportunities, to expand into the world. Creativity gives the animal the capacity to try something out, to see if a certain configuration holds up, if a road that has never been trodden is viable and offers results that are consistent with this need to rejoice in the mere fact of being alive. It is an explosion of vitality that encompasses surprise, affirmation, willpower and above all the pleasure of finding oneself once again in front of the wonder of the world. Consequently, an animal's experience in relation to phenomena resembles, with due adjustments, the way in which a scientist excitedly performs their experiments, *handling* the situation, testing and trying out the formulas of possibility. The concept of experience clarifies the term *subjective* in the sense of "proper of the subject", that is capable of producing a psychological condition or an inner relationship with the world. If experience is an investigative dialogue also and above all about possibilities—about what does not appear and what is not even formally present—we must recognise that experience is the act of *intus legere* the real, that is, of creatively applying the tools one has.

This brings us back to the singular character of subjectivity. We are used to looking for animal intelligence in extraordinary behaviour, the kind that surprises us because it resembles how we humans deal with reality. This is a strange way of getting out of anthropocentrism: instead of recognising our animal nature, with its obvious specificities, we seek to humanise other species. Human beings, precisely because they are phylogenetically adapted to a certain niche, cannot set themselves up as the measure of the world and cannot aspire to be universal. Animalising humans means recognising their own specific nature, on a par with other species, rejecting the humanist notion of ranklessness. This is an epistemological prerequisite, as well as an ontological one, in order to look at other species without prejudice. A child's way of observing animals is certainly naive, but an adult's is no better because of the many biases involved. The comparative obsession, resulting from our need to differentiate ourselves and stand out, prevents us from seeing that subjectivity is manifested in difference and not in standardisation. The personalisation of the gaze lies at the heart of experience: it is the subject's aptitude for making continual course corrections in relation to its own developing needs. Creativity has to do with the subject's existential parable, with the open future which is the cause of joy.

The joy of living stems from this openness to experiential indeterminacy, from the ability to go beyond what has been entrusted to us by phylogenetic heritage or what we have already experienced. It is undoubtedly difficult to understand the origin of this joy, which mixes the surprise of finding one's own expectations matched in the world with the wonder at the whirlwind of possibilities. The joy of living is undoubtedly about astonishment, but I don't think it can be attributed simply to unexpectedness, as if the subject were overwhelmed by a flood of unknown things: rather, it has to do with the feeling of finding oneself again, having the world as a stage on which to express oneself. As I said, the apex of creativity lies at that point where attempts and competence reach a kind of balance between openness to the indefinite and full participation in the feast. In order to understand the creative impulse, one cannot, therefore, ignore this condition of exuberance, which has been spontaneously manifested in every animal since its first appearance in the world. This feeling pervades the subject completely and involves the entire psychological spectrum: the joy of finding something in the world that is already present in the subject, the state of having-been-there-before, and at the same time feeling that the future is open, available and just-right, as if the world had been purposely prepared for the individual.

Joy is expansion into the world, it is surprise and curiosity that suddenly merge into excitement vibrating in every inch of the body and it is the will that meets power, the longing that makes the future full of opportunities and consequences. This eagerness to live takes the most varied directions and expressions inspired by each animal's nature: some we find endearing, others make us smile, others yet terrify or disgust us. This is not the point. Creativity is a way of giving a new voice to the ghosts of the phylogenetic past and the resonances of experience, a way of arranging the clues of the world in a configuration that was not there before. Therefore, it does not mean to ignore but to rewrite the coordinates of orientation; it is a kind of excitement where the subject is caught up in euphoria, like an explorer landing in an unknown continent.

In order to understand this power that springs from the individual, one cannot ignore the tendency to transform the real into a metaphorical space, capable of recalling something else. Play is the most explicit form of this ability to change the existing rules, the one we perhaps understand best from our mammalian perspective. Play is not possible without the assumption of make-believe, of transforming an object into something else and hybridising different behaviours by extracting them from their functional and adaptive meaning. The capacity to go beyond the given, so evident in the playful dimension of many animals, suggests that creativity is based on the subject's abstraction from the objectivity of the situation and from established meanings. The animal is metaphorical insofar as it is capable of this transfer or allusion to something not present or simply possible. Play is an expression of the imagination and the joy of being in the world with full ownership of one's actions, which is why it eludes any attempt at definitive functional attribution. However, this does not mean that it lacks utility. *Animal Play* (1998), edited by Marc Bekoff and John A. Byers, shows the importance of play in innovation processes and in the experiential constructions of the individual: play is also a prerequisite facilitating the development of highly creative behaviour.

In conclusion, there is no opposition between the creative process and the influences that derive from the species' behavioural repertoire, whatever they may be, since animal creativity is always an imaginative use of these. Animal protagonism always operates within constraints, be they phylogenetic, biographical, physiological or environmental, yet these do not impose a behaviour, but establish ranges of movement within which there is freedom of expression. Obviously, in order to have even a minimum amount of creativity, one must play within a certain perimeter, respecting a space for manoeuvre that, however, acts as an elastic propelling one into an elsewhere. The ability to interpret this space is a prerequisite for creativity to emerge. In order to go beyond the already given, trying to unveil new possibilities, it is indispensable to be present in the here and now. Interpreting does not mean merely expressing an evaluation of what is present, but being able to imagine what *could be*. In other words, the animal must be able to glimpse a path where there is only beaten grass, a goal yet unpursued, a course yet unset, a strategy yet untried. In this sense, the interpretation of the here and now means thinking-beyond and not stopping at presences, enabling the openness that intoxicates the subject.

For creativity to exist, an animal has to experience the joy of living, not intersecting the world in a neutral manner, but adopting an enthusiastic approach to events as it is called upon to expand its existential space. Having a goal is a *conditio sine qua non* for the helmsman: it is not possible to regard the helm as a mere tool to be used flexibly. Being oriented removes neutrality or indifference from the course of navigation, because everything is read as favourable or hostile to enthusiasm. The animal, therefore, is a protagonist because it is not indifferent, because its whole being is projected towards something—an interest that arises from the combination of every psychological variable and that produces the copulative need in which we recognise animality as a dialogue with the world.

This joy is where we find the most explicit manifestation of the animal condition, unlike the disinterest shown by the machine despite its performative complexity. An animal always *cares* about something: it is always involved in the situation because there is always something at stake that concerns it.

Works Cited

Agamben, Giorgio. 2004. *The Open: Man and Animals*. Stanford, CA: Stanford University Press.

Bekoff, Marc, and John A. Byers. 1998. *Animal Play*. Cambridge, MA: Cambridge University Press.

Bergson, Henri. 1907. *L'évolution Creatrice*. Paris: Presses Universitaires France.

Bonnet, Charles. 1760. *Essai analytique sur les facultés de l'âme*. Copenaghen: Chez le freres Cl. & Ant. Philibert.

Boulliere, David Renaud. 1937. *Essai philosophique sur l'âme des bêtes*. Amsterdam: Chez François Ghanguion.

Burghardt, Gordon. 1991. Cognitive Ethology and Critical Anthropomorphism. A Snake with Two Heads and Hognose Snakes That Play Dead. In *Cognitive Ethology: The Minds of Other Animals*, ed. Carolin A. Ristau, 53–90. Hillsdale: Erlbaum.

Cuvier, Baron George. 1836. *Le règne animal*. Bruxelles: Société Belge de Libraire.

Darwin, Charles. 1871. *The Descent of Man and Selection in Relation to Sex*. New York: Merrill and Baker.

De Condillac, Etienne Bonnot. 1755. *Traité des animaux*. Amsterdam: Charles-Antoine Jombert.

Derrida, Jacques. 2008. *The Animals That Therefore I Am*. New York: Fordham University Press.

Descartes, René. 1970. *Meditations metaphysiques*. Paris: Presses Universitaire de France.

Golding, William. 1959. *Lord of the Flies*. Oakville: Capricorn Books.

Heidegger, Martin. 1927. *Sein und Zeit*. Tubinga: M. Niemeyer.

Minelli, Alessandro. 2007. *Forme del divenire*. Torino: Einaudi.

Nagel, Thomas. 1974. What Is It Like to Be a Bat? *Philosophical Review* LXXIII (4): 435–450.

Pico della Mirandola, Giovanni. 1942. *Oratio de hominis dignitate*. Firenze: Vallecchi.

Spinoza, Baruch. 1951. *Ethics*. New York: Dover.

Uexküll, Baron J., and Georg Kriszat. 1934. *Streifzüge durch Umwelten von Tieren und Menschen*. Heidelberg: Springer Berlin.

Index

R. Marchesini, *The Creative Animal*,
https://doi.org/10.1007/978-3-031-07414-1

The manufacturer's authorised representative in the EU is Springer Nature Customer Service Centre GmbH, Europaplatz 3, 69115 Heidelberg, Germany. If you have any concerns regarding our products, please contact ProductSafety@springernature.com

Printed and bound by CPI Group (UK) Ltd, Croydon, CR0 4YY
30/09/2026
02225490-0016